students, and its remarkably devoted alumni. Norberg artfully captures the excitement and progress of the time period after 2000 that I have personally experienced at Purdue. This wonderful book is a must-read for all who love this great University."

—*Martin C. Jischke, President Emeritus, Purdue University*

* * *

"With captivating storytelling, drama, and wit, John Norberg has compiled the most comprehensive history of Purdue University to date. His account brings history to life by focusing on the people who shaped Purdue. Written as a series of vignettes, each chapter illuminates a major event or tradition, always with the human element at the forefront. Through diaries, interviews, and other rare sources, Norberg offers a variety of perspectives that define the identity and character of the institution via the personalities involved. Even the most die-hard Purdue fan will find something new to learn and savor in *Ever True*."

—*Sammie L. Morris, University Archivist, Purdue University*

* * *

"*Ever True* is a very good read! Anyone who approaches this book believing the old myth that history is dull and uninteresting will be pleasantly surprised. This book is a gift of great value to the Purdue community—now and as long as there is a Purdue University."

—*Betty M. Nelson, Dean of Students Emerita, Purdue University*

* * *

"Norberg skillfully walks readers through time in this fascinating, comprehensive history of Purdue. His relatable descriptions contain something for everyone to learn. We as students should be proud to be part of this rich history, and Ever True provides us with a sense of what it means to be a Boilermaker: a hardworking, humble individual who has the potential to make an enormous impact. I am ever grateful to the Boilermakers in this book who have made Purdue what it is today. As a trustee, I feel even more compelled after reading *Ever True* to ensure that Purdue's future is as fruitful as its 150-year history. Hail Purdue."

—*Daniel Romary, Student Trustee, 2017–2019, Purdue University*

EVER TRUE

The Founders Series

150 Years of Giant Leaps at
PURDUE UNIVERSITY

JOHN NORBERG

Purdue University Press, West Lafayette, Indiana

Cataloging-in-Publication data is on file at the Library of Congress.
Print ISBN: 978-155753-822-2
ePDF ISBN: 978-161249-543-9
ePUB ISBN: 978-161249-544-6

To all the voices singing "Hail Purdue"

who shaped its past and herald its future

ever grateful, ever true

Teaching Us How to Think

"I remember going to aerodynamics class taught by professor Merrill Shanks. In the first day of class he gave an assignment to read pages 62 to 69 which was Bernoulli's Law and the assignment was to criticize it. It was right then that I realized they weren't teaching us facts and information. They were teaching us how to think. And that turned out to be extremely important in many of the situations I encountered through my life."

—*Neil Armstrong, Aeronautical Engineering '55, fall of 2004 interview with Raymond Cubberley, broadcast media senior producer, Purdue Marketing Photo and Video*

Contents

Part Two 1900 to 1971

Part Three 1972 to 2018

Giant Leaps

Foreword

Those among us of a certain age will never forget the moment Neil Armstrong stepped onto the moon. July 20, 1969, is a day the world remembers and to which every Boilermaker feels a special connection.

The iconic footprints Neil left endure as symbols of human advancement, exploration, and scientific and technological progress. Images of his steps are recognizable across cultures and generations in a way unsurpassed in human history.

As we celebrate that history and the 50th anniversary of that day, we cannot forget that he was not unique in taking, as he described it, a "giant leap for mankind." For the last 150 years, countless Boilermakers have made their own marks, often in the form of scientific discoveries, best-selling books, works of art, patents, new companies, World Food Prizes, Nobel Prizes, or other great achievements. There will only ever be one first lunar landing, but the impact other Boilermakers have made over the last century and a half is no less real.

As with Neil's, these moments didn't just happen. They came only after hours in the lab, office, or library, often after many setbacks and failures, and coaching by great mentors. More often than not, they started at Purdue.

The book you hold tells that story. It documents the history of a university that has shaped countless lives and prepared generations of Boilermakers to create their own success. It is a history of the place where giant leaps begin.

The first of so many such leaps began a century and a half ago, when John Purdue made a substantial gift to create a university in Tippecanoe County, Indiana. Few would have appreciated at the time what this small step would come to mean for our state and the advancement of knowledge across the globe.

As we commemorate this milestone and celebrate our past, we pause to consider the new leaps Boilermakers might achieve in the next 150 years.

Will a project, underway as this volume appears, lead to Purdue being known as the birthplace of the first scalable quantum computer? Who among us will cure diseases, develop lifesaving medicines, or find better ways to feed our growing population? How many future Nobel Prize winners or World Food Prize winners are presently young Boilermaker scholars?

In the next 150 years, will Purdue become known for institutional leaps as well? We have entered an era of great transition in higher education. Questions about the future of the sector abound: Which schools will survive the growing popularity of online education and employers advancing their own programs? What should a land-grant university, with a mission to serve all Americans, do about the disturbingly large portion of the country not currently reaping the benefits of a college degree? How can we reverse the recent spike in pessimism among the public about the value and relevance of a college education and earn back America's trust? (And, of course, why does it cost so much?)

To prepare for an uncertain future, Purdue has taken several initial steps, which have generated their own questions.

Will we maintain our unequivocal commitment to true freedom of speech and inquiry in an era when a new authoritarianism is at least temporarily on the rise? Somehow, we must ensure that in the decades to come, Purdue will always be a place where independence of mind and critical thinking are cultivated rather than stifled.

Can we continue to disprove the common belief that a university like ours cannot both invest in excellence and hold the line on costs? We aim to offer higher education at the highest proven value, and this requires progress in both quality and affordability.

Can we take full advantage of the recent conversion of a leading online university into the Purdue system to equip our residential campuses with the skills and technology they may need to thrive in the digital era? Can we utilize this new addition to expand access to higher education for the millions who are unable to come to West Lafayette? Can we evolve to be not only a top-ranked residential university but a provider of learning throughout a Boilermaker's lifetime? In the twenty-first century, any land-grant university that fails to address these needs is unworthy of its history.

Constantly looking to the future and asking such questions is part of what it means to be a Boilermaker. Our faculty and administrators had the future in mind when in 1962 they created the country's first computer science department. When Purdue's sixth president, Edward Elliott, hired Amelia Earhart to become one of the nation's very first career counselors for women, he had the future in mind. And

Purdue's many feats in aeronautics, engineering, agriculture, drug discovery, and so many other areas have been by definition driven by thoughts and dreams of the future.

Among the dominant lessons of Purdue history is the principle that great histories begin with great aspirations. May this chronicle of a century and a half's small steps and giant leaps provide inspiration to you and future generations. *Hail Purdue!*

Mitchell E. Daniels, Jr.
President of Purdue University
2018

Preface

A full-body portrait of John Purdue hangs on the back wall of the reading room at the University's Virginia Kelly Karnes Archives and Special Collections Research Center. The painting stretches floor to ceiling and captures the nineteenth-century man in a black suit, white shirt, and black tie, his hand on the white knob of a black cane. The only color comes from the pink of his hands and face and his blue eyes that watch people sitting at desks in the Archives reading room.

Working on *Ever True: 150 Years of Giant Leaps at Purdue University*, I've spent many long hours in the Archives with John Purdue staring over my shoulders. Sometimes I even hear him speak to me: "No, no, you've misunderstood. You have it all wrong." That's when I think maybe it's time to take a break. So I get up, walk back to the portrait, and look into those clear blue eyes, searching for answers, listening for voices that speak through the centuries to tell us their stories.

The Virginia Kelly Karnes Archives and Special Collections Research Center is a portal through which the past speaks to us, where people long gone reach out to tell us who they were, how they lived, what they thought, and what they did. Every box from the Archives is opened with anticipation of the wonderful things to be found inside.

The John Purdue death mask is in the Archives along with the china dishware he presented to celebrate a friend's wedding. Some of his letters are there, showing his struggles with grammar and the brevity of a businessman who had no time or patience for small talk. He did math on the back of used envelopes rather than waste fresh paper.

Called the founder of the University, John Purdue didn't create it. The U.S. Morrill Land-Grant Act of 1862, and a decision by the governor and the Indiana General Assembly to take part in it, created the University. Purdue was the founding benefactor. What he did was provide the additional money needed to get the University started, and in return he stipulated he would serve on the board and that it would be named for him and located where he lived. Without his gift the state's engineering and agricultural college would still exist today. But it would be part of Indiana University in Bloomington, or perhaps placed at the center of the state's government and business in Indianapolis. Or maybe it would be in Battle Ground, about eight miles outside Lafayette, and that historic town would be a very different place.

John Purdue was not a complicated person. Born in poverty, he was a self-taught businessman who worked methodically and amassed a fortune. He lived simply, found fulfillment in making money, not possessing it, and gave it away prudently where he believed it would make a difference. He never married. He had no children. He wanted to be remembered. He fathered a university.

Personal recollections of John Purdue in the Archives provide a picture of a man who, while known in his later years for being difficult, was actually kind and generous. He wasn't always "early to bed and early to rise"—although he mostly was. He was also described as a bon vivant who sometimes loved being with friends, eating large quantities of oysters, even sipping the occasional glass of aged brandy. John Purdue comes back to life in the Archives. And that's as it should be. His spirit is alive on our campus today. "The past is never dead," William Faulkner wrote in *Requiem for a Nun*. "It's not even past."

As Purdue University reached its 150th anniversary in 2019, this book was authorized to tell the stories of people and events that shaped it from nothing on flat, treeless farm fields into one of the greatest institutions of learning, discovery, and engagement in the world. In the twenty-first century Purdue faculty and alumni reach into every corner of the globe, impacting life, business, engineering, agriculture, health, science, art, theater, literature, education, government, and much, much more—building, developing, creating, leading the way into the future. It's far different from the University at its meager beginning, when thirty-nine students showed up and only thirteen of them were qualified for college studies. The rest were placed in a preparatory program where the six members of the learned faculty, including the president, in addition to teaching the college curriculum did double duty conducting classes for students who were hardly even qualified for high school.

The history of Purdue University is the story of people. They aren't flat and lifeless, one-dimensional figures staring at us from paintings and black-and-white photographs. They are people who lived and breathed, laughed and cried. They succeeded and they failed, and to understand what they did for Purdue and why requires knowing them as friends, not historical data. *Ever True* tells the stories of Purdue through its people—presidents, trustees, faculty, students, alumni, and more. It provides glimpses of how the University looked and what students and faculty experienced through the decades. What was it like to be a student at Purdue in 1874, 1917, 1945, 1969?

- Who were the Dorm Devils?
- Why was Harvey Wiley, one of the original and best members of the Purdue faculty, publicly reprimanded by the board of trustees?
- Who was Amos Heavilon?
- What university launched the state's first medical school?
- Which president said he would never spend "one damn penny" on music at Purdue?
- Why did Purdue and Indiana Universities create regional campus systems?
- Was the Chocolate Shop a speakeasy during Prohibition?

This book describes the culture of Purdue at a time when nineteenth-century couples planned clandestine meetings at the Old Pump; when male students ambushed their friends on their way to those romantic encounters and cooled them off under the waterspout, once accidentally drenching a University president. *Ever True* tells stories of fraternities serenading sororities, soldiers home from war living in rows of bunk beds crowded into campus attics, student protesters holding a "live-in" at the Memorial Union, and lots more that happened in between.

The book notes the progress of African American and female students as well as Jewish people who were subjected to discrimination and anti-Semitism but persevered not only to achieve their own goals, but to open the doors of opportunity for those who followed. Among the heroes of Purdue are people who saw wrong and righted it. Their stories are an integral part of the University today.

Ever True was a challenging project. History has been made at Purdue every day for the past 150 years. Hundreds of thousands of people have passed through. They impacted Purdue and Purdue impacted them. Their stories are fascinating and insightful, and they all deserve to be part of this narrative. But if every person who played an important role in the story of Purdue were mentioned in this book, there would be space for nothing more than a list of names. Placing 150 years of Purdue into a single book is like trying to fill a thimble with water gushing from a fire hydrant.

There is a wealth of information about Purdue and its people in other books that tell its history, many with a more narrow and detailed focus. Robert Topping wrote the last comprehensive history of Purdue, published in 1988. He also wrote books on Purdue president Frederick L. Hovde, the Purdue trustees, and popcorn king and Purdue alumnus Orville Redenbacher. Robert Kriebel wrote many books and *Journal and Courier* newspaper columns on Purdue and Tippecanoe County people and history, including the lives of John Purdue, David Ross and George Ade, John Stein, and many more. Angie Klink provides amazing insights into Purdue, and more than any other author and historian, she tells the story of women at Purdue through the changing times. Fred Whitford, Andrew G. Martin, and Phyllis Mattheis have written important books on people in Purdue agriculture. Irena McCammon Scott has written a fascinating book that focuses on her relative John Purdue, the family stories about him, and her search to learn more. University Relations Vice President Emeritus Joseph L. Bennett wrote a book about Purdue Musical Organizations. There is much more, all footnoted and included in the bibliography of *Ever True*. But with all of this, there are no books on the five most recent Purdue presidents, so chapters 29 through 44 in part 3 have a strong focus on them and their accomplishments.

On May 6, 2019, Purdue will be 150 years old. The University is celebrating its sesquicentennial from Homecoming 2018 to Homecoming 2019. Also in 2019 Purdue will celebrate the fiftieth anniversary of alumnus Neil Armstrong placing the first human footprint on the moon. Linking the two celebrations, the University's sesqui-centennial theme comprises two of Armstrong's famous words as he first stepped on the moon: *Giant Leaps*. These words have been incorporated into the title of this book. Continuing the theme, *Ever True* includes more than forty profiles of Purdue people who have made giant leaps in research performed at Purdue and in life. The profiles appear between the chapters. This is not an exhaustive list. Thousands of people from today and from the past 150 years could have been profiled. The purpose is to give readers an inkling of the amazing accomplishments of this land-grant university and the impact its people have made on our nation and world.

Many people contributed to this work. Thank you to everyone in Purdue University Archives and Special Collections, with special mention of University Archivist Sammie Morris, Barron Hilton Archivist for Flight and Space Exploration Tracy Grimm, Archivist Stephanie Schmitz, Digital Archivist Neal Harmeyer, and Archivist for University History Adriana Harmeyer. Thanks also to Kelly Lippie and all the people in the Tippecanoe County Historical Association for their assistance, including bringing out a pair of John Purdue's trousers to determine his height and weight. Purdue Black Cultural Center Director Renee Thomas also played an important role in this work, along with alumni James Bly, Leroy Keyes, and Eric McCaskill.

Thank you to Jeanne Norberg, my wife, for support, advice, suggestions, her historical knowledge of Purdue, and help with research including genealogy; David Hovde, whose knowledge of Purdue history is as rich as the history itself; Katherine Purple, Bryan Shaffer, Lindsey Organ, and Kelley Kimm, along with everyone at Purdue University Press who played an important role in this book.

Alumni Association Senior Advisor John Sautter, Dean of Students Emerita Betty M. Nelson, and Head of the Purdue Department of History Doug Hurt, along with David Hovde, Joseph L. Bennett, Fredrick Ford, and Angie Klink, provided wonderful suggestions for this book. Purdue Broadcast Media Senior Producer Raymond Cubberley made important contributions, including the quote from Neil Armstrong at the beginning of the book from one of his interviews. College of Agriculture photographer Tom Campbell and Mark Simons, senior photographer, Marketing and Media, along with several other photographers, provided assistance and great photos for the book.

Vice President for Public Affairs Julie Griffith, Executive Vice President for Communication Dan Hasler, Vice President for Development Amy Noah, Purdue Alumni Association President and CEO Ralph Amos, and Community Relations Director Mike Piggott supported this project, along with many others at Purdue, including Deputy Chief of Staff and Senior Communications Strategist Spencer Deery.

Tippecanoe County Recorder Shannon Withers and her staff went out of their way to assist me with research. My special thanks go to President Mitchell E. Daniels, Jr., for his support and participation in this project and his vision for the Giant Leaps profiles, along with his chief of staff, Gina DelSanto.

As a matter of transparency, I have personally known the last six presidents of Purdue and many other people at the University. As a reporter and columnist for the Lafayette *Journal and Courier* beginning in 1972, I enjoyed opportunities to talk with Hovde. I knew Art Hansen and I know Steve Beering very well. I was employed at Purdue in University Relations and Development from 2000 until retirement in 2013, and I worked closely with Martin Jischke and France Córdova, and briefly with Mitch Daniels. I have great respect for them all. Much of the history in part 3 of this book I witnessed firsthand.

A university is a very special place. It's where we go when we leave home and venture into the world for the first time, pursuing our dreams. It's a place where we mature, establish values and beliefs, make lifelong friendships, fail, succeed, fall in and out of love and back in love again. And most of all, it's a place where we discover how to learn and to think.

The common thread that runs through 150 years of Purdue University is the love and loyalty of its people, who from the beginning through today have always been ever grateful for the incredible opportunities they received. As the words to "Hail Purdue" proclaim, they remain—and always will be—*Ever True.*

John Norberg
2019

Part One

1869 to 1900

1

First Commencement

June 1875

We don't get on very nicely.

—JOHN PURDUE

It should have been the greatest day of his life.

It wasn't.

There was no hint of happiness on the round, clean-shaven face of John Purdue as he rose from his chair and prepared to speak at the first commencement of the University that shared his name. He was a bachelor who had never married. The nine-month-old University and its students would be his only progeny. And he knew it.

It was June 17, 1875, and Purdue had not yet reached his seventy-third birthday. Standing about five feet, seven inches tall and weighing some 240 pounds,[1] he was an imposing presence, not so much for his stature as for his prominence.

One of the wealthiest businessmen in Indiana, Purdue was dressed that day in his standard attire—a black bow tie and a white shirt. His vest, suit coat, and pants were black. Only in the warmth of midsummer did he occasionally switch to stylish white linen trousers purchased in New York City. He often carried a black cane with a white knob as a fashion statement, not as a necessity for walking. His graying hair was short for the time, parted on one side, combed back and across the other. His eyes were clear blue. He often cocked his head to one side as he spoke and listened.

Everyone in Purdue's hometown of Lafayette knew him, if not in fact, by reputation. They called him "Johnny," but not to his face. He was sometimes referred to as "Judge Purdue," or "Esquire," although he was not an officer of the court. "He was clean [honest] after his time and kind," Chase Osborn, a student from 1875, remembered. The University had "a worthy founder."[2]

There was a rainstorm the morning of commencement.[3] Weather cut into the number of people attending the ceremony on the treeless campus located on a sparsely populated bluff across the Wabash River from Lafayette, where many of the coun-

John Purdue.

ty's thirty-four thousand residents lived. But all the people who were required or expected to attend were there. The University's president, Abraham Shortridge, who had just arrived on campus nine months earlier, was present, but not scheduled to speak beyond introductions. Also attending was Indiana governor Thomas Hendricks, a former U.S. congressman, a former U.S. senator, and a future vice president of the United States. Members of the Purdue University Board of Trustees attended, along with all five members of the faculty in addition to Shortridge, who taught moral science and psychology. A small group of students and a smattering of business and civic leaders from the community attended. In addition, "quite a number of our most prominent ladies braved the storm and graced the occasion," the *Lafayette Daily Courier* reported in its evening edition.[4]

Two wooden covered bridges that connected downtown Lafayette to the west bank of the Wabash momentarily spared people traveling to commencement from the downpour. When they reached the west bank, horse hooves and buggy wheels threw mud into the air as travelers proceeded away from the river that in those days was described as "clear and translucent."[5]

There was great local pride in the growing community. The *Daily Courier* described Lafayette as a "jewel" set in "a grand amphitheater," its hills carved out by the Wabash, "a beautiful stream [that] for miles above and below Lafayette abounds in inexhaustible resources." Eight railroad lines cut through Lafayette, north, south, east, and west, allowing boot and shoe manufacturers, woolen mills, and builders of agricultural equipment to ship their goods around the state and nation. "In the manufacture of sash, blinds, doors and moldings we compete with Toledo and Chicago," the *Courier* boasted. "In cooperage [barrel making] we have the largest establishments in the state. Large quantities of pork and beef are packed here. Three large flouring mills send their products to Baltimore and the seaboard. A single distillery is said to buy more corn than is raised in the whole state of New Jersey."[6]

There were grand homes such as the Gothic Revival mansion built by John Purdue's former business partner, Moses Fowler. "Nothing seems to be wanting to make Lafayette a desirable place of residence and every element seems combined for her material prosperity," the *Courier* said. "We have the best Opera House in the west. Our public schools are splendid specimens of architecture."[7]

In addition to church-affiliated schools, there were five public schools in Lafayette, along with a segregated "colored school" at 156 Ferry Street.

Lafayette was home to twelve banks, twelve barbers, five bakers and confectioners, fifteen blacksmiths, forty retailers, twenty-one cigar and tobacco manufacturers and dealers, five dentists, twelve dressmakers, sixty-four retail grocers, eighteen meat markets, forty-four lawyers, fifteen hotels, ten restaurants, twenty-five churches, and sixty-three saloons.[8] Among the churches was an African Methodist Episcopal congregation located on Ferry Street between Eighth and Ninth. It had been established in a wood frame building in 1849. Temple Israel, at 17 South Seventh Street, served the Jewish community. It had also been formed in 1849.

There were four much disparaged "music saloons" in Lafayette where women danced on four-by-eight-foot stages while customers sat and drank at tables. Adjacent to the dancing rooms were "wine rooms" where men could meet the dancers between acts. The women were paid according to how much the patrons drank.[9]

A University rule required students to have permission before visiting Lafayette in the evening—a virtual prohibition. Saloons were off-limits, drinking alcohol forbidden.

As visitors on what later would be named State Street approached the University that first commencement day, they saw a white fence along the north side of the road running the length of the campus. The buildings were few.

A boarding house for faculty and their families faced south and was located just west of where Stone Hall would later be built. It had a 120-foot front and a depth of 68 feet. It was an Italianate structure with two beautiful cupola towers.[10]

On November 11, 1874, the *Lafayette Daily Journal* said, "This building presents a very cheerful and home-like appearance. The hall and stairways are elegantly carpeted and all the appointments are exceedingly neat and tasteful."[11] It included a kitchen, a laundry, a dining room, and an icehouse.

Boarding house for faculty and their families. Beginning in 1875, it was also used for female students. Later it was named Ladies Hall.

Other campus buildings were rather plain in design. Directly north of the boarding house was a four-story dormitory built to accommodate 120 young men. The campus was all male in the spring of 1875. Plans called for two boys to share a suite of three rooms—a study about sixteen feet square and two bedrooms about eight by twelve feet. The building was divided in two by a firewall and included eight bathrooms with zinc-lined tubs and hot water. The rooms were steam heated and lighted by gas.[12] The accommodations were better than some of the boys had at home.

Between and behind the dormitory and boarding house was a laboratory building for scientific research and teaching. Next to it the Power Plant provided steam heat, gas, and water for the campus. It also housed a tower with a bell that woke the students in the morning, signaled lamps off in the evening, and marked the start and end of classes and chapel. Military Hall, where commencement was taking place, was a one-story wooden building on the north edge of the University. A workshop that was also a horse barn and contained water closets was between the Power Plant and Military Hall. From the higher elevations of Lafayette, John Purdue could see his campus buildings.

There was only one graduate at that first commencement, John Bradford Harper, a chemistry student who had transferred from North Western Christian University in Indianapolis, soon to be renamed Butler University.

The formal program for the 1875 commencement listed only a narration by Harper, "The Search for Truth," and several songs by an Indianapolis Glee Club.[13]

Harper's talk concluded that people had historically placed boundaries on human knowledge, only to see those boundaries surpassed. "Who dares to place a limit on human intellect?" he asked. "Who dares to say when the search for truth in this world shall end?"[14]

Following those lines, Governor Hendricks came forward and presented Harper with his diploma. "This is just one step in your career," Hendricks told him, unable to resist an opportunity to lecture a youth. "If you stop here and content yourself with this achievement, your life will be a failure. You must go on."[15]

Near the end of the commencement program, Shortridge introduced John Purdue, who rose stiffly and delivered what appeared to be unprepared remarks lasting about three minutes. As was his custom, he spoke what was on his mind:

> I do not intend to make an address. . . . I merely desire to say a few words. This institution is still in its infancy. I hope that it will grow to become a man. Universities to educate the people, the youth of the people, are very necessary. It is necessary that the people be educated. I found, on looking back to the time of Moses, that education did not do much good because there was little of it. But when the printing press was established and schools, colleges, and universities sprang up, the scales fell from our eyes. And today man is clear of all those evils. Man is on a higher plane. To me the future looks cheerful.
>
> This institution has had a small beginning. My purpose is to educate. I looked over the country in different places with a view to locating a university. I finally concluded that no place needed educational advantages worse than they do just here. The state has named this child after me and the state will take care of it and cherish it.
>
> As the institution has grown, certain evils have had to be overcome. It has been organized in a hurry. The trustees and professors have been selected in a hurry and, of course, they have made some blunders. Those who have, I expect, will leave us. And even if there has been a bad set of men this year there can be good ones the next.
>
> The laws governing the university are imperfect. In most institutions the duties of the officers and trustees are laid down. It ought to be so with this institution. Rules to protect the morals of the students should also be made. The Board of Trustees will perhaps do better next year and remove all the evils that exist.
>
> We don't get on very nicely.[16]

With that, Purdue stopped speaking. Either overcome with emotion or intending to end on that sour note, he stopped talking and returned to his seat.[17]

No one with knowledge of the inner workings of the University was surprised. It was well known that Shortridge and Purdue did not get along. Purdue did not get along with some of his fellow board members. The faculty had split into factions, and the president and the faculty had disagreements with one another and with the students, who felt some of their professors were not qualified.[18] The faculty, their wives, and their children, all living in the single boarding house where they worked and ate together and even slept in close proximity, day in, day out, had taken to bickering.[19]

Things were not going well.

But much would change—and soon.

John Bradford Harper
(right) and Douglas
Graham near the Zuni Dam
*(Photo from the National
Museum of the American
Indian, Smithsonian
Institution [N26712])*

"Who dares to place a limit on human intellect," John Bradford Harper said in his 1875 commencement address when he became the first graduate of Purdue University.

Thirty-three years later in his obituary, the *Los Angeles Times* noted Harper's lifetime "monument" was a dam he had engineered and just completed in New Mexico after four years of work.[20] A college chemistry major who made his career as an engineer, he placed no limits on his intellect and abilities.

Harper was born in 1856 in Fort Madison, Iowa. After the Civil War his family moved to Indianapolis, where he enrolled at North Western University (Butler) to study chemistry. The family rented a room to Professor Harvey Wiley, who taught chemistry at North Western.

In the fall of 1874, when Wiley was hired among the first faculty at Purdue, Harper followed him to the new campus, completed his senior year, and became the sole member of its first graduating class. He was a founding member of Sigma Chi, the first fraternity at Purdue, and played baseball on campus. According to one story, a baseball thrown by Wiley hit Harper in the face and drove a cigar down his throat.[21] He survived, but mostly likely never again smoked while playing baseball.

After graduation Harper became an engineer for railroads. He worked in mining and hydraulic engineering, according to David M. Hovde, retired Purdue associate professor of library science. He served on the board of a Durango, Colorado, company that operated streetcars, became an engineer for the U.S. Department of the Interior, was named superintendent of irrigation general, and finally designed and oversaw construction of the hydraulic earth-fill Zuni Dam on a Native American reservation in New Mexico—at the time the biggest enterprise of its type in the nation.[22]

Near the end of the project, Harper fell ill with pneumonia. He never fully recovered and refused to leave his work for recuperation until it was completed. "Few men have accomplished a task of such magnitude under conditions so unfavorable," the *Los Angeles Times* said in his obituary.[23] While praised for his work, the dam was not as beneficial as hoped.

With completion of the dam, Harper moved to be with family in Los Angeles and died there six weeks later, on March 25, 1908. He was fifty-two years old.

A plaque in his honor placed at the structure he built, now known as the Black Rock Dam, remains in the twenty-first century.[24]

2

On the Cusp of Greatness: A City and a Man

1802 to 1869

Rising with the sun and toiling with a persistent and
resolute spirit, each day was a step forward.

—ELLA WALLACE, ABOUT HER FRIEND JOHN PURDUE

The dissension and displeasure John Purdue expressed at commencement in June
1875 was far removed from the emotions he had experienced just six years earlier
as he pondered the idea and details of bringing Indiana's land-grant university to his
home county.

In March 1869 when the General Assembly grappled with the issue of where
to locate the new school, Purdue was sixty-six years old and lived alone in his Lahr
House suite on Fifth Street in downtown Lafayette. Noisy, smoke-belching trains
passed daily on railroad tracks that ran in the middle of the wide street in front of
his hotel rooms.

Purdue's residence reflected his conservative, bachelor lifestyle—comfortable,
but not fancy, and filled with three hundred books. His apartment was not nearly
as grand as the nearby mansion built by his one-time protégé and former business
partner, Moses Fowler. But then, Purdue never felt the need to ostentatiously display

his prosperity. Everyone knew he was wealthy and successful. What did he have to prove at this stage of his life?

He had operated a dry goods store in Lafayette and owned commercial property. He owned farmland and marketed hogs. He had spent most of the Civil War in New York City supplying pork for Union Army troops. He was a banker and he had owned a newspaper. He had run unsuccessfully for the U.S. Congress and was president and director of the Lafayette Agricultural Works. He was involved in railroads. His hand and his money were in many of the improvements in his community. He was successful, accustomed to being in charge, and used to having his way.

Late into the evening hours during the winter months of 1869, Purdue sat alone in his Lahr House rooms, dimly lighted by gas lamps, reviewing his life, everything he had accomplished, and what would become of his hard-earned success.

Purdue was firm in his opinions and confident in himself. This above all else he believed: He had never made a bad business decision in his life. And he was on the verge of making the biggest, boldest, and most important of them all. As he considered what he would do, rather than waste clean, unused paper, he scratched numbers—adding, subtracting, and multiplying—on the back of opened envelopes as ideas and plans evolved in his head. On those cold winter nights in early 1869 he made very careful calculations.

By early March, Purdue had reached his decision. He would offer a large sum of money to create the new university with two stipulations. First, it would be named for him. Second, it would be located in Tippecanoe County. And not just anywhere in Tippecanoe County. The location would be about eight miles north of Lafayette in the little town of Battle Ground.

John Purdue in Pennsylvania and Ohio

In a sense, whatever else he was doing, Purdue's entire life had focused on education. It was a life that had started in a 360-square-foot log cabin where he lived with his parents and nine sisters, one of whom died in infancy and another as a young woman moving west on a hard trail from Pennsylvania to Ohio.

No one could have seen what lay ahead for Purdue on October 31, 1802, when his life began in Germany Valley, Pennsylvania, deep in the Appalachian Mountains near Shirleysburg, west of Harrisburg.

The United States was a young nation when Purdue was born. The Revolutionary War had ended only nineteen years earlier. In 1802 Indiana was a territory that extended from the border of Ohio west to the Mississippi, south to the Ohio River, and north to Canada. There were only about fifty-seven hundred people of European ancestry living in the Territory.

John Purdue started school at age eight but had been taught at home before that. He was bright, inquisitive, and loved to read books when he could find them. Education at the one-room schoolhouse was not free. His sisters worked to help pay the bills.[1]

Purdue's father, Charles, was an iron smelt worker and farmer. His son worked with him, and when the boy was old enough, he was hired out to work on nearby farms. His mother, Mary, kept the home, raised the children, and cooked the meals. Whatever money was made one day was needed to get through the next. The work was hard and dirty; the days were long, but the family's Dunkard Brethren religious faith kept them going. Purdue's early years in a strict, pious family influenced his later life, when he was known for honesty and discipline.[2]

Parts of John Purdue's boyhood always stayed with him, including an accent that was common in Germany Valley, which he pronounced *Chermany*. His writing and sometimes his speaking style were awkward, and in later years his political foes ridiculed him for it.[3]

If his character emerged from his Pennsylvania youth, his future had its foundation in Ohio, where most of the Purdue family relocated in the 1820s. In Ohio, Purdue struck out on his own, first as a teacher, then as a farmer, and then as a merchant building wealth that would take him farther west in search of more opportunity.

He taught in the same type of one-room schoolhouses he had attended in Germany Valley. He earned barely enough to feed himself. But Purdue would later recall his time teaching before he acquired wealth as the happiest days of his life. It was also the time when he met Moses Fowler, a farm boy who grew up near Circleville, Ohio. Fowler was thirteen years younger than Purdue.[4]

Purdue lived prudently, saved his money, and by 1831 had bought and sold farmland for a profit. He became a drover, taking hogs to market for a commission. Purdue discovered a talent, and it wasn't farming. His talent was in making money. He was a businessman, and he set out on a lifetime of buying and selling—and doing quite well at it.

Purdue Arrives in Lafayette

In 1833 at the age of thirty-one Purdue opened a general mercantile business in Adelphi, Ohio, and brought in his favorite former student, Fowler, then eighteen years old, to help. Their business prospered. They kept meticulous records of all their transactions, including December 9, 1834, when Purdue purchased 240 acres in Tippecanoe County, Indiana, at the northeast corner of what became Creasy and McCarty Lanes on the east side of Lafayette. He purchased the land for $850 without having seen it. In August 1837 Purdue wrote in a letter that he had closed his business in Ohio and was moving to Lafayette. He brought Fowler with him.[5]

Purdue's decision to relocate to Lafayette was another of the great business decisions of his life. Lafayette had only been founded in 1825, and early on it was a rough river town. In his book *A Century and Beyond: The History of Purdue University*, Robert Topping said Lafayette in the late 1830s "was a settlement of mostly saloons, a brawling river village where pigs freely rooted in the dirt streets or cooled themselves in the frequent mud wallows."[6]

Purdue saw something else in Lafayette. He saw potential. It was potential created by its location on the Wabash River at the most northerly point that could be reliably navigated by steamboats. In addition, a canal was coming through. Construction on the Wabash and Erie Canal in Indiana at Fort Wayne and Huntington was already underway when, in 1836, the Indiana General Assembly approved the Mammoth Internal Improvement Act. It was mammoth. It was also very risky. Among its many provisions was extension of the Wabash and Erie Canal from Lafayette to Terre Haute. A financial panic in 1837 led to a depression that extended into the 1840s. Indiana was unable to pay its debts and stopped work on most parts of the Mammoth Internal Improvement Act. The state spent many years recovering.[7]

But not everything about the project was a failure. The Wabash and Erie Canal reached Lafayette in the early 1840s, connecting the community with Lake Erie at Toledo, Ohio. Ships traveling through the Great Lakes could then connect with the Erie Canal at Buffalo, New York, and on to New York City. By 1848 the canal reached Terre Haute, and it was completed to Evansville by 1853. All this led to rapid economic growth in Lafayette, as John Purdue anticipated.[8] He caught the crest of the wave.

In 1851 Lafayette Wabash River docks welcomed thirty-one steamboats going back and forth from Evansville, another twenty-five from Cincinnati, and twenty from Pittsburgh. Others came from Louisville, St. Louis, and New Orleans. At its peak Lafayette exported more than forty thousand barrels of flour and pork every year in addition to one million bushels of corn, four million pounds of bacon, lard, and much more.[9] Completion of the canal to Evansville increased commerce.

There was a downside to all this as well, according to a Lafayette newspaper article published years later: "In the winter, commerce and travel were alike at a stand-still," the article said. "The boats were tied up and the crews went into winter quarters and sought diversions for the cold, idle months. They were a rough and ribald set. Every bar, it is said, had a fighting dog, a fighting rooster, and a fighting man and their brawls frequently made the days and nights hideous. Lafayette was their favorite stopping place and they gave it the reputation of being the roughest place on the Wabash."[10]

When Purdue moved to Lafayette he placed himself on the ground floor of the economic explosion that was about to take place. The river and canal made a number of smart, ambitious people such as Purdue wealthy. And even more prosperity came with the arrival of railroads in the mid-1850s. The arrival of railroads that operated all year long also led to a decline in use of the canal and the Wabash River.

Purdue and Fowler initially opened a dry goods operation similar to their store in Ohio. In their Main Street store they sold items such as coffee, tea, tobacco, and clothing. By 1840 they expanded into larger space and added more retailing and commission sales.

The partnership did not last long. Fowler left the business in 1844 and ultimately involved himself in commerce, banking, railroads, and farmland in Benton County, northwest of Lafayette. The county seat in Benton County was named for him. He is reported to have become one of the wealthiest men in the Midwest when he died in 1889.[11]

Purdue became wealthy in the dry goods business, along with his investments in farmland and other ventures including the marketing of hogs. He made frequent trips to New York City to market goods and eventually kept an office there at 391 Pearl Street in lower Manhattan. He remained in close contact with his mother and sisters in Ohio (his father had died) and frequently helped them financially.

In Lafayette he constructed what became known as the John Purdue Block, a collection of twelve connected stores that occupied the entire block on Second Street between South and Columbia. The Wabash and Erie Canal passed about a hundred feet behind the building. At the time it was said to be the largest brick masonry building outside of New York City. The date carved into the building is 1845, but it was not completed until 1847. Purdue moved his own store to the location in 1846.

He used red bricks on his block, reportedly made in his own kiln. Nearly thirty years later the first buildings that went up at Purdue University used the same type of locally produced red bricks that were common in Lafayette, and as the school grew it became known as "the redbrick campus."

Purdue launched into philanthropy and civic activities soon after he arrived in Lafayette. In their book *Purdue University: Fifty Years of Progress,* celebrating the first half-century of the institution, librarian William Murray Hepburn and history professor Louis Martin Sears wrote, "Purdue at once became a leading citizen. He took an active part in the life of the community. Few enterprises there were of private gain or civic betterment but solicited and received Purdue's support. His purse was always open."[12]

One of the men he worked with was Lazarus Maxwell Brown, who was known as "Mack." In 1856 Maxwell married Mary Wallace of Lafayette, who years later remembered Purdue as generous, hardworking, and a bon vivant who loved walnut pickles, brandied peaches, mince pie, oysters shipped in by the barrel from the East Coast, the occasional glass of fine brandy, and funny stories her husband would not repeat to her.[13]

Ella Wallace, another friend, also knew Purdue and talked about him just eleven years after his death. She said, "He kept his eye fixed on his object, never losing sight of it, though often compelled to pursue it through a crowd of distracting cares and perplexities. The card party, the wine room, the promiscuous dance never entrapped him. Rising with the sun and toiling with a persistent and resolute spirit, each day was a step forward. Instead of patronizing the horserace, or racking his brain [about] how he could make a fortune in an hour, he toiled daily and hopefully, in the line of legitimate business." He was "lucky," Wallace noted. "But it was luck which fortune always gives to those who obey her laws."[14]

Although he was not a member of a Lafayette congregation and did not attend services, Purdue frequently donated to churches that were raising money for specific projects. While he was considered generous, he was also called vain. He gave Second Presbyterian Church $1,000 for a building. He was invited to the dedication service and entered the sanctuary a bit late, just as the congregation was rising to sing the first hymn. Purdue naturally thought they were standing to honor his arrival. "Keep your seats, ladies and gentlemen," he bellowed, pounding his cane on the church aisle floor. "Don't mind me."[15]

Among Purdue's greatest philanthropic passions was education, which was lacking in Lafayette and all of Indiana. In 1840, 75 percent of Indiana children aged five to fifteen did not attend school. By 1850 among the state's more than six colleges and universities, total enrollment was 337.[16]

Purdue's wealth increased as rapidly as his town of Lafayette grew. In New York City he was known as "the King of Produce" and "Mr. Pork."[17] In 1865 Purdue's taxable earnings were reported to be $90,000—the highest in the county.[18]

But like many American entrepreneurs, his joy was in making money, not possessing it, and he found great satisfaction in giving it away. The *Lafayette Daily Courier* noticed this. In 1851, still early in his career, the newspaper commented: "In many . . . ways Mr. Purdue has manifested a commendable town pride with a view to the improvement of the city. Mr. Purdue is a bachelor with no wife or child on whom he can lavish his fortunes and affections. In lieu, therefore, we would just hint to Mr. Purdue the idea of constructing those public improvements which the town so much needs, christening them with his name, and thus they would become enduring monuments of his provident regard for his fellow citizens and they would also perpetuate his name as a model citizen."[19]

Donating his fortune for community improvements christened with his name? John Purdue read that article. A seed was planted. And in 1869 when the Indiana General Assembly was short of funds and deadlocked on where to locate its new land-grant university, the seed blossomed.

3

Bidding for a University

1869

I desire to render a testimonial to the county in which I
have spent thirty years of the ripeness of my life and also to
manifest my interest in the cause of collegiate education.

—JOHN PURDUE

Indiana had a problem. And at the opening of the General Assembly in January
1869, Republican governor Conrad Baker laid it on the line.

The state had participated in the national Morrill Land-Grant Act of 1862 that
provided funds from the sale of public land to create new universities. Indiana received more than $200,000 from the land sale and it was neatly invested and earning
interest. But, Baker estimated, it would take at least $200,000 more from the state to
get a new university up and running. And he said the state was still reeling from the
Mammoth Internal Improvement Act that lured John Purdue from Ohio to Lafayette.

"It would be unwise at this time to make such appropriations as the establishment
of a new college would involve," Baker told the General Assembly. "The public burdens
ought to be diminished rather than increased."[1]

The Morrill Land-Grant Act was named for its sponsor, U.S. representative
(and later senator) Justin Morrill from Vermont. It was approved to help states create

universities open to large numbers of people—not just the wealthy elite. The education was to be practical, focusing on agriculture and mechanics, which came to be called engineering.

There were three major goals of the legislation. The first was to increase the number of students attending college and to open higher education to "working" people. At the time only about 1 percent of the population attended college. They came mostly from wealthy families and were often preparing for the clergy. They studied philosophy and dead languages.[2] The Morrill Land-Grant Act would open the doors of higher education to people from throughout the nation, many of them living in rural areas, and the curriculum would be career-focused.

The second goal was to use these educated students to help build and advance communities, states, and the nation. An industrial revolution had changed the economy and educated people were needed to construct and run factories. The nation had grown rapidly and there was a need for engineers to build roads, bridges, and infrastructure. There were military schools to educate young men in warfare, but what about educating them in creating a better life? Morrill said: "We have schools to teach the art of man-slaying and make masters of deep throated engines of war; and shall we not have schools to teach men the way to feed, clothe and enlighten the great brotherhood of man?"[3]

The third goal was to advance agricultural production. Morrill had noticed that Europeans were applying science to agriculture with great results—a square mile of farming in Belgium supported 336 people. In Virginia a square mile fed 23 people.[4] The United States had fallen behind.

The Land-Grant Act had been approved by Congress in 1859 but vetoed by President James Buchanan. In 1862 Morrill introduced the bill again with slight changes. The Civil War had begun and Southern states that opposed the bill had seceded. The proposal was approved by Congress and signed by President Abraham Lincoln on July 2, 1862.

Among Indiana's two U.S. senators and eleven representatives, only Representative Albert S. White voted for the Morrill Land-Grant Act. He was from Stockwell in Tippecanoe County and had formerly lived in Lafayette. He had no idea how great an impact his vote would have on his home county.

Under the terms of the act, each state was entitled to thirty thousand acres of federal, public land for each of its senators and representatives in Congress. Since Indiana, like many other states, did not have that much public land within its borders, the secretary of the interior issued certificates called *scrip*. States sold the scrip at auction and purchasers then redeemed them for parcels of public land west of the Mississippi River.

While states could use up to 10 percent of the money raised from the sale to purchase land for the new universities, the rest had to be invested and only the interest made available to pay for faculty and materials. The states had to pay for buildings and infrastructure.

On March 6, 1865, Indiana formally decided to participate in the Land-Grant Act. Two years later, on April 9, 1867, the state sold scrip in various sized parcels for 390,000 acres of land and received $212,238.50. Indiana invested the money while the General Assembly debated where to locate the new school and how to pay for it. The governor appointed trustees for the university that did not yet exist.

Indiana Debt Too Large to Start a New University

In his message to the General Assembly on January 8, 1869, Baker said the investment from the land sale with interest had grown to $238,249.90. He noted that by law the state had to establish its land-grant college by 1871 or the money would go back to the federal government. Time was running out. "Hence, it is important that some definite action should be taken on the subject by the General Assembly at its present session," he said.[5]

Not at all sure that land-grant universities were necessary or that they would succeed, Baker believed there was a greater need for public elementary education than for colleges.

In previous sessions the General Assembly had appeared to favor three options. First, create a new agricultural and mechanical university. Second, attach the programs to the existing "State University," which had been founded in 1820—Indiana University (IU). Third, divide the money and curriculum equally between Indiana University and the private, church-affiliated colleges and universities throughout the state, an unpopular choice.

Among the three options, Baker favored creating an independent university but said it would cost too much and the state didn't have the funds to do it. Therefore, he supported attaching the agricultural and mechanical programs to Indiana University with the caveat that at some future date the General Assembly could break the programs off into a new institution.[6]

There was a good argument to be made for placing the land-grant college at Indiana University. IU already had buildings, land, an endowment, faculty, a board of trustees, and classes that the land-grant students would need in addition to agriculture and engineering—English, history, foreign languages, science, and more.

Near the end of the 1865 General Assembly session the Senate actually approved a bill to place the land-grant school with IU in Bloomington. But it reached the House too late in the session for consideration. With the state short of needed funds, cities throughout Indiana began bidding for the new university, offering money to help get the school started if it were located in their community.

During the legislative session of 1869, when an agreement had to be reached, bidding for the new university by communities became intense. An offer from Monroe County and Indiana University continued to receive serious consideration. So did a bid from Marion County that included North Western Christian University. The third bid receiving serious consideration came from Tippecanoe County. The Battle Ground Institute, founded by the Methodist Episcopal Church, offered its buildings, which could accommodate up to six hundred students, and about forty-eight acres of land, together valued at $100,000. Additionally, the Tippecanoe commissioners pledged $50,000 if the university were located within its borders.

The Monroe County commissioners also offered $50,000 and the Marion County commissioners (Indianapolis) offered $100,000. There were more offers, but the contest came down to Tippecanoe, Marion, and Monroe Counties. No proposal had enough support to win when all the other interests voted against it. Moreover, IU was in a dispute with the state and the city of Indianapolis over the ownership of land within the city. The land dispute cost Monroe County support in the legislature. Still, a Senate bill again supporting the Bloomington location was sent to committee on March 1, 1869, and it appeared to be gaining favor.[7]

John Purdue Makes an Offer

The clock was ticking. The session was nearing an end. The Indiana Constitution required that work had to finish on March 8. And that's when John Purdue stepped in. On March 2, 1869, he sent a message with his offer to state senator John Stein of Lafayette and awoke him at two o'clock in the morning with the news.

Purdue's offer was blunt and to the point, as always in his business dealings. He would provide $100,000 cash ($10,000 a year for ten years) for the new school provided it was located in Battle Ground and named for him. Purdue was familiar with Battle Ground. He had been a stockholder and trustee of the Battle Ground Institute in 1857. Stein announced Purdue's offer on the floor of the Senate the very afternoon he received it. He was urged to put it into writing and bring it forth for a vote as quickly as possible.

Timing is everything. And on this occasion the timing was terrible. During the final days of the session the General Assembly was scheduled to consider ratification of the Fifteenth Amendment to the U.S. Constitution. It gave African American males the right to vote.

Indiana's 1851 constitution banned slavery, but it also banned African Americans from entering the state. Indiana had approved the Thirteenth and Fourteenth Amendments to the U.S. Constitution abolishing slavery and giving citizenship to African Americans. But voting rights went too far for some, and to avoid considering the amendment, seventeen Indiana senators and thirty-seven representatives—all Democrats—resigned from the General Assembly. Lacking a quorum, there was no vote on the constitutional amendment. There was also no vote to accept John Purdue's gift, an offer made only two days earlier. The General Assembly adjourned on March 8 with much work left undone.

Had the Senate and House voted on Purdue's original proposal before the resignations, the University would have been located in Battle Ground.

Governor Baker immediately ordered a special election to be held on March 23 (almost everyone was reelected) and called the General Assembly back into special session on April 8.

With a month's break between Purdue's offer and the start of the special session, Monroe and Marion Counties had time to improve their bids, so John Purdue worked throughout that time to raise more money for Tippecanoe County. On March 24 he was present at a meeting in Battle Ground and urged the people of the community to add to the institute's offer.[8] But no additional funds were found.

An April 2 meeting in the Tippecanoe County Courthouse attended by a large number of people was a turning point. At the meeting Stein explained what had taken place during the legislative session and disclosed the 2 a.m. message from Purdue. He said the Senate was prepared to attach the new university to IU, but everyone was so impressed with Purdue's offer that a majority was ready to accept it before the resignations and adjournment. Stein said in the intervening month between legislative sessions opposing forces were gathering, Marion and Monroe Counties would continue to compete, and the Tippecanoe County bid needed more support. Purdue said the bid should total $500,000.[9]

Privately, Purdue and Stein had already agreed to changes in his offer made in March. Stein added a provision that Purdue be made a permanent trustee of the university, and Purdue dropped the requirement that the school be located in Battle Ground. The new proposal only stipulated that the university be located somewhere in Tippecanoe County. That did not include or exclude Battle Ground. But Purdue hoped it would have the impact of attracting investors from Lafayette who might want the university within or close to the city limits.

On April 7 the *Lafayette Daily Journal* reported additional donations were not forthcoming. It said Purdue was then opposed to the Battle Ground location and if someone from Tippecanoe County did not come forward with additional funds, he would open his offer to the entire state and give his donation to the highest bidder.[10]

"Now one thing is certain," the *Journal* said. "If the men of wealth and owners of property here desire the location within two miles of the corporate limits of Lafayette, let them come up to the scratch."[11]

They didn't. So John Purdue did.

On April 15 Purdue set forth his new proposal in a letter to Governor Baker. He wrote: "I desire to avail myself of the opportunity to render a testimonial to the county in which I have spent thirty years of the ripeness of my life and also to manifest my interest in the cause of collegiate education." This time Purdue offered $150,000 and stipulated that the University be named for him, that it be located in Tippecanoe County, and that he be placed on the board of trustees.[12]

Other bids kept arriving. On April 22, 1869, the governor received an offer from Jessie Meharry of Tippecanoe County. Meharry said he had "read with much pleasure" the offer from Purdue and went on to offer 320 acres of his farmland at Shawnee Mound in southern Tippecanoe County for the agricultural college. He said the land was worth $30,000 and his neighbors would pledge another $50,000 cash if the Shawnee Mound location were chosen. His only provision was that no alcoholic beverages would ever be sold on the property.

The Marion County commissioners increased their bid to $175,000. Meanwhile the House of Representatives on April 28 took a series of votes concerning what city should be home to the land-grant college. The final vote resulted in eight ballots for Indianapolis, twenty-seven for Bloomington, and fifty-two for Battle Ground.

To further put off contenders, John Purdue added one hundred acres of land to his bid. Purdue did not say he would donate his personal land or how he would acquire the one hundred acres. He just promised one hundred acres for the new university—at no cost to the state.

The people in Bloomington made one last attempt and proposed that the state accept John Purdue's offer and open the land-grant university in Tippecanoe County, but under the control of the Indiana University trustees. It was not accepted.

On May 4 the Senate approved John Purdue's offer and sent it to the House, but only after some members complained that his demand that the university be named for him was vanity.

In a speech on the Senate floor Stein defended his friend. "It strikes me as vanity worthy of all honor and imitation," he said. "His is the vanity of all the genuine philanthropists."[13]

Thanks to the convincing talk by Stein, Purdue is one of the few public land-grant universities in the nation named for an individual rather than a state.

On May 6 the Indiana House approved Purdue's offer. On the same day Baker signed Senate Bill 156 establishing May 6, 1869, as the official birth date of Purdue University.

The decision set Lafayette on a course for prosperity. On the other hand, it would be one hundred years before Indianapolis was home to a major, state university. And the decision changed the future of IU. With the creation of a separate school focused on agriculture and mechanics, those fields were excluded from the Indiana University curriculum. According to IU historian Thomas D. Clark, "its program was largely restricted to the liberal arts. Thus it was to become increasingly difficult to convince legislators and people that the University served the state in any practical way. Internally, the University administration was forced to seek support to permit the organization of professional courses which would serve nonagriculture and non-engineering constituents."[14]

After members of the General Assembly spent six years debating what county should be the home of the new agricultural school, the Purdue trustees spent the next seven months settling on an exact location.

Purdue University was still a long way from opening its doors.

4

Making of a University

1869 to 1874

His first desire was to be virtuous, his second to be wise.

—Epitaph of Richard Owen, Purdue's first president

In May 1869 the state's new land-grant university had a name. It had a board of trustees. It had John Purdue's pledge of $150,000 and his offer to find one hundred acres of land. It had money from the sale of public lands and $10,000 a year for five years from the Tippecanoe County commissioners.

But it had no buildings, no land, no administrators, no faculty, no library, no books, no classes, no academic standards, no curriculum, no departments, no staff, no rules, and most of all—no students. As for the trustees, two were lawyers, the rest were businessmen and politicians, and they had no idea how to run a university, much less how to start one from nothing.

Since John Purdue was providing the money to get this school going and he had promised to find the one hundred acres, he became the driving force in locating the University somewhere in Tippecanoe County, along with fellow trustee Henry Taylor—a Lafayette lawyer, businessman, and friend of Purdue.

Five sites in Tippecanoe County competed: the Battle Ground Institute, Shawnee Mound, an educational institute in Stockwell, an area called "the Heights" on the east side of Lafayette (later the Perrin neighborhood and Murdock Park), and a

"Second Bank" on the west side of the Wabash River directly across from downtown Lafayette.

On July 20, the Purdue trustees rejected Battle Ground, Stockwell, and Shawnee Mound and decided that the University would be located within two and one-half miles of the courthouse in Lafayette. That left the Heights and the Second Bank as the final contenders.

The state gave the trustees until January 1, 1870, to locate the school. On December 22, 1869, at a board of trustees meeting in the governor's office in Indianapolis, Purdue announced he had options to purchase one hundred acres of land on the Second Bank about a quarter mile west of the little town of Chauncey and within the required distance from the courthouse.

The land came in three parts: fifty-one and one-quarter acres from John and Catherine Opp and Nicholas and Elizabeth Marsteller for $2,750; thirty-eight and three-quarters acres from Silas and Mary Steely for $4,000; and ten acres on the far west end of the tract from Rachel and Hiram Russell for $1. The first two sales were said to be for about half the value of the land. All three sales carried a statement that the sales were in consideration of cash and the location of Purdue University.

The property was roughly what became the Purdue south campus. It ran from State Street on the north to Harrison Street on the south; from Marsteller Street on the east to a point midway between Martin Jischke Drive and what was later named Airport Road on the west. John Purdue did not pay for the land himself—at least not all of it. He raised all or most of the money from donors.

Chauncey

The town of Chauncey had been platted only nine years earlier and it would never be the same. At the time Purdue acquired the nearby land, Chauncey was home to about fifty families and 197 people. As it struggled to establish itself, it sought respectability. In 1868, the town council approved fines for prostitution, along with penalties for shooting a firearm, playing ball, and engaging in public amusements on Sundays. In 1871, the council made it illegal to bathe naked in the Wabash River during daylight hours.[1]

The location of the University beside the little town was going to be an economic boom for Chauncey. But it also brought costs, such as infrastructure, police, fire protection, and much more. In 1871, Chauncey sought help and petitioned to be annexed by the city of Lafayette. Lafayette officials were worried about costs themselves and rejected the petition. In 1888, Chauncey changed its name to West Lafayette. But

Purdue University did not use Chauncey or West Lafayette as its legal address. Its official address was Lafayette until 1972.[2]

The Trustees Struggle to Create a University

With the decision to locate the University on treeless farm fields on the west bank of the Wabash, John Purdue and other members of the board of trustees began the work of creating a university—deciding on buildings, their size and location, hiring faculty, hiring a president, and much more. Because they had no experience at this, making these decisions was difficult, and they traveled to universities as far away as the East Coast looking for ideas and suggestions.

From 1870 when this process began and continuing for the rest of his life, John Purdue was involved in endless disagreements with other trustees and with administrators of the University. More than differences of opinion on specific issues, these discussions involved different viewpoints on who was in charge. Essentially, Purdue believed he had agreed to spend $150,000 and he would build a university for the state. The state believed Purdue had given it $150,000 so that it could build the school. The state was correct, and Purdue had difficulty accepting that.

Among the first disagreements was where to place the buildings. The board began to favor a location on the east side of the one hundred–acre tract, closest to Chauncey and the city of Lafayette. Purdue argued in favor of the west side. In exchange for their property he had promised the Russells that the school would be located near their other land along the western boundary.

In his history of Purdue, *A Century and Beyond,* Robert Topping said the trustees met twenty-three times before deciding where to place the first building, and Purdue still dissented.

On August 9, 1871, the board of trustees finally broke ground. The foundation was started on land where Smith Hall would be located some forty years later. John Purdue did not attend the ceremony.

But it didn't matter where construction began because the work soon stopped, and on January 30, 1872, the trustees told John Purdue to purchase up to two hundred additional acres of land in Tippecanoe County for the University. The trustees were not specific about what land they wanted, fearful that the price would increase if word got out.

But it did get out. John Purdue was looking to purchase the land directly to the north on State Street and the price shot up. He bought eighty-four acres for $24,000 and that was credited against his $150,000 gift.

Using twenty-first-century street names, the area included in the second purchase was roughly from State Street on the south, to Grant Street and Northwestern Avenue on the east, Stadium Avenue to the north, and just east of University Street on the west.

Ultimately buildings for the new university went up north of State Street, and the one hundred acres to the south were considered a good site for the future University farm. John Purdue recycled the stone, brick, and lumber from the work that had begun on that first foundation, and the materials were used for a south campus barn and faculty residence. Buildings went up on the north tract on the west side of what became the Memorial Mall.

Electing a President

On April 23, 1872, the board of trustees hired the University's first president—William S. Clark, president of Massachusetts Agricultural College, today the University of Massachusetts Amherst.

Massachusetts Agricultural College was also a land-grant university that welcomed its first class of students in 1867. Clark had been the third president of the college. He hired its initial faculty, worked on plans for buildings, and was in office when the first students arrived. He was a perfect fit for the needs of Purdue. The board offered Clark a salary of $5,000 per year, a suitable family residence free of charge, and a $1,000 bonus as soon as he located in Lafayette and started work.

On April 26, 1872, Clark accepted the offer and wrote to the board: "You shall have formal response as soon as the Trustees of the College [in Massachusetts] can act upon my resignation."[3] His quick and positive response was not surprising. In Massachusetts, Clark was facing resistance to his dream of an agricultural college.

On May 5 the *New York Times* reported that Clark had accepted the offer from Purdue.[4] And then—"to their astonishment"—the board learned that Clark "had been induced by influences at home to decline the offer of the Presidency."[5]

Trustee John Stein publicly announced Clark's decision at a board meeting on August 13, 1872. The trustees wasted no time finding another man. The same day they held a meeting with Richard Owen and unanimously voted him president at an annual salary of $3,500. They also offered to cover his travel expenses from Bloomington, where he lived and taught at Indiana University.

At a time when whiskers were stylish, Owen was a clean-shaven man with short white hair. He was a geologist, a medical doctor, and a professor at IU. He was born in Scotland in 1810, the son of Robert Owen, who fought against industrial abuses such as child labor during twelve- to fourteen-hour workdays. In 1828 Robert Owen came to the United States and founded the utopian New Harmony community in southern Indiana. He brought three of his sons with him, including Richard. Richard went on to serve with distinction as a colonel in the Civil War, both in battle and as the commanding officer of a Confederate prisoner of war camp in Indianapolis.

For the first nine months after being named president, until May 1873, Owen was not asked to do any work. Finally the trustees told him to develop an organizational plan. They certainly needed one. By federal law, Indiana's land-grant college was supposed to have been up and running by an extended deadline of July 1, 1872. The board had missed that deadline and failure to do so meant potential forfeiture of the land-grant funds. But no action was taken against Indiana at the national level, and in 1873 Congress extended the deadline again. Purdue had until July 1, 1874, to begin classes. They were still not going to make it. The best hopes called for opening the University to residential students in the fall of 1874.

Purdue's first president, Richard Owen.

Owen gave his report to the board of trustees on August 26, 1873. It was sent to committee and copies were ordered printed before it had even been read. It proved to be a disaster and led to Owen's downfall. It was written in three parts: "The Plan of Education in the University," "Recommendations for Faculty and Employees," and "Administration of its Business Affairs."[6]

The education plan opened with the importance of "developing the students' physical and moral faculties, as well as intellectual." Owen recommended "physical training" and "moral instruction."[7]

Concerning administration, Owen went into great detail on ventilating buildings and providing a diet that was "nutritious, palatable and wholesome avoiding the use of pork [John Purdue, "Mr. Pork," could not have approved of that], meats fried in grease, rich pastries and the like." He proposed giving each student a number when he entered the school, "thus readily having a mark by which to designate places at the [dining room] table and avoid confusion or loss by having this number on clothing, napkin rings and other property." Owen proposed the manufacturer and model number for appliances in the kitchen and specified what flour to purchase.[8]

He concluded: "As far as practicable, extensive gravel walks through campus, garden, etc. should afford dry walking even after rains and plants in all directions with shrubbery and flowers, chiefly perennial, should enliven the students' home so that homesickness would be unheard of especially when to these attractive surroundings would be added the kindness which professors, employees, older students, in fact all, should be urged to extend to newcomers."[9]

The printed report was nineteen pages long. Had people of that time been able to read Owen's diary, which went into minutiae such as how many pairs of socks he sent to the laundry, they might not have been surprised at his focus on details. Had they remembered that his father attempted to establish a utopian community in New Harmony, they might not have been surprised at his interest in creating something of a utopian lifestyle at Purdue.

The board was aware of the issues Owen raised. But the report glossed over what they really wanted to know: how to admit students, what classes should be offered, how to hire the faculty. Among those who criticized Owen was Abraham Shortridge, the superintendent of Indianapolis Public Schools. In the January 1874 edition of the *Educationalist*, he said the appointment of Owen was a "mistake" and called for his resignation.

Owen insisted the report was meant only for members of the board—not to be printed and widely distributed. He said it was to provide a starting point for discussion. Instead, it became a starting point for widespread criticism.

On January 8, 1874, the *Lafayette Daily Courier* ran a story that said, "We were under the impression that Professor Owen was the unanimous choice of the Board of Trustees for President. Such was not the case. Judge Purdue was opposed to his selection [although he had voted for Owen]."[10]

At that point Owen realized his short presidency was untenable. On Wednesday, March 4, 1874, he noted in his diary: "I mailed my letter to Lafayette resigning the presidency."[11] The letter was dated March 1 and it was accepted quickly at a meeting of the trustees on March 10.

Sixteen years later, in March 1890, Owen died after drinking from a bottle of embalming fluid mistakenly given to him instead of mineral water. The epitaph on his gravestone in Marble Hill Cemetery, New Harmony, Indiana, reads, "His first desire was to be virtuous, his second to be wise."

Owen was virtuous and wise. Now the Purdue trustees were searching for a new president who would be virtuous, wise—and know something about how to open a university.

5

Presidents Come and Go

1874 to 1876

I knew something of the parts of speech, but I didn't know
whether to say "have seen" or "have saw," "I done" or "I did."

—Worth Reed, Purdue student in 1874

The Purdue University trustees were looking for a new president who was virtuous, wise, and knowledgeable, but they should have added one more trait: an
ability to get along with John Purdue. The first three qualities were much easier to
find than the fourth.

When they accepted the resignation of President Richard Owen on March 10,
1874, an academic session had already begun. Federal land-grant regulations required
that Purdue be up and running by July 1, 1874. So the University held a special session
from March through June for a limited number of students.

The first member of the faculty, John S. Hougham (pronounced *HUFF-um*), hired
in 1872, was the instructor. The courses were offered to students free of charge, but
they had to find a place to live on their own. No one was allowed to live on campus.
On Wednesday, March 4, the *Lafayette Daily Journal* reported twenty-one students
were signed up, fourteen from Lafayette and seven from Chauncey. In addition to
classes, there were also extracurricular activities. On April 20 the newspaper reported

that a Purdue University team played a game of baseball against students from Ford, a public school in Lafayette. Athletics had begun at Purdue.[1]

The first person admitted to the March–June Purdue session was Charles Howard Peirce, of Lafayette, the son of trustee Martin Peirce, who was donating his salary from the University toward the planting of trees on campus. Charles Peirce was also the first to enroll for the regular semester that started the following September, so by all accounts he was the first student at Purdue.

Eight women applied for admittance to the March–June session and were denied because of their gender. On March 12 the *Journal* reported that the women "were very disappointed to learn that for the present they cannot be admitted. The subject will be considered and probably decided upon at the next session of the board. . . . We understand the prevailing sentiment among the members of the board is to admit them, and we earnestly hope that such will be the case."[2]

It did not come up at the next board meeting, and Purdue opened in the fall of 1874 with an all-male enrollment. Indiana University had started admitting women on the same terms as men in 1867.

John Purdue was present at a board meeting on April 16, 1875, when trustee John Sutherland offered a resolution that read: "Resolved that after the end of the present scholastic year ladies be admitted as students in Purdue University on like terms as to qualifications, discipline and deportment as gentlemen."[3] The motion was approved unanimously.

Purdue's Second President

John Purdue supported the admittance of women in the 1875 vote. But the previous year he had been so unhappy with other board actions that he boycotted many of the trustee meetings. Among the meetings he did not attend was one on the afternoon of June 12, 1874, when, at the urging of Governor Thomas Hendricks, the trustees hired Abraham Shortridge as the University's second president—the same Shortridge who six months earlier had said the appointment of Owen was a mistake and called for his resignation.

Shortridge was a balding man with a long white mustache that extended far beyond the corners of his mouth. He had served for eleven years as superintendent of Indianapolis Public Schools. He championed public education for African Americans. The *Journal* praised his selection in its June 13 edition: "Professor Shortridge is an experienced educator and possesses the necessary qualifications for success in his new and arduous position, in an unusual degree. Combined with his ripe scholarship and other qualities necessary to the schoolroom, he is a gentleman of great energy, business tact and administrative talent."[4]

Shortridge was in his prime. He had not yet reached his forty-first birthday. What the newspaper failed to note was that Shortridge was in poor health and nearly blind. The governor had convinced him to accept the Purdue position against the advice of Shortridge's doctor.

Admitting Students

Purdue now had its second president. It had five buildings—a boarding house, science building, men's residence hall, power plant,

Abraham Shortridge, Purdue's second president.

and military hall—and a workshop. It was ready to open, and the board anticipated some two hundred students would show up for the entrance examination, classification, and admission on September 17 and 18, 1874.

As the University prepared to open there was no ceremony and much confusion. The evening before entrance exams were to be given, the faculty discovered no questions had been drafted. They hurried to put an exam together. Further adding to the problems, one of the professors received a telegram informing him of the death of his four-week-old child near Cincinnati, and Shortridge had to leave for Henry County in Indiana due to the death of his brother.[5]

To be admitted to the University that first year, students needed to be at least sixteen years of age. They had to pass an examination in English grammar, including spelling and composition; history of the United States; geography; elements of natural philosophy; physiology; and arithmetic, including the metric system, algebra, and plane geometry.

On opening day, only thirty-nine students showed up to take the test, most of them from Tippecanoe County. Thirteen were found prepared for college study, and they became the first Purdue University students. There was one senior, John Bradford Harper, who transferred from North Western Christian University (Butler University), no juniors, three sophomores, and nine freshmen. The remaining students were placed in a Preparatory Class also called the Academy, where faculty worked to qualify them for university studies.

Students appeared to come and go. A report from the trustees to the governor dated November 1, 1874, stated that forty-six students were in attendance at that time.[6] The official Purdue historical enrollment record for the fall of 1874 lists sixty-five students.

Among the first students in the Preparatory Class was fifteen-year-old Richard Wetherill of Lafayette. He would become a Lafayette physician who studied in Berlin and Vienna as well as in the United States. Following his death in March 1940 his will provided Purdue University with nearly half a million dollars for scholarships and what became Wetherill Laboratory of Chemistry.[7]

The earliest examples of admission examination questions date to 1884:

- Describe, briefly, the aborigines of Northern, Central and Southern North America;
- Describe briefly the time, cause, results of the nullification movement;
- Give a brief account of the annexation of Texas.
- Give an analysis of the following sentence: "Like as a father pitieth his children, so the Lord pitieth them that fear him."
- What is a modifier? How many kinds? Give an example of each?
- What parts of speech have comparison, declension, inflection?
- Math,

$$\text{Given } \begin{cases} \frac{1}{2}x + \frac{1}{3}y + \frac{1}{4}z = 62 \\ \frac{1}{3}x + \frac{1}{4}y + \frac{1}{5}z = 47 \\ \frac{1}{4}x + \frac{1}{5}y + \frac{1}{6}z = 38 \end{cases} \text{ to find } x, y \text{ and } z$$

extract the Square Root of $b^4 + 24a^3b + 10a^2b^2 - 8ab^3 + 9a^4$.[8]

The First Faculty

Six members of the faculty were listed in the 1874–1875 Purdue Annual Register: John Hougham, professor of physics; John Hussey, professor of botany and horticulture; William Morgan, professor of mathematics and engineering; Harvey Wiley, professor of chemistry; and Eli Brown, professor of English literature and drawing. Shortridge was listed in the Purdue Annual Register as teaching moral science and psychology.

When the University opened in 1874 there was no professor of agriculture at the state's agricultural university. But that soon changed. Charles Ingersoll became Purdue's first professor of agriculture in 1879, and eleven agricultural students were listed in the 1879–1880 Purdue Annual Register.

The Curriculum and Fees

Courses of study listed in the 1874–1875 Purdue Annual Register included agriculture (with no faculty) and horticulture, civil engineering, mechanical engineering, mining engineering, architecture, industrial design, natural history, and chemistry. Degrees were civil engineer, mining engineer, bachelor of chemistry and bachelor of science, and doctor of science. Liberal arts courses, including French, German, history, and English literature, were also part of the curriculum from the beginning. Although tuition was free for Indiana students, there was a matriculation fee of $10. The room rent, fuel, and light fee per term was $5, and there was a $5 janitor's fee. Board was $3.50 per week and washing was 75 cents per dozen. There were three terms per academic year.

Student Life

Moving-in day at Purdue in those early years was called "settling." The biggest job during settling was taking a bed tick—a cloth bag—to the horse barn and filling it with freshly threshed oat straw. That was the student's mattress.[9]

Worth Reed, from Covington, who graduated from Purdue in 1880, was admitted to the Purdue Academy in 1874. In an article in the *Alumnus* magazine in 1919 he remembered the campus from its early years:

> I entered Purdue about the middle of September 1874. The only preparation I had was obtained in the country school. Very few applicants, if any, were turned down. At that time I knew something of the parts of speech, but I didn't know whether to say, "have seen" or "have saw," "I done" or "I did." [I was] more apt to say "you was" than "you were."
>
> Professor Hussey was a rare soul. He could teach any subject then taught in the ordinary college. He was a fine teacher when he had good material [students] to work on and with. But alas, his material was very raw. We used to play

the age-old student trick—get him on some other subject when we were not well prepared.

Professor Wiley roomed in the Dormitory the first year or so in order to keep these wild and wooly embryonic students in order. But Wiley was a big, jolly boy himself and so we had a good time. The campus was at that time little better than a pasture field. A few small trees had been placed for future shade.

The young women came the second year and then things moved much better. The young men began wearing collars, blackened their boots and fixing up generally.

We had a good time. When the girls came and when we had become acquainted with some of the Chauncey girls we had our parties, our picnics and other festivities. We came to feel in time that we were students in a real college. That we were green we did not then wholly admit.

The people of Lafayette were tolerant and helped us in many ways. After a period of probation we became fairly acceptable in polite society.[10]

Cornfields surrounded the campus. Trees and bushes were freshly planted, and they were so small students were sometimes disciplined for jumping over or on top of them. While chapel in the upper floor of the laboratory building was mandatory, there were often empty seats. In the winter a canvas canopy was stretched over the chapel balcony floor to save heating expenses.[11]

Controversy

As the first academic year progressed into the second semester, 1875 become "turbulent," according to Purdue historian George Munro, who graduated from the University in 1897 and went on to teach engineering: "There were strained relations in the Board of Trustees, though careful writing of minutes from the meetings largely concealed that fact. The faculty was divided into factions. The president and the faculty had difficulty with the students and with each other and Mr. Purdue naturally opposed a board that had juggled him from a commanding position to one of impotence. Also, he naturally maneuvered to displace a president [Shortridge] who was chosen by the governor and not by himself."[12]

The faculty and Shortridge even argued over the University's horse and spring wagon—a buggy with two rows of seats. It was for use by anyone on the faculty, but Shortridge considered it his and took his family on frequent rides. The wife of at least one professor was verbally upset, stating she had every bit as much right to the wagon as the president.[13]

Wiley said Shortridge had differences with Professors Hussey and Hougham. There was also a major disagreement between Shortridge and the students concerning University rules and discipline. With so much dissension, the board accepted Shortridge's resignation at a meeting on December 8, 1875. Shortridge "cordially" wished "the institution a prosperous future and that it ultimately may be made to serve the purposes for which it was endowed."[14]

The resignation was effective December 31 and Shortridge returned to Indianapolis at the start of 1876. On a September day in 1906, Shortridge, who had by then been totally blind for five years, attempted to board an interurban electric tram near his home, stepped in front of the car, and was struck. He lost a leg. He died October 8, 1919, in Deaconess Hospital, Indianapolis, from congestion of the lungs.

A Third President for Purdue

On February 16, 1876, the Purdue trustees met to name another new president for the University. There were two candidates: Emerson White, who had been superintendent of schools in Ohio and president of the National Education Association, and Daniel Reed, who had served as the president of the University of Missouri since 1866. Missouri had become its state's land-grant institution in 1870 during Reed's presidency, which would end in 1876.

White was in Lafayette for meetings in December 1875 and John Purdue, along with another board member, approached him about his interest in the University presidency. He was interested. The board vote was four to one in favor of White. There is no record of who opposed him. At Purdue, White earned a reputation for seeing the big picture, setting a progressive agenda, keeping his eyes on details—and getting into a controversy that went all the way to the governor and General Assembly.

Purdue president Emerson White.

"White [was] rigorous, even austere, in his thought and kept a tight control over university activities both large and small," H. B. Knoll wrote in *The Story of Purdue Engineering.* "He insisted, in a small matter, in having a key for every lock on the campus, and if there were only one key for a case of scientific specimens, the professor in charge had to go to White to get it."[15] White had a high forehead, dark wavy hair, and a full, graying beard.

The First Female Students

White's term began shortly after the first female students had been admitted. Shortridge had a long history of support for women in education, but White did not.[16] The Purdue Annual Register for 1875 stated:

> The university opens all of its courses of study to young women as well as to young men. No distinction will be made in examinations, expenses or classes. They will be furnished dormitory privileges in the boarding house [where the faculty families lived], equal in comfort and convenience to the quarters furnished to young men in the dormitory proper. No separate courses have been provided for women. It is expected they will pursue the courses as already arranged subject to any reasonable modification. It is believe that the work as presented in horticulture, natural history and industrial design will prove attractive to them.[17]

While the University and all its courses were open to women, the Register encouraged them toward certain fields of study and, more directly, away from engineering and agriculture.

Eight women enrolled in the Academy in the fall of 1875, but none qualified for the University. Two in the senior preparatory class were Lora Rosser, sixteen, of Battle Ground, and Hattie Taylor, seventeen, of Lafayette.

The first female faculty member was Sarah Oren. She was first employed in July 1875 "as female teacher of the university" with a salary of $1,000 per year. Shortridge soon changed her title to "assistant professor of mathematics." By August the board had changed her title to professor of botany and raised her salary to $1,500. The male members of the faculty each received $2,000 with the exception of Brown (professor of English literature and drawing), who, like Oren, received $1,500.

According to the Register, by the fall of 1876 the first six women had been admitted to university studies as opposed to the preparatory school. Hattie Taylor was a freshman along with Florence Taylor, and twins Mable and Maude Miller.

Their sister, Eulora Miller, was a sophomore, and Hattie Mercie Brown was a junior. Twenty-eight women were in the preparatory school.

Eulora Miller became the first woman to graduate from Purdue, in 1878. In February 1920 in a *Purdue Alumnus* article, Wiley called Miller "one of the most brilliant and charming girls I have ever known."[18]

Near what later became State Street, about fifty yards east of the Boarding House, was a well dating back to the farm that had previously occupied the land before it was sold to the University. Female students lived in the Boarding House. A walk was always maintained between the building and the well, protected by shrubs that screened it from public view. It became a lover's lane and a place where male and female students met in the evening by "chance." Female students had strict "hours" when they were required to be in the Boarding House and males were prohibited from visiting. But the young women were allowed out to the well even after "hours," and if a man happened to be there when she arrived—well, stretching the rules has always been part of student life.

Derexa Morey Errant enrolled at Purdue and was a member of the class of 1879. Along with many other students, she lived in Lafayette. Years later she remembered her experiences, especially traveling to and from campus:

> In fair weather the walk morning and evening was both a benefit and pleasure. In wet weather it was another matter. At first there were no walks and no protection after leaving the bridge [over the Wabash] and the mud on the levee and at the foot of the hill was often ankle deep. Later a plank was put down, starting from the bridge, but extending only about half way along the levee. Some of the hardy ones walked in all seasons. Dr. Wiley was one of them. Most of us took to vehicles of some kind when bad weather set in. My first winter I was one of a group who hired an old-fashioned omnibus to carry us back and forth. These trips together were one of the joys of our life and never to be forgotten.[19]

On June 16, 1876, Purdue held its second commencement. White used the occasion to deliver a lengthy address in which he said the University must lead in the fields of engineering and agriculture and never follow. He reorganized the curriculum.

Three months later he would give a very different speech. It was a somber speech that would mark the end of one era of the University and the beginning of another.

Eulora Miller Jennings

Eulora Miller was not yet sixteen years old in 1875 when she began daily travel from her Lafayette home to Purdue University, over the Wabash River, and through the west bank levee that "was so rich in a coating of mud that it was about like a pancake batter."[20]

Three years later she became the first woman to graduate from Purdue.

Miller said the arrival of women on campus met with some teasing and gender competition.[21]

"One time I was the only one in the class able to answer one of Professor Thompson's questions," Miller remembered in a 1924 interview. "He wrote on the board in Latin, 'a woman the leader.' This aroused the boys considerably, but helped establish the girls' work."[22]

Miller also remembered a surveying course where women had to tromp through farm fields and climb fences wearing train dresses and hoop skirts.

She entered Purdue with her twin sisters, Mable and Maude, who did not graduate. Their parents were John and Amanda Miller, early Lafayette settlers from Pennsylvania. John Miller was an attorney and from 1869 to 1877 Lafayette postmaster.[23] He believed in education. His son, Melville, graduated from Asbury College (DePauw University) and served as U.S. assistant secretary of the interior under President Theodore Roosevelt.[24]

A fourth daughter, Katherine, married Oliver Peirce, grandson of Martin Peirce, an early Lafayette settler, wealthy businessman, Purdue trustee, and friend of John Purdue. Oliver and Katherine Peirce had a daughter who married Burr Swezey Sr., a 1913 Purdue graduate and founder of Lafayette National Bank in 1934.

After graduation, Eulora Miller became librarian at Purdue, did graduate work at the University and Columbia University, and wrote four short plays for performances at women's club meetings.

In 1890 she married Rufus Jennings, of Milwaukee. They moving to Berkeley, California, where he was a successful businessman and promoter. Following the 1906 San Francisco earthquake the mayor named him secretary of the Committee of Twelve charged with city rehabilitation.[25] She taught metalworking at the California College of Arts and Crafts.[26]

Eulora Jennings served the First Church of Christ, Scientist, in Berkeley for forty years. She was head of the church's building plan committee and largely responsible for selecting renowned architect Bernard Maybeck, who designed the building in 1910.[27] She fabricated large bowl reflector lamps to hang over the center aisle.

The church is considered an American masterpiece.[28] It continues in use and has been designated a National Historic Landmark.

Chase Osborn

In 1860 Chase Osborn was born in a one-room Indiana log cabin. He died eighty-nine years later in an "old fashioned cracker farm house"[29] in Possum Poke, Georgia.

In between he was governor of Michigan and regent for the University of Michigan. He amassed a fortune in timber and iron ore, giving most of the money away to schools, charities, and communities. "He could have lived in a palace," people said. "But he preferred a cabin."[30] He loved nature, sometimes slept in a lean-to in Michigan's Upper Peninsula, and addressed God in daily prayer as "father and mother of earth."[31]

An April 8, 1919, edition of the *Saturday Evening Post* even credited him with discovering the source of light for the firefly.[32]

Osborn was a boy when his family moved from the log cabin in northeast Indiana to a three-room house in Lafayette. The seventh of ten children, he mostly fended for himself, selling rags, scrap iron, and newspapers on city streets.[33]

He described his boyhood in Lafayette as running away from school, fighting, and swimming in the Wabash River.[34]

Fourteen years old when Purdue opened in September of 1874, he managed to enroll in the lowest level of the preparatory school and remained on campus until 1877, completing his freshman year.

"There was no organization, no classification, no standards, no anything at the beginning," he said of those early days on campus. "I seldom wore socks and never any underclothes because I didn't have any to wear."[35]

Leaving Purdue, he set out on a lifetime that included work as a newsboy, a typesetter, a newspaper publisher, an author, a millhand, a bellhop, and a coal wagon driver. As governor of Michigan from 1911 to 1912, he authored one of the nation's first workers' compensation laws and pulled the state out of deep debt. He was friend to nine U.S. presidents.[36]

While he spent his winters in Georgia, he summered on Duck Island, Sault Ste. Marie, Michigan, in a log cabin where his great-grandson still remembers visiting. "He would give us brand new dollar bills, and that was a lot of money to us," George "Nick" Pratt said. "He would ask us questions and listen carefully to the answers."[37]

Osborn always valued his years at Purdue.

"I love the place," he wrote in 1919. "I love every Purdue woman. I love every Purdue man. Sometimes [in my imagination] I steal out of the early shadows and into the warmth and glory of Purdue's noon-day sun."[38]

6

His Spirit Departs

1876

He who writes upon the tablet of the human soul
does that which no time can efface.

—Purdue president Emerson White

September 12, 1876, was a particularly beautiful late summer day. The August humidity had disappeared and there was a hint of autumn chill in the morning air. It marked the opening day of classes at Purdue and the beginning of the University's third year.

John Purdue was fast approaching his seventy-fourth birthday. He was feeling quite well. Health concerns over the past several months, including chills and vertigo, had sent him from the downtown Lahr House to the Lafayette Hygienic Institute at 360 South Seventeenth Street. The institute offered Purdue rest, a good diet, and hot baths. The treatment seemed to be working.

On the morning of September 12 Purdue sent word to his nephew, Thomas Park, who lived in Lafayette, that he wanted to take a ride. Park picked him up at the institute in a horse-drawn black buggy and drove him to Purdue's Agricultural Works implement factory. Purdue inspected the factory and next visited his university across a covered bridge over the Wabash River.

On campus, Purdue found faculty and students hurrying to start their day and excited about the start of a new academic year. But they took time to stop and chat with the founding benefactor of their school.

Purdue walked to a site at the center of campus and inspected the foundation of University Hall, a $30,550 building that was just getting started. When it opened the next fall it would be the centerpiece of campus, the heart of the University. Construction had been stalled for years by disagreements concerning its location and size, but now all was going well.

Purdue still had some energy left. He set out on a stroll through campus, along paths that carried students and faculty among the five completed buildings plus the workshop. The day was passing and it was time for him to move on. Back in his buggy, Park delivered Purdue to his rooms at the Lahr House.

Two days later Purdue returned to his university across the same covered bridge. This time he was carried in a black hearse drawn by horses, followed by a military band, members of the Lafayette Guards, and a long line of carriages carrying many of the citizens and all of the prominent people of his town. There were 105 carriages in the funeral procession. Most of the stores in town were closed.

John Purdue's Declining Physical Health

As classes began in September 1876, Purdue had been showing signs of aging. His business interests, which he had handled so carefully throughout his life, were in disarray. He had been having health problems for at least four years.

On May 15, 1872, the *Lafayette Daily Journal* noted: "The Hon. John Purdue took his departure the day before yesterday morning for the West Baden Springs, in Orange County, this state. Many friends join him in the earnest hope that he may find a speedy restoration from his present shattered state of health."[1] He was there for two weeks. On March 4, 1873, the *Journal* reported: "Hon. John Purdue is around attending to business as bright as a new dollar, and none the worse apparently for his severe fall on the Opera House steps the other night."[2] In the winter of 1876 he had another fall, this time in front of the Lahr House.

On May 6, 1876, shortly before he attended Purdue's second commencement, the *Courier* said: "Judge Purdue has removed bag and baggage from the Lahr House to the Hygienic Institute where under the skillful treatment of the doctor in charge he hopes to rebuild his shattered health."[3] He recovered and during the summer kept a busy schedule, making trips to New York, Boston, and Cleveland.

In early September 1876 he visited Indianapolis, where he again slipped and fell down long iron steps at the Indianapolis Savings Bank, escaping serious injury but no doubt badly bruised and sore. On September 11 he was taken back to the Hygienic Institute, and the *Journal* reported that he was "quite ill."[4]

But on the September 12, the day he visited campus, "he was feeling quite well and shaking off the burdens of his cares," according to the *Courier*.[5] When he returned to the Lahr House after his visit to campus, "he sat down in the public room for awhile and, complaining of chilling sensations, was induced by Mr. O. K. Weakly (proprietor of the Lahr) who had watched over him and waited on him tenderly for years, to lie down for a rest."[6]

He was taken to his rooms in the hotel where a log was burned in his fireplace. He appeared comfortable and asked for oyster stew. He ate it all and appeared ready for a nap after this hearty, favorite meal. When Weakly returned a half hour later, Purdue was nowhere to be found. Purdue had sent for Park, who worked at the *Journal*.

Park drove Purdue to the Hygienic Institute and returned to his job, which indicated there was nothing that appeared critical about Purdue's condition. In fact, at the Institute, Purdue had tea, talked with people in a common area, and walked around the garden before going to his room. A short time later, two attendants arrived to check on him. It was 5 p.m. They entered and found Purdue lying face down on the floor near the door, as if he were trying to reach it to summon help. The attendants rolled him on his back. The Hon. Judge John Purdue gasped, took his last breath, and died.

Declining Mental Health

"A horse was mounted in hot haste and a messenger dispatched to the city," the *Courier* reported. "Weakly, John Sample, and other cherished friends were brought quickly and all that medical aid could do was on hand. But his spirit had departed. . . . No citizen of Lafayette was so widely known. From his early years until his death he had been prominently and honorably identified with the best interests of the city. He cherished an honest pride in Lafayette. . . . But his enduring monument is Purdue University."[7]

Since Purdue was known for his abilities in business, people were shocked when they discovered he died with no will. In the University's Second Report to the Governor, White said Purdue still owed $65,000 of his $150,000 donation to the University, and he had secured that through a Warren County farm named Walnut Grove. It had been his favorite place to visit.

In the days, weeks, and months following his death, Purdue's finances were found to be in disarray. His declining mental as well as physical health was discussed in the newspapers.

On September 15, the *Courier* ran a story under the headline "Judge Purdue's Mental Condition." It read:

> The other day, and in perfect propriety, was alluded to the mental depression of Mr. Purdue growing out of the financial complications. It was a matter of common notoriety and has been on everyone's lips for months. Mr. Purdue's mental condition has been for some time a subject of serious solicitude to his friends—with all his life long and abundant caution he has become personally liable for an immense sum of money. When some weeks since he was asked how much he had become liable for he answered with the simplicity of a child. "I don't know. I signed everything they brought to me."[8]

The article said that the day before his death he had argued with the driver of his buggy about the location of his own business, the Agricultural Works. Purdue insisted on being taken to an Eleventh Street location. The driver took him where he requested. But the Agricultural Works was not there.

During the last six years of his life, Purdue's temperament had changed. He had become increasingly more argumentative and difficult. He had several serious falls on steps. He was confused about the location of his own business in his hometown. He was making very poor financial decisions. His handwriting had deteriorated considerably. Today doctors might determine he had dementia, but no such diagnosis was made then. His cause of death was listed as "apoplexy," a stroke.

The morning after his death a group of prominent citizens, including Moses Fowler, who had spoken very little with Purdue since their business partnership ended thirty-two years earlier, met and made decisions about the funeral. Purdue president Emerson White joined the meeting in progress. They decided to go as a group to the Hygienic Institute at 3 p.m. to take the body to the home of James Spears, at the southeast corner of Fifth and South Streets. An honor guard of students watched over the remains through the night, and a military detail from the Lafayette Guards also stood by. A large photograph of Purdue was placed in the room along with black mourning drapes. Tuberoses forming a cross were place at the head of the casket on an alabaster column. Lying on the caskets were symbols of the cross, an anchor, and a heart that represented faith, hope, and charity. A religious service led by three Lafayette Christian ministers, which included the singing of hymns, was planned at the Spears home for 2 p.m. on September 14. There would then be a procession to the University, where Purdue had said he wanted to be buried. Three of his sisters and a number of nieces and nephews came for the funeral. The religious service at the Spears home was canceled. For unknown reasons the body was not embalmed, and at the appointed time it was decided to proceed immediately to burial. Christian clergy took part in the graveside ceremony.

A committee chose the burial spot directly east of University Hall. They thought that in time they would put a monument at his grave, but no one ever did. It was decided that the University was his monument and nothing else was needed.

President Emerson White delivered the eulogy.

> The life of the deceased was not devoted to amassing wealth for selfish ends. He used his wealth in a large-hearted manner and churches and schools and every other agency for the good of the community received his support. . . . In the last few years of Mr. Purdue's long life there have been conflicts and I cannot be misunderstood when I say that whatever has disturbed the serenity of his old age has been a result of a breaking down of his mental powers. If in the future it should turn out that his later business ventures were not judicious, it will be attributed to the infirmities of age. His errors in business were, in my judgment, largely due to the failure of the brain to do its part. In the five months of my acquaintance with him I have wondered that he attempted to do business, that he did not throw off all business cares and live in peace. But his early habits of active industry urged him forward in every duty. It is believed that Judge Purdue's gift to assist in founding the University that bears his name was the chief joy of his declining years and the one satisfaction that kept him serene in the last struggles of life.[9]

White concluded:

> The gift of our friend of $150,000 to aid in the establishment of this institution will make the name of John Purdue live just as long as learning lives and people keep their civilization. . . . This institution will be a perpetual monument to his memory. . . . While it was the job of his life, these buildings will crumble to dust, the very ground on which we stand may be a wilderness and the owls hoot in the branches above this place, but every truth and good impulse given here will never die, but live forever in the hearts of the students here instructed. . . . We may engrave upon brass and rear temples, but they will crumble into dust.

But he who writes upon the tablet of the human soul does that which no time can efface—which will grow brighter throughout the age of eternity.[10]

In the years to come, many people from the University would make contributions to life and society that would make the name of Purdue grow brighter still.

And the first among them was Harvey Wiley.

7

Old Borax

1880 to 1906

Imagine my feelings and those of other members of the
board on seeing one [of] our professors dressed up like a
monkey astride a cartwheel riding along our streets.

—Purdue trustee John Colcord Dobelbower

They were embarrassed. They considered it undignified. And the gentlemen on the Purdue University Board of Trustees simply would not stand for it.

So Professor Harvey Wiley—one of the first people hired by Purdue, one of the most respected and loved by students and faculty alike, and the institution's most successful researcher—was called to a meeting of the board.

He had no idea what was coming. He actually expected a commendation, maybe even a raise. Instead, he became the first member of the Purdue faculty to be publicly reprimanded by the trustees.

By his own account, in 1880 Wiley, a medical doctor and Purdue's professor of chemistry, had "scandalized the faculty and community"[1] when he purchased a nickel-plated Harvard roadster bicycle. Bicycling had become popular in the United States and Europe in the 1880s, and a Bicycle Club of Lafayette had been organized. But it's safe to say the Purdue trustees were not members.

Wiley's bicycle had a huge front wheel and a much smaller one in back. It was very difficult to ride because the passenger sat so high. The front wheel diameter was fifty-two inches, and the back wheel eighteen inches. The seat was fifty-three inches above the ground.[2] He often rode wearing knickers.

It's not surprising that Wiley would be interested in bicycling. He enjoyed sports and had helped to organize Purdue's first baseball team. He even played baseball with the students, which some considered undignified for a member of the faculty. He lived in Lafayette and walked daily to campus, so another, faster means of transportation interested him. And he liked to try things that were new and different. In the early twentieth century, while living in Washington, D.C., he would claim he owned the third automobile ever taken out on the capital streets. And, he said, he was the first person in Washington to have an accident driving a car.

Wiley described the experience of pedaling his bike as like "riding through thin air. Striking a stone or stump meant disaster, for an undignified fall was the result. I acquired a bicycle uniform with knee breeches and I rode daily through the streets of Lafayette, over the bridge across the Wabash and up to the university, frightening horses, attracting attention and grieving the hearts of the staid president and professors as well as members of the Board of Trustees."[3]

In *The Story of Purdue Engineering*, Professor H. B. Knoll said Wiley learned how to ride the bicycle by practicing in the center of campus, in front of University Hall and John Purdue's grave. He picked this spot because he needed the help of students to get him on, get him started, stop him, and get him off until he could figure out how to do it alone. Sometimes, sensing danger, he gave up and threw himself into the shrubbery that had been so carefully planted to beautify the campus.

It wasn't long before the board members reached the limit of their patience. That's when Wiley received a notice to appear at their next board meeting.

"When I entered the room I was impressed with the density of the silence and the straight faces of the mourners," Wiley remembered. "For some time nothing was said. I sat in mute expectancy."[4] Finally trustee John Colcord Dobelbower told Wiley why he had been summoned. Dobelbower, a Lafayette newspaper editor and publisher, was bearded and graying, with a stern look that came naturally to him. Seated at the board table, he did not address Wiley directly.

The disagreeable duty has been assigned to me to tell Professor Wiley the cause of his appearance before us. We have been greatly pleased with the excellence of his instruction and are pleased with the popularity he enjoys among his pupils. We are deeply grieved, however, at his conduct. He has put on a uniform and played baseball with the boys, much to the discredit of the dignity of a professor.

Harvey Wiley.

But the most grave offense of all has lately come to our attention. Professor Wiley has bought a bicycle. Imagine my feelings and those of other members of the board on seeing one [of] our professors dressed up like a monkey astride a cartwheel riding along our streets. Imagine my feelings when some astonished observer says to me, "Who is that?" And I am compelled to say, "He is a professor in our university." It is with the greatest pain that I feel it my duty to make these statements in his presence and before the board.[5]

At this point, giving up all hope for a raise in pay, Wiley asked for pen and paper. He resigned from the University on the spot and left the room. It's not known if he arrived and left on his bicycle. The next day he received a letter from the board: "Dear Sir, The board of Trustees unanimously refuses to accept your resignation. Respectfully, John A. Stein, Secretary."

In the years that followed, Wiley never mentioned if he continued riding his bicycle after the reprimand or if he continued playing baseball and wearing knickers. He probably did. But it didn't matter. He would not stay much longer at Purdue before heading to Washington, D.C., where he became the "father" of the U.S. Pure Food and Drug Act, passed by Congress in 1906. It was one of the most important pieces of legislation in U.S. history. His research that led up to the passage of the act earned him the nickname "Old Borax."

Like many prominent men of his time, Wiley was born in a two-room, dirt floor log cabin on October 18, 1844, near Kent in Jefferson County, Indiana. His father was a farmer, a carpenter, and a plasterer, and for fifty years served as an elder in his church. His theology was stern and he abstained from all social events. He did not celebrate Christmas, and while it's clear from his writings that Wiley loved and respected his father, they were quite different.

In 1867, after serving in the Union Army during the Civil War, Wiley received a bachelor's degree from Hanover College in Indiana, along the Ohio River. He earned his master's degree there in 1870 and a medical degree in 1871 from the Medical College of Indiana in Indianapolis. In 1873 he received a BS degree from Harvard University after only several months of study.

Thirty years old when Purdue opened in 1874, Wiley was the youngest member of the first faculty, so he was chosen to live and keep order and decorum in the male students' boarding house. But as the students said, Wiley claimed to be "blind and deaf" to their antics.[6]

He said in the earliest days that many of the Purdue students were not well behaved and the courses were crude. "Most of the boys who came from the country to join the Purdue Agricultural School were not even well enough trained to enter high school," Wiley said. "In addition to this, the bad boys of the city who were expelled from the high schools sought a more congenial environment at Purdue. We had, for that reason, a number of boys from the city who modestly [were] described as being 'tough.'"[7]

Wiley was popular with everyone, including Governor Thomas Hendricks, who was a member of the board of trustees. In 1875, when it was clear to everyone that Abraham Shortridge would have to leave, Wiley was called to a meeting with Hendricks.

According to Wiley, the governor said: "The only professor who has given entire satisfaction during the past year is yourself. If you were a little older we would make you president of the school today."[8]

In addition to his dorm responsibilities, military training was assigned to Wiley. He was a corporal during the Civil War and served only one hundred days—not an extensive military background. But he accepted his Purdue assignment with typical enthusiasm and took the students on their first camping experience. "We marched to a point on the Wabash from eight to ten miles northeast of Lafayette and pitched our camp," Wiley wrote in his autobiography. "All the strict rules of camp life were enforced, including the guard house. There were some evidences of depredations in the nearby chicken coops and I found one morning a very attractive dead rooster under my pillow. In addition to our military exercises we indulged freely in bathing in the Wabash."[9] He took the same easygoing attitude to military training that he did to policing the dormitory.

In the summer of 1876 Wiley went to the U.S. centennial celebration of the Declaration of Independence in Philadelphia. He spent six weeks there studying exhibits, among them two Gramme dynamos—electric generators that produced direct current. He purchased the smaller of the two and had it sent back to Purdue. When it arrived, he made a lamp and put it in the tower of a campus building. He claimed it was the first electric light produced by a dynamo west of the Allegheny Mountains.

Early commencements at Purdue continued to be interesting. Wiley recalled graduation day in 1877 when newly elected governor James "Blue Jeans" Williams spoke.

Purdue Hall, men's residence hall, an original campus building.

Williams, a southern Indiana farmer and a Democrat, was plainspoken. His nickname came from the denim clothes he always wore, including suits sewn by his wife.

Wiley wrote of that 1877 commencement:

I shall never forget the impression he made on me and on others. He was dressed in his traditional blue jeans suit, which evidently had not been cut by a tailor. As he sat on the platform in the place of honor at the right of the president of the university and looked out over the vast audience in the gymnasium, there was no sign of expression . . . in his countenance. Finally the time came when it was customary for the governor of the state to make his address. Governor Williams was a very tall man, said to resemble in his physical contour Abraham Lincoln. He got up apparently by sections and without any unique haste. He stood for a moment or two perfectly silent, looking over the audience but betraying no emotion. Finally he began his address. I shall never forget the first sentence: "Edicate a boy, and he won't work." This was his theme and he launched forth into a bitter denunciation of higher education for a boy, affirming that it would unfit him for any useful vocation.[10]

Williams was typical of rural people in Indiana and elsewhere in the mid-1870s who did not understand why they should send their sons to college to learn how to farm and do mechanical work. But in time the governor became a strong supporter of Purdue, and two of his grandsons attended the University.

In 1878 Wiley initiated the first significant research at Purdue. He examined sugars and syrup being sold in Indiana and found that glucose was being used to adulterate honey, syrup, and molasses. He was not opposed to glucose. But he was opposed to the adulteration of products and their mislabeling.

President White, Success, and Controversy

While Wiley was well on his way to success, Purdue president White also had many accomplishments. He recruited W. F. M. Goss to Purdue from MIT. In 1879 Goss founded the Purdue School of Mechanics, and in 1882 he developed a full course of studies in mechanical engineering. In 1900 Goss was named dean of engineering, a position he held until leaving the University in 1907. In *A Century and Beyond,* Topping said Goss "shaped Purdue undergraduate engineering from literally nothing in an unimproved basement in old Science Hall into a national educational treasure."[11]

In addition to Goss, in 1879 White hired Charles Ingersoll—Purdue's first professor of agriculture. Ingersoll left in 1882, the same year Purdue awarded its first bachelor's degree in agriculture. He was replaced by Professor William C. Latta. Frederick Whitford and Andrew Martin wrote in a biography of Latta that "until his death in 1935 he helped shape what eventually would become a major education institution known worldwide for its diverse student body, quality education and wide assortment of agricultural majors and specializations.

William C. Latta, director of the Agricultural Experiment Station, with Purdue's first female trustee, Virginia Meredith.

He would live to see how his contributions helped to dramatically change the way Indiana's farming community viewed a Purdue education in agriculture."[12] During his tenure Latta hired additional faculty and expanded the agriculture curriculum from three to four years.

In 1881 White recruited Annie Peck to Purdue. She taught Latin and was the matron for the female students. Male and female students alike regarded her as among the best of the Purdue faculty. She also had an unusual side interest for a woman of her day—she climbed mountains. At the age of eighty-four she listed "mountain climber" as her occupation in *Who's Who in America*.[13] When Peck resigned from Purdue in 1883, White eliminated Latin from the curriculum, believing it did not fit his vision of a university focused on agriculture, engineering, and science.

But with all his successes, like Owen and Shortridge, White's term as president was marked by disagreements. His downfall was a fraternity.

Sigma Chi became the first Greek letter society at Purdue in 1875. Shortridge did not approve of fraternities, but enrollment at Purdue was weak and he did not think he was in a position to fight against them. White took a harder approach. In 1877 he instituted a rule prohibiting new students from joining Greek fraternities. He believed fraternities were immoral, unsuited for the industrial classes doing technical studies, and interfered with academics.

Without the ability to recruit new members, Sigma Chi continued on the Purdue campus but underground, not disclosing its membership or activities and initiating new members when classes were not in session. Over time several members of the fraternity were expelled or suspended from the University, and the issue of whether or not a fraternity could exist on the Purdue campus went to Tippecanoe Circuit Court.

In 1881 Judge D. P. Vinton ruled for the University. He didn't know that two of his sons were Sigma Chis. The fraternity lawyers appealed to the Indiana Supreme Court, which ruled for Sigma Chi, allowing it on campus.

White countered with a rule that said fraternity members could not receive honors or take part in other activities. In 1883 the issue hit the Indiana General Assembly, where the governor and lieutenant governor were both Sigma Chis and many legislators supported Greek societies. In the General Assembly, a rider was attached to the two-year appropriation bill funding Purdue, stating that the University could not receive its funds unless restrictions against fraternities were rescinded. Purdue backed off its hard stance against fraternities, but it was too late. The session ended without an appropriation bill being approved, and there was no state money for Purdue during the next two years.

At a board of trustees meeting on March 19, White resigned, believing that he had become the focus of the controversy. His resignation was accepted by a 4–2 vote of the board. White stayed through the summer of 1883 during the transition to a new president. He then moved to Cincinnati, Ohio, and remained active in educational issues. He also visited Purdue from time to time, where all was forgotten and forgiven, and he was very well received. His last visit was in 1894. He died in Columbus, Ohio, on October 21, 1902.[14]

Wiley never forgot the Sigma Chi controversy:

> Why a Greek letter fraternity wanted to have a society made up of boys who could not tell the Greek alphabet from a roving electron, I do not know. But the boys took to it like a duck to water, among them my most promising pupils. I was opposed to the secret societies but I believed in the boys that belonged to them. I did not believe in excluding a boy from the institution because he joined a society that I did not like. [But] from my point of view, I would not allow anyone to join a [Greek] letter society who could not repeat the *Iliad* in the original from memory.[15]

A Fourth President

With White's resignation, some of the students wanted the new president to be Wiley. According to Wiley, they campaigned to have him elected against his wishes. He said, "I discouraged in every possible way their efforts, but their petitions were circulated extensively and taken to members of the legislature, then in session, for the signatures of many of these gentlemen."[16]

In *Health of a Nation*, Wiley biographer Oscar Anderson told a different story: "Since late February [1883] Wiley had been at work to mobilize support behind his candidacy [for the Purdue presidency]. His campaign gathered momentum and by the middle of April he was confident that he had the backing of the legislature, the board of agriculture, the agricultural press, the students, and graduates, three-fourths of the faculty and nine-tenths of the people."[17]

The board met on June 7, 1883, to elect a new president. Before voting on the new leader, the board unanimously accepted Wiley's second resignation from Purdue. This time he was definitely leaving. He had already taken the oath of office to become chief of the Division of Chemistry in the U.S. Department of Agriculture.

The two main candidates for the fourth Purdue presidency in nine years were John C. Ridpath and James Smart. Ridpath was vice president of Indiana Asbury University in Greencastle, Indiana—soon to be renamed DePauw University. Smart, a native of New Hampshire, had been Indiana state superintendent of public instruction, president of the National Education Association, and a member of the National Association of School Superintendents. In addition to serving as a principal in Toledo, Ohio, he had been superintendent of Fort Wayne, Indiana, schools and served on the State Board of Education, so he was well known in Indiana. While he had not attended college, he was given honorary degrees from Dartmouth College and Indiana University.

When the board met to elect the new president, Smart won on the third ballet by a vote of 4–2. It was the second (with White) and last time a Purdue president was elected without a unanimous vote. The minutes did not record who cast the negative votes, but they might have been the same two trustees who opposed White's resignation.

U.S. Pure Food and Drug Act

Wiley's future was in Washington, D.C. There he began a lengthy crusade to change the food industry. Before the Civil War most people grew food for their own consumption. But the industrialization and concentration of people in urban areas that followed the war changed the nation's food supply. Preservation techniques were needed and the least expensive methods were chosen. The result was adulteration of food products that were not properly labeled. At the same time, "medicines" were being advertised in newspapers and other periodicals promising incredible cures. In truth, the "medicine" was often nothing more than alcohol. Some of them, including some marketed for children and infants, contained cocaine and other drugs. For example, Mrs. Winslow's Soothing Syrup, "mother's friend" for teething children, contained morphine. There was nothing on the labels to tell consumers what they were actually using.

Wiley not only had to convince manufacturers to change their ways and label their products correctly, but he also had to convince consumers about the dangers of the products they were consuming. He believed in what he called "trying it on the dog." In 1902 he received authority to test various food preservatives by serving them to young men. Twelve male volunteers ate their specially prepared meals together in the basement of the U.S. Department of Agriculture. Wiley understood how to use the media to his advantage, and he opened the doors of his work to reporters. He called his project hygienic table trials. But a *Washington Post* reporter named the young men "The Poison Squad," and that caught on with the public.

Members of the Poison Squad would eat meals with gradually increasing amounts of commonly used additives. The processes stopped as the young men started to get sick. One of the first preservatives they were given was borax, which was used to give products the appearance of freshness. It was used in the food industry to doctor decomposing meat. When Wiley himself was dubbed "Old Borax" by the press, he did not discourage the name.

Other additives tested included sulfuric acid, saltpeter, formaldehyde, and copper sulfate. Wiley's efforts, the impact of additives on the Poison Squad, and coverage in the news media led to Congress passing the Federal Meat Inspection Act and the Pure Food and Drug Act in 1906. The central purpose of the Pure Food and Drug Act was to ban adulterated and mislabeled food and drugs. Wiley was named the "father of the Pure Food and Drug Act," and the legislation led to the creation of the Food and Drug Administration.

When he left the government, Wiley worked with *Good Housekeeping* magazine. He wrote on the dangers of cigarettes, linking smoking to ill health forty years before the Surgeon General's report that linked smoking to cancer.

Wiley died June 30, 1930. It was the twenty-fourth anniversary of the signing of the Pure Food and Drug Act by President Theodore Roosevelt.

Annie Peck

When Annie Peck died in 1935 at the age of eighty-four, the *New York Times* declared her "the most famous of all women mountain climbers."[18]

She wouldn't have liked that. "I have climbed 1,500 feet higher than any man in the United States," she had said. "Don't call me a woman mountain climber."[19] She was a mountain climber. Period.

Peck was born in Providence, Rhode Island, in 1850. She was a twenty-first-century woman locked in a nineteenth- and early twentieth-century culture that gave women more restrictions than rights.

During her lifetime she reached all the highest mountain peaks in Central and South America, Mt. McKinley in the United States, and others in Europe, including the Matterhorn, but was widely criticized for climbing in pants rather than a long skirt or dress, as was expected of women in her time.[20]

Peck taught Latin, elocution, and German at Purdue University from 1881 to 1883 and was considered among the best teachers. But her path to higher education had been filled with roadblocks.

Her family tried to discourage her from entering a coeducational university. When she wrote Brown University president Ezekiel G. Robinson asking for admission, he replied: "Women are not encouraged to seek higher education."[21]

University of Michigan president James B. Angell disagreed. He accepted Peck and encouraged her to enroll in the classics curriculum.[22] She graduated in 1878, earned a master's degree there in 1881, and began teaching at Purdue.

While her teaching went well, Peck called living in Lafayette "wretched." Costs for merchandise were high and the weather was extreme, with a horrible winter in 1882 followed by months of rain. The Wabash River flooded three times.[23]

After leaving Purdue she studied in Hanover, Germany, and Athens, Greece, where she learned to speak French, Spanish, and Portuguese.

Her love for mountain climbing began in 1885 and continued throughout her life. She never married and was an active suffragette. At the top of one mountain she left a sign: "Votes for Women."

Peck made her living lecturing about her mountain climbing. She authored four books and climbed her last mountain in New Hampshire at the age of eighty-two.

She was honored throughout South America and her fame spread throughout the United States and Europe. After one spectacular climb it was said, "She has done all that a man could, if not more."[24]

The inscription on her gravestone states: "You have brought uncommon glory to women of all time."

8

Turbulence, Enthusiasm, and Accomplishments

1883 to 1893

About the most unchangeable thing in
the world is a college student.

—Purdue Dean Stanley Coulter

The Civil War had been finished for less than a year when George Ade was born
in Kentland, Indiana—the youngest of seven children and the one who would
baffle his banker father the most.

Kentland, north of Lafayette, was an agricultural community, but George had no
interest in farming, nor, seemingly, in anything else. While college was considered
a waste of time in their community, the Kentland school superintendent noted that
George, while an average student, perhaps should apply for the one county scholarship to Purdue. Ade's father agreed. He won the scholarship. He was the only one
who applied.

On a late summer day in September 1883, George Ade said goodbye to his parents and boarded a train for the short ride south to Purdue. It was a defining year
for Ade, who would go on to be among the most celebrated American writers of the

early twentieth century. In 1904 he had three productions on Broadway in New York City at the same time.

It was also a defining year for Purdue University with the arrival of a new president—James Smart. There were 214 students at Purdue, including the Academy, when Ade arrived. His freshman class consisted of thirty students and only eight would graduate, among them Ade.[1] It was a small school about to get larger with Smart breathing fresh air into the campus after the stern leadership of Emerson White.

The Engineer's President

In *The Story of Purdue Engineering,* Professor H. B. Knoll wrote, "With the inauguration of Smart as president a period opened in the story of Purdue that was notable for its turbulence, its enthusiasm, and its record of incredible accomplishment. Perhaps in no other period in [University] history were the students so happy, rowdy and creative and the staff so badly paid, ambitious and inspired. This was the period when new wonders were welcomed with loud student yells, when new schools were born, when exciting equipment was installed in the new engineering laboratories."[2]

And if all of that wasn't enough, Smart had the coolest beard of any Purdue president before or since. It started at his sideburns, which were trimmed short and

Purdue president James Smart.

grew longer halfway down his face. The area beneath his lower lip, along with his throat, was clean-shaven. His mustache and beard, which covered his cheeks, hung freely down below his jaw. The hair on his head was neat and parted on the side.

Purdue University in the early 1880s was still tied to its earlier days when students were generally unsophisticated. Ade recalled that his first impression was that male students wore ascots or ready-made cravats—wide ties that were stuck inside a shirt or vest. Some had horseshoe tiepins along with derby hats that had wide brims and low crowns. Male students wore bell-bottom pants that were only called trousers by those putting on airs. "The only suit of evening clothes in the dormitory belonged to a sophomore," Ade would later remember. "When its owner wore the suit other students leaned out of the window to hoot."[3] Women wore corsets and dresses that extended from their chin down to the floor. They wore their hair "up."

For entertainment, young men sometimes had dorm room parties featuring raspberry punch. They played penny-ante poker and when they could sneak the opportunity, they smoked Richmond Straight Cut and Sweet Caporal cigarettes.[4] Smoking was banned on campus, but they leaned over a fence, placing their head and cigarettes off University property, and got away with it.

Smart came to be remembered more for engineering than agriculture, and was called the "engineer's president," because people in Indiana still did not see the need to send their sons to college to learn how to farm. Smart was waiting until they changed their opinion.

Agriculture

While Smart is most remembered for his work in engineering, his presidency was a very important time in agriculture. In 1887 Congress approved the Hatch Act, providing land-grant universities with federal funds to create agricultural experiment stations to investigate improvements in food production and pass that information to farmers.

In 1889 the Indiana General Assembly passed the Farmers' Institute Act and placed it under the control of Purdue. The institutes had actually begun several years earlier, sponsored by the State Board of Agriculture. William Latta, hired by former president Emerson White, ran the program from the onset. It marked the beginning of an extension service at Purdue. From 1889 to 1891, Farmers' Institutes took place in about half of Indiana's counties, but by 1893 all ninety-two counties were taking part.[5] Topics included farm profits, sheep, dairy, pigs, horses, insects, and much more. The institutes were not just for men; they also offered help for women. Latta believed in education for females.

One of the first speakers Latta involved in the institutes was Virginia Claypool Meredith. They helped launch her career. She remained an active part of the institutes until 1920, with the exception of six years when she worked out of state. Born in Fayette County in 1848, she graduated from Glendale Female College, in Ohio, in 1866 and married Henry Clay Meredith four years later. He died in 1882 and, in an unusual decision for a woman of that time, she took over the operation of their farm herself. She gained national fame in farming and raising livestock and became known as the "Queen of American Agriculture."

Her institute talks taught men how to farm. The men were initially skeptical that a woman could teach them anything. But her knowledge of farming was immediately clear, and she was a powerful and inspirational speaker. According to an article in the *Indianapolis News,* she had an "individuality about her, an atmosphere of inexhaustible strength, of calm confidence in herself arising no doubt from her long habit of self-reliance, but never merging into arrogance or egotism."[6]

The first bachelor's degree in agriculture at Purdue was awarded in 1882 while White was president, and by 1887 about twenty-eight men were enrolled in the curriculum. The year 1887 also marked the start of the Purdue Winter Courses, providing agricultural education from eight to eleven weeks at a time of the year when farm work was less busy. While it did not fulfill requirements for a degree, it went a long way toward convincing farmers to apply the lessons of science to food production. Short courses lasting several days or weeks were also offered.

Engineering

According to Knoll, Smart and W. F. M. Goss combined to lead Purdue engineering into twentieth-century prominence. Goss, Knoll said, "was alive with ideas and lofty in aspirations. Goss and Smart worked in harmony, forming a team that would not let any obstacle, large or small, stand in the way of engineering progress."

The School of Civil Engineering, which had been a department, was founded in 1887 and the School of Electrical Engineering in 1888. Civil Engineering shared facilities in a mechanical building, but Electrical Engineering got its own facility.

School of Domestic Economy

At the same time that Purdue launched Civil Engineering, it also began instruction in veterinary science (but not medicine) and launched a School of Domestic Economy. Domestic Economy was canceled after two years because of inadequate enrollment.

The 1887–1889 Purdue Annual Register stated that Emma Ewing, professor of domestic economy, would conduct a series of lectures, most of them once a week. Ewing was one of the significant early women in the history of Purdue. A native of New York State, she gained national recognition as an instructor of cooking following the Civil War. In 1882 she had established the Chicago School of Cookery, and she authored several cookbooks. She came to Purdue in 1887, but left in 1891 when University women failed to show an interest in her instructions. "It is not the mission of every woman to do housework," she said. "All women are not called housekeepers.

But it is of paramount importance that every woman who attempts to preside over a household should be thoroughly instructed in the domestic arts."[7]

Stanley Coulter

Although Goss and Smart were filled with energy and vision for engineering, they were short on money. In addition to the state funding shortages that emerged from the fraternity controversy, the Indiana legislature was still not completely behind the idea of public higher education for the masses of people, and Smart kept the University running on "faith, hope and credit."[8] In 1887 Purdue again received no state appropriation. Wealthy Lafayette citizens provided the University with some cash and merchants extended credit.

Smart was in Europe during the summer of 1887 when members of his faculty became concerned about the University's ability to pay their salaries. Some of them left for more stable positions, and Smart returned to discover he needed to hire new professors. He remembered an acquaintance, Stanley Coulter. They had met in 1880 when Smart was state superintendent of public instruction and Coulter was the principal at Logansport High School. In the summer of 1887 Coulter was fishing on Lake Maxinkuckee in Culver, Indiana, when a young man rowed out to bring him a telegram. It was from Smart, asking him to take a position at Purdue. Coulter read it, put it in his pocket, and continued fishing and smoking his pipe.[9] That fall he started at Purdue as a professor of zoology and assistant principal of the preparatory school, beginning thirty-nine years of service to the University, including being named dean of science in 1907 and dean of men in 1919. In 1917 the Biology Building opened, and in 1960 it was named in his honor.

Stanley Coulter.

Coulter remembered his early days on campus in an article for the 1924 *Debris* yearbook:

Dorm Devils drop water on people below from Purdue Hall.

Purdue Hall was the dormitory for men [who were] familiarly known as Dorm Devils. Their favorite indoor and outdoor sport was throwing water upon unwary passers. No wise professor ever passed near enough to the dorm windows to come within range of one of those paper bags filled with water.

Students who first came to land grant colleges were not quite of the same type as those who registered in the old-line colleges. They were less conventional, less adept in social affairs perhaps, but just as eager in extra-curricular activities and in attempting to establish traditions. I doubt if there was a dress suit in the student body. I know that there was none in the faculty. Not ten percent of the students knew how to dance and those who did generally concealed the fact. They had just as effective ways of wasting time as

Purdue's original campus buildings in 1876. In 1877 University Hall opened. By the mid-1880s trees and shrubbery had been added.

do the students today. They cut classes, skimped work, followed the team. About the most unchangeable thing in the world is a college student.[10]

School of Pharmacy

The first big move in Smart's presidency came with the establishment of a pharmacy program in 1884. Rapid developments were taking place in the industry. Colonel Eli Lilly had started his company in Indianapolis in 1876, dedicated to using science to produce medicines that would help people, pushing aside decades of elixirs with extravagant claims being promoted to the public by other drug producers.

Among those in Indiana advocating for professional training of pharmacists was John Newell Hurty, an Indianapolis drugstore owner. Hurty had been trained at the Philadelphia College of Pharmacy, the first of its kind in the nation, and he apprenticed under Lilly. In talks with the new president of Purdue, Hurty encouraged Smart to start a pharmacy program, believing it fit into the land-grant mission of the University. Smart agreed, but with one stipulation: Hurty would have to lead the program for at least two years. The board of trustees approved it in December 1883, and it began in 1884 with seven students and four professors. It became the third state-supported school of pharmacy in the nation.

Smart's Many Responsibilities

Smart served as director of admissions and dean of students in addition to his presidential duties. His attitude toward discipline seemed to side with Harvey Wiley. He let the students have their fun—up to a point. Students somehow were able to get cows up on the roofs of buildings. Roosters were placed in the chapel organ. One morning Smart arrived at chapel and saw that all the students had arrived early. He then stopped and stared "in blank amazement" at a seven-foot-high pile of railroad ties. The pulpit and Bible were on top of the ties. As the faculty filed in and observed what had happened, Smart sent a professor to find another Bible so the service could take place. He turned to the students, congratulated them on their efforts, and said they would have a great future in a mechanical line. Campus lore claimed Smart opened his songbook and made a selection at random. But he probably knew what he was doing when he announced they would sing hymn number 256. The pianist played, and the students and faculty sang "Blest Be the Tie That Binds."[11]

Another evening students hid in the bushes by the Old Pump with plans to capture the first man who walked by, drag him to the pump, and soak him in water.

The Old Pump.

It was a common prank to "cool off" male students heading to the pump to meet a young lady. But the first man who walked by was Smart. In the darkness and in their haste, the students didn't recognize the president, and he got soaked. No one was punished in spite of the fact that the students were certain he recognized their voices and knew who they were.

Emma Montgomery McRae.

Mother McRae

In 1887 Smart hired Emma Montgomery McRae as professor of English literature and "lady principal." She would come to have a lasting impact on Purdue, influencing the lives and careers of women who would play leadership roles at the University and elsewhere in the twentieth century.

Born in 1848 in Loveland, Ohio, she was the daughter of a Methodist minister and moved to Indiana with her parents when she was five years old. With a master of arts degree from Wooster College in

Ohio, she was better educated than some of the men on the faculty. She had taught school in Vevay and Muncie, Indiana, where she served as high school principal from 1867 to 1888. McRae, a widow, would spend twenty-six years at Purdue, retiring in 1913. Women students called her "mother," and she fulfilled the role of a dean of women, although she was never given that title. As more young women came to Purdue, the faculty moved out of the boarding house, and it was renamed Ladies Hall for the female students with Mother McRae in charge. In some black-and-white photos of the time McRae had a stern look on her face. But she was said to be kind and friendly.

"Socially she occupied a very high position," according to her biography in Purdue's Virginia Kelly Karnes Archives and Special Collections Research Center. "She was a delightful hostess and was never so happy as when her apartments in the Ladies Hall were the scene of some social gathering. Her annual Easter Monday reception was always brilliant and the invitation list was always large. As a public speaker, she won not only by the power of her thoughts and the eloquence of her manner, but also by the sweetness of her soul."[12]

John McCutcheon and George Ade

In addition to being one of the great writers of his time, Ade, along with his friend John McCutcheon, became among the great characters in the history of Purdue. After graduation McCutcheon became a Pulitzer Prize–winning political cartoonist whose commentaries ran on the front page of the *Chicago Tribune*. He was called "the dean of American political cartoonists." Another cartoonist would graduate from Purdue in 1917. His name was Harold Lincoln Gray, an orphan from West Lafayette, and his cartoon was *Little Orphan Annie*.

McCutcheon grew up in Lafayette and lived at home while attending Purdue, a three-mile walk away. He had to reach the University for classes at 8 a.m. "We either walked or drove in a buggy or else did the trip in a herdie—a kind of horse drawn bus that held twelve or so shivering students putting their bodily warmth against the wintery blasts that whistled through every crack," McCutcheon wrote in his autobiography, *Drawn from Memory*.[13]

By his own account, Ade was more interested in theater at the Lafayette Opera House than in his studies. During his sophomore year he joined other Republicans on a trip to Lafayette, where they shouted down Senator Daniel Voorhees, a Democrat, as he tried to address a crowd gathered at the Tippecanoe County Courthouse. In later years Ade would host huge political rallies at his Brook home, including the kickoff of William Howard Taft's 1908 presidential campaign.

Novelist Booth Tarkington attended Purdue from the fall of 1890 to the spring of 1891. He went on to receive two Pulitzer Prizes.

A Boilermaker his entire life, Pulitzer Prize–winning political cartoonist John McCutcheon.

Young men playing pranks on one another are not unique to modern times. McCutcheon recalled taking Fanny McGrath to an evening reception at Purdue, driving her to the University from Lafayette in his family buggy. He hitched the horse to a post outside Ladies Hall. When they left the reception three hours later, the horse was still hitched to the post. But the buggy was gone and he had to walk

Early campus photo showing University Hall (center), sometimes called Old Main in the early days, Ladies Hall (left), Purdue Hall (next to University Hall), and the first Mechanical Engineering Building (far right).

the horse and Fanny home. It took three days to find where his "buddies" had hidden it, miles away in a cornfield.[14]

Ade and McCutcheon got into a lot of mischief, but they pushed Smart too far when they attended a ladies' literary society meeting without asking permission. They were called to his office and reprimanded.

Student Organizations and Publications

It was during the Smart years that student organizations began giving the University traditions that continued into the twenty-first century. Purdue began to look and "feel" like a university.

In 1886 a student drum and bugle corps emerged and played during military training. It would evolve into Purdue Bands—an organization started by students who came to Purdue with a love for playing musical instruments.

A student newspaper, the *Purdue Exponent,* was founded in 1889. The Purdue yearbook, the *Debris,* also began in 1889. McCutcheon chose the name because he

Cartoon by alumnus John McCutcheon says it all.

believed, incorrectly, that it was a French word meaning "a collection of works." It continued for 120 years into the age of the Internet, when sales declined. A glee club started in 1893, directed by Cyrus Dadswell, an organist in Lafayette. There were eleven members.

Intercollegiate Athletics

The first athletic teams at Purdue were baseball and track. Football began in 1887 and basketball in 1900. The first collegiate football game in the United States was in November 1869 when the College of New Jersey (later Princeton) played Rutgers. By the time Purdue introduced the sport, Princeton was famous for having the best team in the nation. J. B. Burris, '88, captain of the first Purdue team, believed they should use Princeton's colors, black and orange. To be distinctive, Purdue picked black and old gold.

In 1891 Purdue defeated Wabash 44–0. The Crawfordsville *Daily Argus News* headlined a story about the game: "Slaughter of Innocence, Wabash Snowed Completely Under by Boiler Makers from Purdue." *Boiler makers* was a name for men who made or worked on boilers or craftsmen who worked with steel. It was also a nineteenth-century term used for industrial blacksmiths. All were large, strong men. Perhaps the author of the headline was using "boiler makers" to reference Purdue engineers, who were sometimes confused with blacksmiths by people who didn't understand the profession. Whatever the purpose, it was considered derisive and an accusation that Purdue had used nonstudents on its football team. But in Lafayette, the fans loved it. The city newspapers picked up the name, and by 1892 the *Purdue Exponent* was using it.[15] Purdue athletic teams have been known as the Boilermakers ever since.

Smart loved football and supported the University's athletic programs. In 1895 he initiated efforts to create the Western Conference, which later became the Big Ten—the oldest Division One conference in the nation. According to Knoll, "in the stands Smart fervently rooted the players along, and when they scored he yelled, danced, and hugged his fellow spectators in the classic manner of the rabid football fan."[16]

Corliss Steam Engine and the Schenectady

In 1890 Smart and Goss changed engineering at Purdue forever when they brought a Corliss stationary steam engine to campus. The engine, used for research, gave Purdue something that could only be equaled by MIT.[17]

Corliss steam engine.

Purdue next became a leading university in the world in locomotive testing. Goss is the man who had the idea, in his own words, "to mount a locomotive in the laboratory in such a manner that it might be operated as if upon the road, under conditions which would permit its performance to be determined with a degree of accuracy hitherto unknown."[18]

The Schenectady.

Goss brought a train engine named Schenectady to campus. The Schenectady Locomotive Works in Schenectady, New York, built the engine for $4,000.[19]

By the fall of 1891 the Schenectady was in place on a treadmill. The train arrived on a railroad line south of campus. Smart called an all-university holiday so students and faculty could help to move the eighty-five-thousand-pound locomotive to a research facility a mile and a half away. It took the young men and three teams of horses eight days.

"The locomotive was the showpiece, the distinguishing mark, the thundering reason why engineering students were glad to be at Purdue," Knoll said. "It became the symbol of the energy and vitality of the schools of engineering."[20] In 1893 engineers and railroad businessmen who visited the World's Columbian Exposition in Chicago traveled south to West Lafayette to observe its locomotive research.

In time a Schenectady II and III and an engine called Vauclain, Purdue No. 4, would replace the first locomotive. Among the advances made through research at Purdue was an airbrake system that was later required on all trains in the United States. The University also had a Railroad Museum on campus filled with historical items that had been donated.

Purdue locomotive research continued until 1938, when railroad companies stopped funding the work and the Vauclain was declared unsafe. In 1944 it was sold for scrap. Some of Purdue's railroad equipment was sent to museums.

The first years of Smart's presidency had been an incredible success. But nothing he accomplished would compare to what came next. And when faced with tragedy, he would muster a Purdue spirit that has inspired and rallied the University and its people ever since.

GIANT LEAPS IN RESEARCH

William Freeman Myrick
(W. F. M.) Goss

It was A. A. Potter, Purdue's dean of engineering from 1920 to 1953, who best summarized the accomplishments of his predecessor William Freeman Myrick (W. F. M.) Goss.

Goss, Potter said, was Purdue's first professor of practical mechanics, its first and only professor of experimental engineering, its first dean of engineering, the first to promote, on a large scale, close relations between Purdue and industry, and the first professor at the University to conduct applied research outside of the field of agriculture.

Goss set the standard for everything that was to come in Purdue engineering.

He came from simple, hardworking people. Born in 1859, his father was owner and editor of the *Barnstable Patriot* newspaper and collector of customs in Cape Cod, Massachusetts. Even as a boy Goss was interested in mechanics. He built a model steam engine from waste material in his father's shop. At the age of seventeen he installed a steam engine in a boat.[21]

In 1877 he began a two-year program in mechanics at M.I.T. and upon completion took a position with Purdue as an instructor. Potter wrote: "Under his leadership Purdue became the leader west of the Hudson in the field of practical mechanics."[22]

In 1888 Goss was granted a six-month leave for study and visiting factories. When he returned his title was changed to professor of experimental echanics.

Among his first moves was acquiring a train engine, "a steaming, smoking, fire-breathing monster to be studied, teamed, and before long revered as a symbol of power and romance," according to Professor H. B. Knoll in his 1963 book, *The Story of Purdue Engineering.*[23]

The engine was named the Schenectady and was hitched to a post with its wheels on a treadmill while students and faculty studied its performance. It could pull thirty thousand pounds, but if a student placed his hand in front of the locomotive it would register a change in stress. With this, Goss started "a type of experimental engineering which contributed greatly to the progress of locomotive and other railroad equipment design in Europe as well as in this country," Potter said.[24]

For many years Purdue was the only "neutral" site where railroad companies could take their equipment for testing, and they made great use of the Goss lab.

In the mid-1920s, long after Goss was gone, the program he established developed an air brake system that became standard equipment on American railroads. And a train engine would eventually become the official Purdue mascot.

9

One Brick Higher

1890 to 1895

We can only help others permanently by helping them
in a way that prepares them to help themselves.

—Purdue donor Amos Heavilon

October 31, 1892, was a day of celebration at Purdue University—a day of rejoicing like nothing that had ever been seen on campus before. If John Purdue had
lived to see that day, it would have been his ninetieth birthday. But no one was focused
on the University's founding benefactor on this day. The nearly seven hundred students and forty-nine faculty members assembled at 10:15 a.m. chapel in the redbrick
University Hall weren't thinking about the past. They were cheering for the future.

Purdue president James Smart addressed the assembly and briefly talked about
the needs of the University as it prepared to enter a new century. Smart had been
struggling to build Purdue into his vision for its future. His dreams always exceeded
the University's finances, but he never backed away. "It has been our ambition to
make Purdue one of the most thorough and best equipped technical schools in the
country," he had said. "The state of Indiana can afford nothing less."[1]

In 1890 Smart asked the Indiana General Assembly for $60,000 to build a new
engineering laboratory. The total value of the entire campus including land, buildings,

and fixtures was less than $500,000.[2] Now Smart was asking for a building that would equal 12 percent of the value of the entire University!

He didn't get it. The state appropriated $12,000 for the building. But, ever the optimist, Smart moved ahead with his plans, using the funds to build the first phase of what he hoped would one day be the best academic engineering facility in the nation. The modest building opened in January 1892 and became home to the Schenectady and the Corliss engine. It was a start.

On Monday morning, October 31, 1892, at chapel, Smart told the students and faculty how he would accomplish the next step in his vision. There was a man with them that morning, he said, who was the first major benefactor to the University since John Purdue. He was giving nearly half his hard-earned fortune to the University, and it would be used to expand the mechanical laboratory.

Smart then introduced Amos Heavilon of Clinton County, southeast of Lafayette. An editorial in the November 1, 1892, *Purdue Exponent* described the reception Heavilon received: "Never has the old chapel echoed back such enthusiastic yells as those that broke from the throats of the happy students."[3]

That evening, alone in his home, Heavilon shared his feelings with his diary. "Self went to Lafayette and closed up the donation to Purdue," he wrote. "I was introduced to six or seven hundred students and the cheers I got was deafening. The only time in my life that I could do anything that appeared to be appreciated. It was a surprise for me. My natural timidity prevented me from enjoying it to its full extent, yet it was the happiest day of my life, I think."[4]

The *Exponent* described Heavilon as a "quiet and unassuming man." He never married and lived modestly with his parents until they died. "By business-like management of his farming interests and judicious investments in real estate he became the owner of extensive farm lands."[5] In 1890 he helped organize the Clinton County Bank.

His diary entries portray a man who was lonely, depressed, and in poor health, but was encouraged by his gift to Purdue. On Tuesday, September 22, he wrote that he had visited the University. "Self went to Lafayette and Purdue and found the university a big thing, nearly seven hundred students and with the appliances and their mode of teaching certainly learns the young men and women very rapidly to be useful men and women. I think the institution will do much good to a class of young people that are worthy. And most need help, as the patrons are mostly poor young people and think I could not do better with some of my means than to help them some."[6]

On Wednesday, October 26, he had completed the agreement with the University. "Done the biggest day's work on my life," he said of the donation. "Don't know whether or not I have done a good thing, but there is much less for to fuss over when I am gone."[7]

On November 3 he was happy with "many flattering notices in the papers of what I had done for Purdue University." But by December 9 he was tired of hearing from people who wanted money: "What I done for Purdue University has caused the people to think I am so generous that all that ask will receive, especially the women. I have got a great many letters. The statement in connection with that gift in the papers that I was a bachelor did it. If I answered the mail I think there would soon be a perceptible increase in Uncle Sam's post office business." A month later he wrote, "Have letters from nearly every part of the U.S. and today got one from Mexico, most of them begging for money. I thought I was important enough before the donation to Purdue but that increased it one hundred fold."

At the end of 1892 he summed up the year, saying it had been the most important of his life. "I gave almost one half of my whole life's accumulations to an industrial school at Lafayette, Indiana, not because I thought less of my relatives (but, must, in truth, say have been losing confidence in most of them), but am thoroughly convinced that we can only help others permanently by helping them in a way that prepares them to help themselves."[8]

Heavilon's gift totaled $35,000. In early 1893 the Indiana General Assembly responded to it by providing an additional $50,000 for the new building. Smart pulled together more money for the project. According to William Hepburn and Louis Sears in their book *Purdue University: 50 Years of Progress,* "In order to hasten completion of the building it became necessary to borrow money on the security of the Heavilon gift and on the strength of the legislative appropriation and to issue warrants to professors and others willing to accept them in lieu of cash. Thus, the resources of the university were severely taxed."[9] Construction on the building, to be named Heavilon Hall, began in March 1893.

The most impressive feature was a tower that rose 140 feet above its entrance. The tower was built for a museum, drafting rooms, and recitation rooms as well as offices. But even more importantly, the tower served as an exclamation point proclaiming to all who saw it that Purdue had arrived among the great universities of the nation and world.

Charles Stuart, a Lafayette attorney and president of the Purdue University Board of Trustees, did much of the work of obtaining the gift from Heavilon. Stuart had been appointed to the board in 1885. He was elected president in 1888 and served in that position until his death in 1899. Throughout his tenure he worked with and supported Smart's vision for the campus and helped create the Lafayette law firm that became Stuart and Branigin.

Heavilon died of pneumonia on November 18, 1893, before the building was completed. He had given the University a gift that would transform it. It was an incredible moment for Purdue. It had also been an incredible moment for Heavilon, the highlight of his life.

The building was dedicated on Friday, January 19, 1894, amid much fanfare. During an afternoon ceremony Smart presented Heavilon Hall to Indiana governor Claude Matthews. The dedication was "a gala day at old Purdue," the *Lafayette Morning Journal* reported on January 20. The 3 p.m. dedication ceremony in the University Hall Chapel was filled with "well dressed, intelligent looking, decorous people," the newspaper said. They had climbed three flights of stairs to reach the chapel. The Trinity Methodist Church choir sang.[10]

The evening was highlighted by a reception in Heavilon Hall. People walked through the building and marveled at all it offered. In a large room on the second floor, an orchestra played and guests danced. "Life and laugher pervaded every hall, recess and room," the *Journal* reported.[11]

Charles Benedict Stuart, Purdue trustee from 1885 to 1899, president of the board the final eleven years. Stuart did most of the work securing the gift from Amos Heavilon that resulted in Heavilon Hall.

Fifteen hundred people were invited. Palms, ferns, and evergreens provided the decorations along with yellow, white, and pale green muslins draped around doorways. Purdue's color of black was probably avoided because of its Victorian association with mourning. There were carnations from the Purdue Greenhouse. There had never been an evening like it in Tippecanoe County.

Four days later, on Tuesday, January 23, at 8:30 p.m., there was an explosion in the Heavilon Hall boiler room. A fire started, and soon there was a second explosion that blew out the walls of the room. The fire quickly lit up the nighttime sky. Children in Lafayette ran through the streets shouting, "Purdue is on fire!"[12] Students rushed to the building and ran inside where they could. They carried out equipment that could be saved.

"The news of the conflagration spread with astonishing rapidity and by 9 o'clock the campus was dotted with spectators," the *Journal* reported the morning of January 24. "They came from every direction and by every method of locomotion. People left the opera house to cross the river to see the fire. Ladies came by the score and stood in the snow and slush for hours. The heat from the fire melted the snow for some distance around the building and people who went that close to the shops were compelled to wade through water and slush."[13]

Firemen hurried to the scene from Lafayette. Horses had to haul firefighting equipment a long distance and were exhausted trying to get up State Street Hill in West Lafayette. When firemen arrived, they discovered their hoses were not long enough to extend from cisterns at the far west end of campus to the burning building. They had to send for more hoses from Lafayette.

"President Smart arrived at the campus [from his Columbia Street home in Lafayette] after the fire had gained such a headway that hope had been abandoned," the *Morning Journal* reported. "He did not remain long. The sight was too much for him. He had devoted much time and energy to the construction of the building. He had achieved the greatest purpose of his life when it was completed. . . . To see months of toil rendered useless in an hour, to see the University crippled and expectations crushed was too much for even a man of his energy and pluck and perseverance." Smart cried as he watched the fire consume his greatest achievement.[14]

By 9:40 p.m. the roof of the main building was engulfed in flames. A reporter for the *Courier* described the beauty amidst the horror:

> Nothing could check the flames. All the efforts of the firemen and the helping students were in vain. It was simply a case of stand away and see the building go. No painter could have transferred that scene of fire to his canvas, no artist could have formed colors to depict the flow of living light that made the scene so gorgeous. The people who went to the fire from the city saw something that no one had ever seen in Lafayette before. It was Lafayette's greatest fire in many ways—in beauty, in awfulness, in destruction and in wretchedness. . . . Let it simply be said that the reader must imagine a glow of red, yellow, blue and green, a brilliancy indescribable, a horrible, seething mass, an illumination that could be seen for miles.[15]

There were some people in Lafayette who said the fire was God's retribution because people had sinned by dancing during the dedication celebration.[16]

The following morning, even as people throughout the community were reading about the fire in their newspapers, Purdue students, faculty, and staff gathered in the chapel in University Hall. Outside in the cold winter air Heavilon Hall was still smoldering, smoke rising from the ruins. The chapel itself smelled of smoke rising from the clothing of people who had spent the night watching the pride of the campus consumed by fire. Students slouched in their seats. Some leaned forward with hands on their knees, their heads down. They believed they had lost everything. On

Heavilon Hall after the fire.

the stage an equally grief-stricken President Smart rose and faced the assembly. He gave the greatest speech of his life and closed with words that came to personify the spirit of Purdue. His words were extemporaneous. No one wrote them down exactly as he said them. Recollections differ only slightly, a word here, a word there. But the essence of what everyone heard was the same.

"We are looking this morning to the future, not to the past. I am thankful no one was injured," Smart said. And then he paused. He straightened his posture and his voice became strong, no longer grieving for what had been lost. He was now ready to lead the University in building out of the ashes something even greater than had been done before. "But I tell you young men that tower shall go up one brick higher."[17]

Smart's words are remembered in the history and folklore of the campus and in awards presented for excellence. They symbolize a determination to always do better than anything that has been done before and to face setbacks with the determination not just to go on, but also to excel. Smart's speech in chapel that Wednesday morning ended with loud cheering by the students.

The building was valued at about $175,000. Insurance and salvaged equipment totaled $100,000. In 1895 the Indiana General Assembly appropriated $36,000 for rebuilding and another $25,000 to replace money drawn out of the University's general fund for the project. Manufacturers rebuilt machinery and donated or discounted equipment. Students reconstructed apparatus. This time, the new tower included a

Second Heavilon Hall.

clock and chimes, thanks to gifts totaling $800 from the Ladies' Matinee Musical of Lafayette, the class of 1895, and others. The Schenectady was shipped to Indianapolis for repairs and returned on a new campus railroad line.

When completed, the facility was officially known as the Mechanical Engineering Building. It wasn't until April 9, 1930, that the board of trustees named it Heavilon Hall. At the time, a new facility under construction on the Purdue Mall was named the Mechanical Engineering Building

The second Heavilon Hall with its tower, clock, and bells stood in the same spot as the first until 1956, when it was taken down amidst some protests. A third Heavilon Hall opened at the same location in 1959. The clock and bells from the second Heavilon Hall tower were placed in storage. In 1995 the bells were placed in the 160-foot-tall Purdue Bell Tower. In 2011 the Heavilon clock was placed in the Roger B. Gatewood Wing of Mechanical Engineering, where it continues to keep perfect time.

But what of Smart's promise that the tower would be rebuilt one brick higher? Brilliant as he was, this time Smart was off the mark. On December 4, 1895, the building reopened. It wasn't one brick higher than the Heavilon Hall tower that had been destroyed by fire nearly two years earlier.

It was nine bricks higher.

GIANT LEAPS IN LIFE

Elwood Mead

In 1924 the Purdue *Debris* listed twenty-five prominent alumni. The first featured was Elwood Mead.[18]

Mead and two classmates became the University's first graduates in agriculture in 1882. His career took him around the world and to a dam that was the largest concrete structure ever attempted at the time. Purdue recognized him for serving "the industrial and agricultural development of the entire world."[19]

Born in 1858 in Patriot, a southern Indiana town of several hundred people, Mead grew up on a farm. He enrolled at Purdue on a scholarship and worked for his meals, receiving a bachelor of science degree. In 1883 he received a civil engineering degree from Iowa State and came back to Purdue for an engineering master's degree in 1888.

That same year he was named the first state engineer of Wyoming and developed the Elwood Mead code to resolve controversies in irrigation farming. It became a model for many western U.S. states and some nations.

In 1897 he became chief of irrigation investigations in the U.S. Department of Agriculture and taught at the University of California.

He was involved in a traffic accident in 1901 and his right arm was surgically removed. Three years later when he received an honorary doctorate in engineering from Purdue—the first honorary degree in engineering awarded by the University—he kept the empty sleeve of his suit tucked inside his right coat pocket.

In 1907 he accepted a position with Australia as an advisor in the development of agriculture by irrigation. His recommendations marked a turning point in Australia's agriculture, becoming a model for the world.[20]

Mead went on to advise governments in New South Wales, Canada, Hawaii, Haiti, Java, South Africa, and Palestine.

He became special advisor on reclamation in the U.S. Department of the Interior in 1923, and one year later President Herbert Hoover named him commissioner of reclamation. In that position, he became the chief engineer of the Boulder Dam on the Colorado River, later named the Hoover Dam.

Mead died in January of 1936 at the age of seventy-eight shortly after the dam's completion. Secretary of the Interior Harold Ickes said: "Perhaps no man contributed more to the planning of Boulder Dam and certainly no one had a more important part in the actual construction of it than Dr. Mead."[21]

After his death its reservoir—one of the largest man-made lakes in the world—was named for the Purdue graduate from Patriot, Indiana: Lake Mead.

10

The Look and Feel
of a University

1890 to 1900

Purdue, Purdue, Rah, Rah, Rah, Rah! Purdue, Purdue, Rah,
Rah, Rah, Rah! Hurrah, Hurrah! Bully for old Purdue!

—Purdue Cheer, 1892–1893

I n 1893, the Columbian Exposition world's fair opened in Chicago, lighted at
night by nearly one hundred thousand incandescent lamps. It celebrated the four
hundredth anniversary of the voyage to a "new world" by Christopher Columbus.
But even more, it celebrated all that humankind had accomplished and the promise
of the twentieth century.

Among the organizers of the exposition was Virginia Claypool Meredith from
Cambridge City, Indiana, who was named to the Board of Lady Managers at the
fair and elected vice chairman of its executive committee. She worked closely with
Bertha Palmer of Chicago, who was chairman of the executive committee and president of the board, and played a huge role in highlighting the interests and successes
of women around the world.[1]

Meredith, already proclaimed the "Queen of American Agriculture," played a
huge role in the history of Purdue, speaking at Farmers' Institutes, advocating for a

Memorial Fountain near John Purdue's grave, Electrical
Engineering Building in background.

department of home economics, and much more. In 1921 she would become the first female member of the board of trustees. (More on Meredith in part 2.)

The 1890s were an exhilarating time at Purdue. The University had only been open for sixteen years when the decade began, and it had shaken off its early awkward, unworldly, unsophisticated aura.

By 1899 more than seven hundred students studied at Purdue. In 1894 President James Smart eliminated the old Academy for unqualified students that sometimes made the campus look and feel more like a glorified high school than a major teaching and research university.

Many of the extracurricular activities associated with modern universities emerged at Purdue during this period, along with academic and research programs that were drawing national and even international attention.

It Was a Wonderful Time to Be a Boilermaker

Students in the 1890s were closely associated with their class—freshman, sophomore, junior, or senior—and their graduation year. One of the first things a class did upon arriving on campus as freshmen was gather to decide on class colors and "yells" that would be shouted in unison at football games and convocations—each class trying to outshout the others: "Rah, rah, rah! Zip boom roar! Boomalack, boomalack, '94!"

Card games like Hearts, New Market, and Whist were all the rage, as was "going out calling," Tulu (chewing gum that "perfumed the breath, cleaned the teeth and aided digestion"), ten-cent shows, Happy Hollow, Tecumseh's Trail, the YMCA, high teas, poker, tennis, and chapel.

Students chewed the rag (talked) and scrooched up to fit one more person in a buggy. They spooned, they bellyached, they liked things with razzmatazz, and, gee, they had a corker of a time, according to the slang of the day. It was a golden age for popular music when Tin Pan Alley formed in New York City and people at Purdue sang with a piano to "The Band Played On," "The Sidewalks of New York," "Sweet Rosie O'Grady," "Oh Promise Me," and "On the Banks of the Wabash," which weren't actually all that far away.

The 1899 *Debris* yearbook included tongue-in-cheek fashion and etiquette suggestions for students:

- "The elite of the dormitory no longer wear their slippers when going to breakfast."
- "It is considered bad form to take a young lady to the opera house when the tickets cost more than fifty cents apiece."
- "A negligee breakfast costume is popular with the waiters at the [Ladies] Boarding Hall."
- "It is good form to throw water on your friends in the dormitory."[2]

"Dorm Devils" who occupied the men's residence hall initiated and terrified new residents, made certain everyone was all wet from time to time, and generally raised havoc whenever the opportunity arose.

Student handbooks were filled with advice for new students, such as these suggestions in 1892–1893:

- "Don't forget the reception for new students. It is always an enjoyable occasion."
- "Supply yourself with college colors, black and old gold."
- "Do not address instructors and fellows as professor."
- "It is customary to salute professors that you know by touching your hat (as you pass on campus). There is no rule on the matter, however."
- "Learn the college yell: Purdue, Purdue, Rah, Rah, Rah, Rah! Purdue, Purdue, Rah, Rah, Rah, Rah! Hurrah, Hurrah! Bully for old Purdue!"

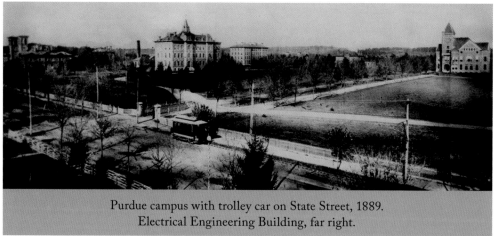

Purdue campus with trolley car on State Street, 1889.
Electrical Engineering Building, far right.

- "You will need a military suit by the middle of the fall term. It will cost you about $15. Arrange your finances accordingly."
- "Calling hours at the Ladies Hall are Friday, Saturday and Sunday afternoons and evenings from 3 to 10 p.m. Remember this."
- "There is no hazing at Purdue. Don't be scared."

No hazing at Purdue? Good luck trying to tell that to the Dorm Devils.

Reginald Fessenden, Brilliant, Absentminded Professor

Among the faculty in 1893 was Reginald Fessenden, a professor of electrical engineering who had worked for Thomas Edison before arriving at Purdue. He was the epitome of the absentminded professor. Students howled at the stories of Fessenden walking into a classroom and beginning his lecture, only to be informed he was in the wrong room. He once poured cream into his napkin ring instead of his coffee, and one morning at his boarding house, he reportedly came down for breakfast wearing the wrong tie.

Faculty meeting, 1896–1897.

His wife sent him upstairs to change. When he didn't return, she went after him and found he had undressed and returned to bed.[3]

Fessenden helped Westinghouse install the lighting for the Columbian Exposition in Chicago and left Purdue in 1893 for Western University of Pennsylvania in Pittsburgh. Seven years later he became the "father of radio," the first person to transmit a human voice without wires. He received hundreds of patents in his life, including some for a device called "television."

African American Students

According to some sources, in 1890 George Washington Lacey was the first African American to graduate from Purdue. His degree was in pharmacy, and he practiced in Chicago on State Street.[4] The University kept no records of students by race at that time and there are no Purdue photographs of him.

David Robert Lewis, one of Purdue's first African American graduates, in 1894.

David Robert Lewis, an engineer, is also sometimes listed as the first African American graduate of Purdue, in 1894. He came from Greensburg, Indiana, a town southeast of Indianapolis, and went by "D. Robert." His civil engineering senior thesis at Purdue was titled "Highway Road Construction" and included a review of European road building techniques.

Lewis was among the nation's first black engineers. He taught mechanical drawing at Virginia's Hampton Institute for twelve years and wrote of his students: "To make each one think, to make him self-reliant, to feel the responsibility of his own effort is a constant aim and endeavor." After leaving the institute he became a real estate broker in Pittsburgh.

Female Students

Young women continued to enroll and graduate from Purdue. They held leadership roles such as editor of the *Debris*. Each yearbook included a report on class activities, and in 1890 men in the Class of 1893 reported as much on the beauty and morals of their female counterparts as their intellect:

The girls in our class, as well as in all others of the University, are in the minority; yet in class-meetings . . . they vote and rise to points of order as well as the boldest men. Their moral training and good qualities have added much to the dignity of the class; while their intellectual abilities and aptness to learn have thrown a halo of intellectual light about the University. To describe the beauty of our girls—to guard a title that was rich before, to guild refined gold, to paint the lily, to throw perfume on the violet, to smooth the ice, or add another hue unto the rainbow, or with taper light to seek the beauteous eye of heaven to garnish—is wasteful and ridiculous excess.[5]

Young women felt there were inequities on campus, and they were not shy about making their opinions known. One even wrote a poem in the 1895 *Debris,* asking Smart why there were no sororities on campus for women.

There were six fraternities listed in the 1895 yearbook: Tau Beta Phi, Sigma Chi, Kappa Sigma, Sigma Nu, Phi Delta Theta, Sigma Alpha Epsilon. The first local sorority at Purdue, Phi Lambda Psi, had been formed in high school and would be recognized at Purdue in 1906. In 1915 it was chartered nationally with Kappa Alpha Theta.

The Purdue Annual Register of 1892–1893 lists University graduates and their occupations at that time. In all, there were fifty-six female graduates listed since the first women enrolled in 1875. Fourteen of them were working outside the home. All of them were teachers, three of them at the university level. The 1900–1901 Annual Catalogue (formerly the Register) lists seventy members of the Purdue faculty. Eight of them were women, including Carolyn Shoemaker, an instructor of English. She would become part-time dean of women and play a major role at the University in the upcoming new century. (See part 2.)

Most women who attended college in the nineteenth century came from the middle class. The poorest families could not afford even the minimal costs of sending a student to college, and since women were only expected to marry, they did not need career training. Women from wealthy families were simply prepared for a life of leisure.[6]

At Purdue in 1890 less than 17 percent of enrolled students were female compared to 36 percent nationally.[7] Nearly all the women were enrolled in the School of Science, which included the liberal arts. The preferred courses of study chosen by women at Purdue were history, biology, literature, and chemistry.[8] Just as Indiana governor James "Blue Jeans" Williams had feared university studies would leave men uninterested in work, a number of people in Indiana feared that four years of college would result in women losing interest in marriage and children.[9]

Many women took courses in the industrial arts program, which included drawing, woodcarving, and china painting. Among the most distinguished members of the industrial arts program at Purdue was Laura Fry, who arrived at the University in 1891. She only stayed one year, but returned in 1893 and stayed at Purdue until 1922. She was head of the art department and under her leadership Purdue developed a reputation in ceramic painting. In 1898 she was a founder of the Lafayette Art League. Fry developed and patented a technique for evenly applying pigment to ceramics.

International Enrollment

Land-grant universities were not created exclusively for students from their state. Purdue quickly enrolled students from Illinois and surrounding states, and international enrollment followed. The first international student known to enroll in the University, listed as a freshman in the 1888–1889 Purdue Annual Register, was Frederic Charles Scheuch, in mechanical engineering, from Barcelona, Spain. He came from Spain but he was an American. His father was in the U.S. diplomatic service. Also in that Register, listed as a special student in agriculture, was Genzo Murata of Yamaguchi, Japan. In the *Debris* 1890 yearbook Murata is listed as a senior studying science, becoming the first known international graduate of Purdue. Another Japanese student graduated in 1894, Seizo Misaki from Hyogo Prefecture, Japan. His degree was in mechanical engineering and he was a member of the Tau Beta Pi fraternity.

The 1900–1901 Purdue Catalogue listed 1,043 students from thirty-five U.S. states and the District of Columbia, along with students from the Philippine Islands, Belgium, Bulgaria, Germany, Ontario, and South America.

Enrollment and Fees at the Dawn of the Twentieth Century

Tuition was free for students who lived in Indiana, but there were annual fees: a $5 entrance fee and a $10 incidental fee. Library and laboratory fees were $12 for freshmen, $16 for sophomores, and $20 for juniors and seniors. In 1901 out-of-state students paid $25 annual tuition in addition to the fees. The 1900–1901 Purdue Catalogue lists the fee for room rent, heat, and light in the men's dorm and Ladies Hall as $1 to $2 per week, and board was $2.50 to $3 per week. With limited space in the men's dorm and Ladies Hall, most of the students boarded in West Lafayette and Lafayette, or lived at home with their parents.

Part Two

1900 to 1971

11

Dawn of a New Century

1900 to 1921

A conscientious New Englander who tried, with little success,
to adapt himself to a somewhat hilarious world.

—Purdue professor George Munro on Winthrop Stone

B orn during the U.S. Civil War, Winthrop Ellsworth Stone was in his prime
on May 4, 1900, when he gave a speech to high school students in Anderson,
Indiana. As young people do, the teenagers in his audience were looking forward.
As older people do, Stone was looking back.

It was the dawn of the twentieth century and as acting president of Purdue
University, Stone told the students he marveled at how far science and technology
had advanced in his lifetime.

"I suppose that when this generation has disappeared and the historians feel free
to speak their minds about us without fear of dispute, they will likely agree that this
last half of the Nineteenth Century was characterized by its wonderful scientific and
industrial development," Stone said. "Men are still living who rode upon the first
railroad train, who read the first electric telegram, who used the first rapid printing
press, who have been active in all the marvelous developments of the iron and steel
industry, who slept in the first sleeping car, who used the first turbine water wheel,
who first used ether to allay pain, who discovered similar drugs and nitroglycerin."[1]

It was an amazing list of rapid accomplishments. But it would pale beside what people would live to see next. There was nothing in Stone's powers of imagination that enabled him to comprehend that some of the teenagers in his audience that day would live to see a man walk on the moon. And that man would be a Purdue University graduate.

Having lost their longest serving, most popular, and most successful president, the Purdue trustees proceeded cautiously as well as swiftly in selecting the next person to lead the University. Their first move after the death of James Smart was to name Stone, a professor of chemistry and vice president of the University, as acting president.

By summer, two men were being considered for the Purdue presidency. One of them was Stone, thirty-eight, who was married and a pious Presbyterian. The other was fifty-five-year-old former Purdue professor Harvey Wiley, who was still single with a well-known reputation for not attending church. Wiley enjoyed his work as chief of the Division of Chemistry in the U.S. Department of Agriculture and said he was not interested in becoming president of Purdue, but that didn't stop the trustees from considering him.

At 2 p.m. on July 6, 1900, the board gathered in Denison House in Indianapolis. The *Lafayette Daily Courier*, an afternoon paper, reported that the meeting to elect a new president had been delayed until 3 p.m. when former U.S. president and current Purdue trustee Benjamin Harrison arrived. He had returned from a trip to Yellowstone Park. In an early edition the newspaper reported, "At present, it looks as if H. W. Wiley, of Washington, D.C. will be appointed president of Purdue, although nothing definite will be given out until the meeting is over." In a whiplash postscript in the same edition the newspaper added, "A later telegram to the *Courier* says Dr. W. E. Stone unanimously elected."[2]

The minutes of the meeting state that Stone was the only person nominated. He was elected with an annual salary of $5,000.

Wiley later said board members favored naming him president and several newspapers reported that. But then Harrison spoke up and said the former Purdue professor was an excellent person but not suitable because he was still a bachelor and he didn't attend church. According to Wiley, the other trustees had so much respect for the former U.S. president that they switched their support to Stone.[3]

At that point, Wiley had been considered for the Purdue presidency in 1875, 1883, and 1900. He would never be a candidate again.

Stone—whose middle name honored Elmer Ephraim Ellsworth, the first Union officer killed in the Civil War—graduated in 1882 from Massachusetts Agricultural College (now the University of Massachusetts Amherst). He studied chemistry and

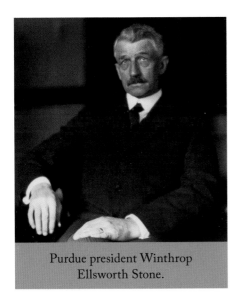
Purdue president Winthrop
Ellsworth Stone.

biology at Boston University before moving to Germany and the University of Göttingen, where he received his PhD in 1888.

In July 1888 Stone returned to the United States, joined by Victoria Heitmuller. They had met in Göttingen and married in Nashville, Tennessee, in 1889, the same year he became a professor of chemistry at Purdue. In time they had two sons, David and Richard.

A Disciplined Leader

Stone came from a distinguished family. His brother Harlan was named chief justice of the U.S. Supreme Court by President Franklin Roosevelt. Stone was the second Purdue president in a row born in New Hampshire. George Munro, a professor of engineering during Stone's tenure, called him "a conscientious New Englander who tried, with little success, to adapt himself to a somewhat hilarious world."[4]

Stone had his thumb on everything that took place on campus. When a department needed a typewriter, he remembered one had been stored in a campus closet and gave it to them. He kept a small black book in his pocket that listed the salary of everyone on the faculty. When the head of a department complained of flies and asked for screens on his office windows, Stone pointed to his own windows: "No screens," he said. And the department head did not get screens.[5]

The Tank Scrap

From early in his presidency Stone encouraged putting an end to a popular student activity called the Tank Scrap, which was more of a brawl than a scrap. In 1894 a large water tank was placed in West Lafayette on Salisbury Street at the southern edge of Grand View Cemetery. It was visible from the University, but it was not on Purdue property. The tank was viewed as an ideal location to paint the graduation year for University classes.

As soon as new students arrived on campus each September, upperclassmen pointed out to them that the numerals painted on the tank by a previous class were highly offensive and a disgrace to the University. Something would have to be done, and it was up to the new freshman class to take action.

Of course, this didn't happen without a fight. In the earliest days it was known to be wide open, no holds barred, and some years there were several battles with the numerals on the tank changing several times. In later years it was decided there would be just one battle. Eventually there were rules that were often ignored. Sometimes the battle took place at 2 a.m., and the number of spectators eventually grew into the thousands.[6]

During the brawls each class took prisoners and the victors paraded the vanquished to downtown Lafayette, where they were publicly ridiculed. They were sometimes left tied up all night or bound to signs that read "For Sale."[7] In later years the defeated were taken to Stuart Field next to the Armory.

West Lafayette water tower with class numerals, once the site of the Tank Scrap.

Stone had warned the students as early as 1904 that the Tank Scrap was "intolerable and indefensible."[8] In September 1913 the inevitable happened. When the fight was finished, a sophomore, Francis Obenchain, of South Whitley, Indiana, lay motionless. He died from a broken neck. The following day the entire Purdue community gathered and took a vote. The decision was unanimous. The Tank Scrap tradition ended forever.

McAnnix Burning

Stone also oversaw the end of another popular student event—McAnnix Burning. It celebrated students finishing their required courses in mechanical engineering and included a mock trial at which "McAnnix" and sometimes "Hy Draulics" were found guilty of various crimes. It ended with a book burning. In 1912 it also featured a lynching of "McAnnix" and "Hy Draulics" and activities involving the occult. That was the end of McAnnix Burning following criticism from faculty and students.

A Buried Bell

When Purdue opened in 1874, a bell was placed in the Power Plant to wake the students, announce classes, and notify everyone that it was 10 p.m. and gaslights had to be turned off. In September 1877, the bell was ruined in a fire that partly destroyed the Power Plant. A new bell was purchased. The plant continued in use until 1903, when it was demolished, and the bell was placed in storage in the locomotive lab. When Purdue defeated Indiana 27–0 in 1904, the students took out the bell, paraded it in downtown Lafayette, and left it on the courthouse steps. After it was left downtown and had to be retrieved several more times following Purdue victories, Stone hid the bell from the students, burying it where Hovde Hall was later built. In 1907 the students found it, dug it up, cleaned it, and built a frame for it. Students, faculty, and alumni donated funds to construct a Bell House. The Victory Bell is still taken to Ross-Ade Stadium and rung after Boilermaker victories.

A President for the School of Agriculture

Stone had a focus on the School of Agriculture. He worked closely with John Skinner, who became Purdue's first dean of agriculture in 1907. Skinner led Purdue agriculture to national prominence in a career that extended to 1939.

During his presidency Smart had waited for Indiana farmers to realize the importance of science to agriculture and come to Purdue. Instead of waiting for farmers to come to Purdue, Stone and Skinner took Purdue to farmers.

Purdue campus surrounded by farm fields. (Photo by President Winthrop Stone)

The federal Smith-Lever Act of 1914 created the system of cooperative extension services, but Purdue had done extension work since the 1880s. The Purdue trustees had created a Department of Agricultural Extension in 1911, and the General Assembly provided it with a small appropriation. While young people came to Purdue to study agriculture, it was decided the best way to reach large numbers of people was to extend the campus into communities throughout the

Dean of Agriculture J. H. Skinner operating a horse-drawn mower on Purdue farm, 1920.

state and teach farmers where they lived and worked. Farmers' Institutes and short courses were part of extension work. Eventually extension would come to include agents working in every county.

In 1905 Professor George I. Christie became part of Purdue agriculture, and in 1906 he became an associate in Purdue extension and later superintendent. Educated at

Students with cow in agricultural class.

Iowa State, he was called "eager, shrewd, aggressive, and a sound teacher," and a showman who knew how to take science to the towns and crossroads of Indiana. As a young man he had actually been a carnival barker. An often-told apocryphal story about Christie addressed his enthusiasm. The story said he stopped at an Indiana farmhouse and found people there waiting for the preacher to lead a funeral service. The preacher was late, so the funeral director asked if someone else could speak. Christie saw an opportunity. "I didn't know the deceased," he said. "But I know the work Purdue is doing in alfalfa and I'd like to tell you all about that."[9]

Educational trains were one of the most successful means of reaching Indiana farm families with programs for both men and women. Among the most popular was the "corn train," where Christie used his incredible teaching skills.

In 1909 the *Indianapolis Star* declared that the state had "forged to the front as producer of the best corn in the world."[10] As proof, the newspaper cited awards at national corn shows in Chicago, Illinois, and Omaha, Nebraska. "In no other state is such [an] effort for the growing of better corn being made as in Indiana through corn trains, lectures, state and local corn shows, the state corn school at Purdue and the State Corn Growers' Association. The gospel of 'better corn' has been preached by experts to 100,000 farmers."[11]

Christie was among those the newspaper credited for the success:

> The fact that Purdue has in its agricultural faculty an enthusiastic corn expert has contributed in no small way to the high attainments of Indiana as a corn state. This man is Professor George I. Christie. . . . It is mainly through his efforts supplemented by the members of an able corps of professors of the Purdue School of Agriculture aided by prominent Hoosier farmers that an increased interest in the growing of corn has taken a genuine hold of the farmers of the state. . . . The work of the experiment station at Purdue has much to do with the improvement of maize in Indiana.[12]

Most ag students were male, but women were admitted. In 1919 Lillian Louise Lamb became the first woman to graduate from Purdue Agriculture and was teased by the male students for her last name so well-suited to agriculture.

Department of Education

In 1907, the Indiana General Assembly passed a law creating qualifications for teachers. Stone told the board of trustees that Purdue should start a program to train teachers, and a department of education began in the School of Science. George

Roberts became the first member of the faculty in 1909. By 1920 there were seven faculty in education, and a graduate school was created in 1929.

Student Traditions

As Indiana public schools improved, more in-state students enrolled at Purdue. In the first years of Stone's presidency the University doubled in enrollment from 932 in 1900 to 1,861 in 1905.

The students of the early twentieth century had great Boilermaker spirit and started traditions that would last for many years. In 1904 several seniors saw corduroy trousers in a Lafayette store and decided they would be perfect for men in their class to wear. That fall half the seniors wore corduroys. The next year's senior classes adopted the trousers as a privilege—no one else could wear them. Seniors used the first home football game of the year to parade onto the field with their "whistling" trousers. At about the same time, juniors adopted the tradition of selecting hats to wear, and in 1907 freshmen were forced to wear green beanies.

Freshmen were also forced to sit in the upper gallery during chapel and convocations. "Freshmen up," was a familiar call from the upperclassmen as students filed in for events. Student handbooks published by the Young Men's Christian Association advised new arrivals on what they would encounter. Since University housing was in

Main entrance to campus at Class of 1897 Gates
off State Street, early twentieth century.

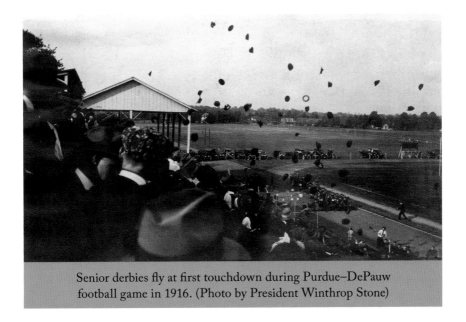

Senior derbies fly at first touchdown during Purdue–DePauw football game in 1916. (Photo by President Winthrop Stone)

short supply, the vast majority of students lived off campus. Students ate at boarding clubs that were not associated with their rooming houses. Prices ranged from $3 to $4 per week.

Ten Olympic Gold Medals

One of the most celebrated athletes in Purdue history was Ray Ewry. He graduated in 1894 with a degree in mechanical engineering and participated on the track team. But his fame began in 1900 when he won three gold medals at the Paris Olympic Games. He won seven more golds in subsequent games for a total of ten. His events were the standing triple jump, the standing long jump, and the standing high jump.

As a boy growing up in Lafayette, Ewry had contracted polio and used a wheelchair. It was feared he would be crippled for life. The doctor recommended a leg exercise: jumping. Years of jumping therapy turned him into an Olympic champion.

African Americans at Purdue

The University struggled with its diversity. African Americans enrolled at Purdue in small numbers at the dawn of the twentieth century, most of them in pharmacy, a profession that was needed in their segregated home communities.

The University was not immune to the racism that was common in the United States. Like many communities, Lafayette was segregated, and African Americans were not permitted to live in West Lafayette. The 1902 *Debris* yearbook, published by the senior class, included a satire about a fictitious "Southern Club" at Purdue. "To become a member of this club," the yearbook said, "the applicant must be white, must live south of the Ohio River, must have witnessed at least one lynching."[13]

The 1904 *Debris* featured a story about a minstrel show that included performers in "black face" and a song titled "Sports from Darktown."[14]

Richard Wirt Smith graduated in pharmacy in 1904. The *Debris* staff commented on all the senior photos. Most of the comments were sarcastic. The comments on African Americans were indicative of attitudes at the time. "Smith is somewhat of a buttinsky but it is partly excusable as he is the only one of his kind in the Pharmacy class," the *Debris* said about Smith.[15] He went on after graduation to own Smith's Pharmacy in Indianapolis.[16]

In the 1905 *Debris,* John Henry Weaver from Marion, Indiana, is listed as a graduate in pharmacy and a member of the Purdue track team and the pharmacy football team. "Weaver is an exception," the *Debris* said. "He bears the distinction of being the black sheep of our bunch and his popularity is due to the fact that he never shows any signs peculiar to the class with horns. He also enjoys the distinction of being the only one of his race who was eligible to wear the letter 'P' at Purdue."[17] A broad jumper, he was the first African American student to receive a varsity letter in Purdue athletics. But Purdue would soon follow a national trend of segregation in athletics in which black students were not allowed to play in varsity competition. Weaver became a food inspector in Chicago.[18]

In 1906 Samuel Saul Dargan tied for first in his class and earned a bachelor of science degree. Albert DeWitt Bailey graduated in 1909 with a bachelor of science degree and was a member of the Athletic Association. He became curator of the Indiana University Law School.[19]

Through much of the decade from 1910 to 1920, the *Debris* reacted positively to African American students, but they were very few in number. David Nelson Crosthwait Jr. enrolled at Purdue in 1909 and received his bachelor of science degree in 1913. He earned a master's degree in mechanical engineering and went on to a remarkable career that ultimately brought him home to Purdue. Born in Nashville, Tennessee, he grew up in Kansas City, Missouri, and attended segregated schools, where he excelled in math and science. He was nicknamed "Crossie" at Purdue, and the 1913 *Debris* commented under his commencement photo: "Dave . . . is a good man. He has successfully supported himself through the entire four years which marks him as a man to be respected and judging from his tenacity of purpose, Crossie will make a success of his future."[20]

During his career Crosthwait received thirty-nine patents for heating systems, vacuum pumps, refrigeration methods, and temperature-regulating devices. He helped design the heating systems in New York City's Rockefeller Center and Radio City Music Hall. At the end of his career he taught at Purdue and was presented with an honorary PhD in 1975. He died the following year.

In 1913, Homer Milton Taylor, of Indianapolis, became the first African American to earn a Purdue degree in chemical engineering. The *Debris* noted: "Taylor wants to become a teacher and with all the earnestness and energy he has shown in his work here, we know for a certainty that he will make his goal." He became director of the Mechanical Department, acting superintendent of buildings and grounds, and school engineer at West Virginia State College.[21]

Elmer James Cheeks in 1914 became the first African American to receive a Purdue degree in electrical engineering. His *Debris* commencement write-up stated:

"Ell" comes to us from the Blue Ridge Mountains of Virginia via the Cleveland Ohio schools. He has the distinction of being not only the only member of his race in our class but also, the first member of his race to take a degree from the school of Electrical Engineering. His chief characteristic is to stick to a thing until he gets it. . . . The fact that he has supported himself through his school and college courses stamps him as a man to be respected. He has made many friends both among the students and faculty many of whom will feel more kindly toward his race for having known him.

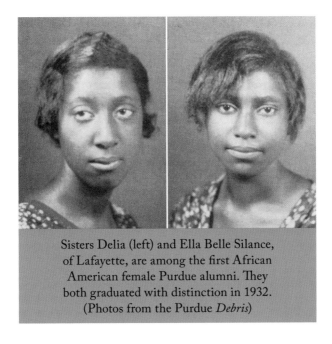

Sisters Delia (left) and Ella Belle Silance, of Lafayette, are among the first African American female Purdue alumni. They both graduated with distinction in 1932. (Photos from the Purdue *Debris*)

He became an electrical engineer for the City of Cleveland, Division of Light and Power.

Since for many years the University did not keep records on race, a Purdue graduate student, Alexandra Cornelius, went through ninety-nine issues of the *Debris* to find information about African Americans and campus attitudes. "There is little doubt that African-American students were making academic progress at Purdue" at the dawn of the twentieth century, she said. "Increased participation in professional and social clubs and organizations indicate that black students were becoming more integrated into Purdue's social life. Yet the campus continued to feature minstrel shows that depicted African-Americans as buffoons."[22]

International Students

In 1909 a Chinese Students' Alliance was founded at Purdue. By 1917 there were fourteen members. The 1905 *Debris* reported on the visit on May 24, 1904, by a Prince Pu Lun, of China, who was described as "a distinguished diplomat and a man of the highest rank in his country."[23]

The 1906 *Debris* lists Felix Edmunds Cuellar of Linares, Mexico, who received a degree in mechanical engineering. "He of the unpronounceable name is going to revolutionize Mexico along engineering lines when he returns to his native home," said the comments with his senior photo in the *Debris*. "He is one of the few foreigners who have stayed with us to the end."[24]

The Purdue Catalogue for 1910–1911 lists 12 students from seven foreign nations along with 361 from states outside of Indiana. Nations represented included Brazil, China, Cuba, Greece, Mexico, and Switzerland. Total enrollment was 1,705. Purdue was tuition-free for Indiana students, $25 a year for everyone else. There were $40 in fees that everyone paid.

"I'll flunk you if I can, and I can if I want to"

A professor who was loved by students of the late nineteenth and early twentieth centuries was Michael Golden. He was also the member of the faculty who most bewildered them.

Golden taught and became director of practical mechanics during a Purdue career that began in 1890 and ended when he retired in 1916. Born in Ireland, Golden grew up in the Boston area. It was home to many Irish immigrants and as a boy, Golden

Michael Golden with his sisters, Katherine and Helen.

worked barefoot in mills. He made spare cash as a boxer and he was good. The son of Irish immigrants, John L. Sullivan, the heavyweight champion of the world, said of him, "He didn't have to be a college professor. If he'd stayed with boxing he'd have gone far in the fancy."

Golden used money he made boxing to attend the Massachusetts Institute of Technology, and after two years Purdue's W. F. M. Goss hired him to teach.

He was serious about his work and dedicated to the University. If anyone questioned his authority, he was quick to respond. If an argumentative student challenged what he said, Golden invited the student to join him outside, where the issue would be settled in a boxing match. He taught at Purdue for twenty-six years. No one accepted the offer.

He delivered his lectures in a rich Irish brogue. In his book *The Story of Purdue Engineering*, H. B. Knoll wrote, "He could be arbitrary and threatening: 'I'll flunk you if I can, and I can if I want to.' 'I'll flunk you just as soon as I can get my pencil out.' 'I don't care whether you are right or wrong. If I say you're wrong, you're wrong.'"[25]

When Golden gave his class a quiz and a baffled student handed in a blank sheet of paper with only his name on it, Golden gave him a minus ten—zero points for the quiz and minus ten points for spoiling a perfectly good piece of paper.

Completed in 1910, Purdue's Practical Mechanics Building was the largest engineering construction project under Stone's leadership. In 1920 the board of trustees,

Michael Golden Laboratories. Students called it "Mike's Castle."

at the urging of alumni, renamed it Michael Golden Laboratories. Students simply called it "Mike's Castle." The main part of the building was taken down in 1982 when Knoy Hall was built, but laboratories attached to the former main building continued in use into the twenty-first century.

Golden had two sisters, Katherine and Helen, who followed him to Purdue, earned bachelor's and master's degrees, and became instructors.

"Hail Purdue"

The song that has tied generations of Boilermakers together is "Hail Purdue." In 1912 during the Stone presidency, Purdue student James Morrison wrote the words and sent them to another student, Edward Wotawa, who wrote the music. Wotawa was a member of the Glee Club, which performed the song at a convocation, and it was an immediate hit. "Hail Purdue" was copyrighted in 1913, and it's been sung countless times since at everything from athletic events, to alumni gatherings around the world, to student demonstrations on campus. Wotawa graduated in 1912 and Morrison in 1915.

Tragedy

Very soon after being named president, Stone led the University through a terrifying event that shook its soul. It happened on Halloween, 1903. There was no University fight song at that time, hailing Purdue and its heroes. But it was a time when Boilermaker men and women became heroes rushing to help injured and dying people, most of them football players, lying amidst twisted metal and splintered wood in a railroad yard in Indianapolis.

GIANT LEAPS
IN RESEARCH

Katherine Golden Bitting

She was the first nationally acclaimed Purdue University female researcher. But whenever her name was mentioned at the school where she got her start, she was identified as the sister of Michael Golden, a fiery, flamboyant, very popular Purdue professor of mechanics.

Katherine Golden Bitting had a distinguished research career in biochemistry and food preservation that began at Purdue and lasted nearly fifty years. Her 1937 obituary in Lafayette's *Journal and Courier* identified her as a "nationally known scientist" and "famous graduate" who had received an honorary PhD from Purdue two years earlier—and the sister of Michael Golden.[26]

One of her first research projects with her husband, Avril Bitting, was to determine how ketchup could be produced commercially without adding preservatives. Preservatives in ketchup were used to hide unsanitary conditions at the factories. In 1906 the use of those preservatives was impacted by former Purdue professor Harvey Wiley and the Pure Food and Drug Act. The Bittings turned their Lafayette home into a model ketchup factory and analyzed more than sixteen hundred bottles of the condiment.[27]

The parents of Michael and Katherine were poor Irish immigrants who arrived in Canada, where Katherine was born, then moved to Lawrence, Massachusetts.

Michael worked as a barefoot boy in factories and earned additional money by boxing to attend a two-year program at M.I.T. In 1884 he was recruited to Purdue by W. F. M. Goss, the University's first teacher of mechanics, who also had a two-year degree from M.I.T.

While teaching at Purdue, Michael Golden earned a bachelor of engineering degree and a professional ME degree. His success opened opportunities for others in his family. Katherine and another sister, Helen, followed him to West Lafayette and received bachelor of science degrees in 1890. Katherine immediately began teaching biology at Purdue and earned a master's in 1892. Helen began teaching in practical mechanics with her brother in 1906.

Purdue launched Michael, Katherine, and Helen Golden, children of poor immigrants, into successful lives that impacted many people through their teaching and research.

Katherine lived with Michael for a time before marrying Bitting, also a member of the Purdue faculty.

The two continued research throughout their lives with the U.S. Department of Agriculture and later with the National Canners Association. They wrote many books and collected more than forty-three hundred volumes of gastronomic literature that were donated to the Library of Congress. And they discovered that adding a precise amount of sugar and vinegar to ketchup kept it in fresh without harmful preservatives.

12

Football Train Wreck: Cheers to Tears

1903

Tomorrow will be the greatest day of my life.

—PURDUE STUDENT CHARLES GRUBE

I t was the most terrible day ever at Purdue University.

It started with all the excitement and anticipation that pulses through college campuses on fall football Saturdays. "Great gridiron struggle today, Lafayette will send an immense delegation to Indianapolis to witness the contest between Purdue and Indiana universities at Washington Park for supremacy in the great college sport," the *Lafayette Morning Journal* proclaimed on Halloween morning—October 31, 1903.[1]

Game time was set for 2:45 p.m. and ten thousand people were expected. Thousands of laughing, cheering football fans gathered in downtown Lafayette at the Big Four Railroad Station, many preparing to board two trains with wooden passenger cars bound for Indianapolis and others just wanting to share in the fun. "The contest will be the greatest to be played this season between Indiana colleges and the outcome practically determines the state championship," the *Journal* noted.[2]

At 3:15 p.m. that day Lafayette's afternoon newspaper, the *Courier*, hit the streets with its headlines and stories. It listed eleven dead Boilermakers following a horrific train crash. There would be more deaths. It took place on what would have been John Purdue's 101st birthday.

There were several big football games for Purdue each season. But the contest against Indiana University had reached epic proportions. Edward "Robbie" Robertson, a 1901 Purdue graduate and football star who set a world record with seven field goals during one game in 1900, returned to campus two weeks before the big contest in 1903. He had taken a leave of absence from work in Indianapolis to help coach the Boilermaker team.

Nearly a thousand people boarded the first train. The football players, coaches, managers, and trainers occupied the lead coach. Members of the band and the faculty with some female guests were in the second, and fans filled the coaches behind them—including Purdue president Winthrop Stone and his wife, Victoria.

It was an early morning departure, about 7:30 a.m., and in the lead passenger car team members and coaches were confident, laughing, talking about the game and their plans for victory. The second train departed ten minutes later. At 9:50 a.m. the first train of coaches entered the north side of Indianapolis and came to a sharp curve along a gravel pit. It was a switching area where trains could be moved from one track to another. At the controls engineer W. H. Shumaker believed he had the full right-of-way to rush through to downtown Indianapolis. Some later reports said he was traveling at twenty-five to forty miles per hour. The engineer and conductor both said the train was moving about twenty miles per hour.

In the switching area, railroad workers and engineer E. J. Smith were moving seven steel coal cars on the main track at about six miles per hour, thinking they had the right-of-way. Coming to the turn, the engineer of the Purdue train, Shumaker, saw the coal cars coming toward him on the same tracks only one block ahead. He hit the air brakes, reversed his engine, and having done all that he could, jumped from the train. Fireman L. E. Irvan also saw the crash coming and moved to the coal car. Both Shumaker and Irvan survived.

When the two trains collided, the engine of the Purdue train was knocked from the track. It lay on its side, "a heap of broken and bent metal and splintered wood . . . enveloped for several minutes in a blinding mist of hot and hissing steam," according to a *Journal* article the following day.[3]

Stone rushed out of his coach, ran forward, and what he saw was unspeakable carnage: body parts, blood, and youth cut short. Stunned, he gathered himself together and joined others helping the injured. Seventeen people died from horrible injuries,

Purdue train wreck.

all of them in the first passenger car occupied by the team and special guests. The *Courier* reported that afternoon:

> The dead are on every hand. One player had his head cut off. Another player was struck in the face with a piece of iron and was cloven to the brain. . . . The arm of a man was fastened beneath the wheel of the second coach, though the body was not found. . . . The horrible screeching of the injured and the mangled remains of the dying struck terror to the hearts of the passengers. . . . The bodies of the football players were suspended from the wreckage. Strong men wept and cried aloud when they looked upon the faces of their dead classmates and fraternity brothers. Many of the girls rolled up their sleeves and knelt at the side of the stricken men and bathed their wounds until the services of surgeons could be secured.[4]

A physician from the University of Chicago was lecturing students in the assembly room at the Medical College of Indiana in downtown Indianapolis when a messenger rushed in with a note that read: "Terrible wreck on the Big Four at Northwestern Avenue. Send as many seniors as possible." Panic erupted as students rushed to the door, all trying to get through at once. Within ten minutes one hundred medical students were helping to treat the injured.[5]

A student was found crushed on the track beneath a heavy timber. It took several men to lift the board, and they asked his name. "Powell of Texas," he said, gasped, fell back, and died.

Meanwhile twelve hundred excited students arrived in Indianapolis on a special train from Bloomington. "They were met with news of the wreck," the *Courier* reported. "Instantly the bands ceased, colors were lowered and tears followed."[6]

Among the dead was Robertson, who had joined the team for just this one game. Also dead was Patrick McClair, the team trainer from Chicago, and Newton Howard, a Lafayette businessman and loyal Boilermaker fan. The other fourteen people who died were members of the football team. At least fifty were injured.

In Lafayette and West Lafayette business ceased at noon when the names of the dead began to arrive by wire. It was difficult to determine who had been injured and killed. At the scene of the accident, people picked up students and rushed them to hospitals without telling anyone who they were taking and where they were going.

The *Indianapolis Journal* reported on the afternoon of the accident:

> Parents of the Lafayette students who were known to be on the train became frantic. The telephone offices were crowded and at the Western Union telegraph office two hundred fifty messages were received in two hours. . . . The Big Four railway offices were besieged by hundreds of men, women, and children, but nothing was given out there, so the crowds returned to the newspaper offices and waited for fresh bulletins to appear. Occasionally a shriek would be heard as some person in the throng saw a name of a relative or close friend added to the list of casualties. Several women were prostrated and were carried home. Mrs. Harry Leslie, mother of [former football player and team manager] Harry G. Leslie, learned at noon that her son had been badly hurt. Then news came that he was dead, only to be followed by another report that he was still alive. She was so shaken that neighbors had to carry her to the bedroom where she raved and cried in agony. Her grief was pitiful to the extreme.[7]

It was later reported that Harry Leslie had been taken to the morgue. But someone there saw him move, and he was rushed to medical care. In 1929 he was elected governor of Indiana.

On Sunday, November 1, the *Indianapolis Journal* provided first-person accounts. I. S. Osborne, captain of the Purdue team, told the newspaper:

> "I was riding backward. . . . My feet shot from under me and I was plunged forward. I have no distinct recollection of what happened, but from the crashing

of glass I must have shot through one of the car windows. It seemed as if the fall would never end. I struck the ground and someone fell on top of me, but who it was I do not know as the air was black with coal dust. And just to think that (Walter) Roush (22 years old) was killed, the man with whom I roomed for three years." And with a sigh he turned on his cot at St. Vincent's Hospital and would say no more.[8]

F. M. Hawthorne, of Wingate, left end on the Purdue team, told the *Journal*:

I guess I was thrown through a window. I only know I felt myself flying through the air. It seemed to me I was five minutes on the way and I could feel myself moving and hear humming in my ears. I didn't see anything. My head was dizzy and it was all dark with flashes now and then. I found myself lying across the ditch when I came to. I can't shut my eyes now without feeling that sickening sensation of going through the air in the darkness with the roaring in my ears and the sparks in my eyes.[9]

Purdue president Stone remained in Indianapolis to find, identify, and then keep track of the injured and in some cases dying students in several Indianapolis hospitals. In the days that followed, as accounts of the tragedy appeared in newspapers throughout the nation, worried parents called Stone's office seeking information on their sons and daughters. Telegrams came with a common closing plea: "answer quick."[10]

A mother in California inquired if her son was injured. A long-distance call asked for information concerning Purdue student Harry Walmesley. Stone responded he had "no reason to believe he was injured." From Veedersburg, Indiana, came a request on the condition of Charles Furr, of Fountain County, a member of the football team. "From facts in our possession we fear the worst. Reports conflicting," Stone responded. His mangled body had been found under the train engine. From Davenport, Iowa, C. W. McManus asked for information on his son. "Your son is hospitalized in Indianapolis. Communicate with Dr. W. Stone, Denison Hotel, Indianapolis," was the response from the president's office. The parents of Gabriel Drollinger, twenty-one, of Mill Creek, Indiana, had already been advised by a member of the Purdue faculty that their son had died. They wanted to know where they could recover the contents of his pockets. Stone told them the items were held at the Indianapolis police headquarters.[11]

There were intense moments at the hospitals. D. M. Hamilton, of Huntington, the father of Jay Quincy Hamilton, wrote to Stone: "I fear that when you met my wife at Indianapolis she used interminate language and accused you unjustly. I know that you are too large a man to notice such expressions under such conditions and yet I want to assure you that the feeling that you were in any degree responsible for that awful catastrophe cannot and does not exist within us." Jay Hamilton's injuries were fatal.[12]

In Beardstown, Illinois, all business was suspended during the funeral for Walter Hamilton. "He was the very soul of fidelity," the *Purdue Exponent* reported. "In his pocket after his death was found a New Testament which had been presented to him by a young lady to whom he was engaged."[13]

The pressure on Stone and the intensity of his grief showed at a memorial service for the students held on campus on November 11. Railroad workers from the coal train in the switching yard said they were not aware of the Purdue train rushing toward them. The engineer for the Purdue train said he had no knowledge of the coal train. Stone didn't buy any excuse:

> The fatal collision was the result of man's foolish or ignorant disregard of the simplest natural law. God did not desire the accident. We cannot conceive that he ever wishes evil to his creatures. But, he has entrusted them with knowledge and with power to guide and utilize these natural forces and when man uses his power foolishly or neglects the responsibilities . . . evil follows. Not even God himself can prevent the evil resulting from a foolish or neglectful human act. We are all responsible for one another. It is said that some of those concerned in the railroad disaster did not know of the special train. It was their duty to know and their ignorance is the cause of this day of mourning. Somewhere the plea of "I did not know" or "I forgot" becomes the admission of fatal guilt.[14]

While an investigation of the accident continued, the railroad reached settlements with the families. Settlements ranged from $1,200 to $3,750. Some families hired attorneys who were paid out of the settlements.

On December 24, 1903, a grand jury investigating the accident decided against issuing any indictments. The jury said workers in the yard should have been notified that the Purdue train was coming through, but they were not. The jury placed the blame on "an imperfect system of train handling."[15]

Purdue fielded a football team the next fall. It went 7–2, including a 27–0 victory over Indiana in Indianapolis. Talk of a memorial on campus began almost immediately after the accident and culminated on May 30, 1908, when the Memorial Gymnasium was dedicated. Fund-raising for the gym had actually begun in 1892, preceding the train wreck, but enthusiasm for the project greatly increased after the accident. Hundreds of people including alumni, students, trustees, faculty, and employees of Purdue donated to help build the gym. The Big Four Railroad also gave money toward construction. At the dedication, Stone said donated funds totaled $54,255.66, with another $5,000 in pledges yet to be collected. The University added $25,000. Stone said:

> Often as I have watched this structure rise under the builders' hand has come to my vision a heart-rending scene which I can never quite banish from my recollection. . . . That dreadful calamity of that beautiful October morning in 1903 when the flower of our athletes met dreadful death. . . . No one could have dared think then that out of that disaster was to spring a new hope, the most fertile imagination could not conceive a vision of this splendid building arising from the depths of misfortune. . . . This building is a monument and memorial to the seventeen young men intimately connected with the life and activities of Purdue University who on the morning of Oct. 31, 1903 suffered death. . . . This building is the embodiment of the great truth that out of disaster comes victory, that in all human experience the day of sorrow and disappointment is not the last day but there surely follows a new dawn where hope revives and life take on new aspects of promise and interest.[16]

John Wooden, who became a legendary coach at UCLA, played basketball for Purdue in the Memorial Gymnasium. He graduated in 1932.

Convocation, Memorial Gymnasium, September 14, 1927.

The front entry of the Purdue gymnasium was built with seventeen steps, one for each person who died. In the twenty-first century the building, which had been converted to classroom use, was renamed for former provost Felix Haas. The seventeen steps remain.

Among the most compelling items found after the train wreck was a note inside the pocket of Charles Grube. The twenty-one-year-old member of the football team had written it the day before the big game.

"Tomorrow will be the greatest day of my life," it read.

It was the last day of his life.

13

The Problem of Purdue's Duty to Women

1904 to 1918

Be a man, Miss Shoemaker. Be a man.

—President Winthrop Stone

Images of Purdue University students lying injured and dying along railroad tracks in an Indianapolis switching yard were still fresh in Winthrop Stone's memory on May 4, 1904, as he stood to deliver an address beside a portrait of the first woman to be memorialized in a building on campus.

The woman was Eliza Fowler, the wife of Moses Fowler, who had come to Lafayette more than sixty years earlier with his business partner, John Purdue. Moses Fowler met and in 1844 married Eliza in Lafayette. Fowler died in 1889, having amassed a multimillion-dollar fortune.

Upon his death Eliza Fowler inherited a large part of his wealth and began to use it for charitable purposes. She donated land for a new Presbyterian church near her home. And in 1901 Fowler, who never studied at Purdue, walked into Stone's office to announce an unexpected gift to the University—$60,000, which was later increased to $70,000. The only condition of the gift was that it be used to build

Purdue campus, 1905: University Hall (background, far left), Fowler Hall, Electrical Engineering Building (center), Power Plant, and Heavilon Hall (right).

an auditorium on the campus. Unlike John Purdue and Moses Fowler, who liked having their names attached to businesses, a town, and a university, Eliza Fowler asked for no recognition.

Purdue named the new building after her anyway—Eliza Fowler Hall. It marked the largest gift to the University since John Purdue launched it in 1869, and it was the first major gift by a woman. Located in the center of campus, Eliza Fowler Hall became the main meeting and administrative building. The seating capacity in the auditorium was more than 1,300.

Taking an idea from Smart when he announced the gift from Amos Heavilon, Stone told the Purdue students, faculty, and staff about the Fowler gift in a morning chapel. "The complete surprise of the students and faculty was followed by a momentary silence and then came an outburst of joy which will never be forgotten by those who heard it," Stone said in his building dedication speech. "For many years the university has possessed no assembly room large enough to contain the student body nor of suitable character for any public occasion. The many embarrassments growing out of these conditions have been at times almost intolerable."[1] From 1895 to 1902, commencement had been moved to a tent on the Oval in front of University Hall. In June 1898 a rainstorm hit, blew down the tent, and left everyone drenched.

In addition to administrative offices and the auditorium, Fowler Hall included a beautiful pipe organ donated by James Fowler, Eliza Fowler's son. The auditorium also had a new grand piano. Stone kept it locked up and he had the only key. Eliza Fowler died in 1902 before her hall was completed.

Auditorium in Eliza Fowler Hall including pipe organ donated by her son, James Fowler.

Home Economics

In September 1905, Purdue began a program in home economics for female students. A majority of women on the campus took part in the full four-year program or at least some of its classes.[2]

"Its creation was an appropriate recognition of the part that women are entitled to in the educational advantages of Indiana," Purdue librarian William Murray Hepburn and history professor Louis Martin Sears wrote in their 1925 book, *Purdue University: Fifty Years of Progress*. "The rising feminist movement possibly may claim some credit for the University's more serious grappling with the problem of its duty toward young women."[3] The Hepburn and Sears book was among the earliest to use the word "problem" in relation to courses of study for women.

Purdue agriculture professor William Latta was an advocate for educating women. As head of the Farmers' Institutes, he included women speakers on domestic and farming issues. Latta advocated for a home economics program at Purdue, and he believed it should be in the School of Agriculture. In addition to Virginia Claypool Meredith, the "Queen of American Agriculture," who spoke on agricultural issues, he recruited other female speakers to attract women to the institute events. Latta and Meredith played important roles in Purdue home economics.

In 1895 Meredith was completing her work with the World's Columbian Exposition and was planning to return to her farm. Life is what happens while people make plans, and Meredith's lifelong love of agriculture and advocacy for women would lead her in 1897 to the University of Minnesota to start a home economics program. Following the death of a friend, she had adopted the woman's two children, a boy, Meredith (named for Virginia), and his sister, Mary Matthews. The children went with her to Minneapolis, and in 1904 Mary became the first woman to receive a bachelor's degree in home economics from the University of Minnesota. Virginia Meredith left Minnesota and returned to Indiana in 1903.

She had continued speaking in Indiana and elsewhere while at Minnesota, but once free from her academic work, she spent even more time talking about farming and homemaking. She resumed her work with the Purdue Farmers' Institutes and the Winter Short Courses. She advocated for creation of a home economics program at Purdue. Meredith's support played an instrumental role in the creation of the Purdue program in 1905. She helped to change the mind of Stone, who initially was not interested in the new department. In the years ahead, Meredith would be an important part of the University in many ways. The first woman on the Purdue Board of Trustees, she served from 1921 until her death in 1936.

The Purdue *Debris* of 1906 said the new department in home economics was created "to give the young women of the University the same opportunity as that

Female students cooking in home economics class.

which is given to the young men, of thorough preparation and training for her inherited profession—that of home-making."[4] Courses included food principles, home and household management, nutrition, hygiene, cooking, and textiles. The classes were held in Ladies Hall, and groups of women took turns living and working in a "practice house."

The first head of Home Economics was Francis Harner. She left Purdue in 1908 and was replaced by Henrietta Calvin. In 1912 Mary Matthews, the adopted daughter of Meredith, became head of the department her mother had worked so hard to create. Matthews was an innovative crusader for home economics programs and education for women. In 1926 she created at Purdue the first nursery school in the history of Indiana. She became the first dean of the School of Home Economics and continued to lead the program until she retired in 1952 with the title dean emerita. For many years she was the only female academic dean on campus.

Mary Matthews.

During Matthews's tenure, Gertrude Sunderlin and her students invented the "Master Mix," which was created in the home economics labs of Purdue. Homemakers used it for quick and easy preparation of baked goods. Matthews was instrumental in the creation of the new Home Economics Building, completed in 1922. In 1958 it was named in her honor.

In 1914 Matthews hired Lella Gaddis to work in the summer school, training home economics vocational teachers. Gaddis remained at the University until her retirement in 1947, and with Matthews she forged opportunities for women at Purdue and throughout the state.

While Matthews focused on home economics, Gaddis became Indiana's first state leader of home demonstration, traveling through rural Indiana teaching women about nutrition, hygiene, safe water, childcare, and more.[5]

Women were already performing important roles on campus even before the home economics program began. President James Smart had hired Emma McRae in 1887. Students knew her as Mother McRae. In 1900 McRae brought Carolyn Shoemaker to Purdue, where she taught English literature. Shoemaker had received a master's

Washing machine demonstration by the Purdue Extension Service in 1921.

degree from Purdue fourteen years earlier and for a time cared for her incapacitated mother. In 1912 McRae retired. In 1913 Stone summoned Shoemaker to his office in Fowler Hall. Shoemaker was described by Professor H. L. Creek, who was head of the English Department where she taught, as a woman who "calm as she might seem, had a deeply emotional life . . . her power as a teacher and woman lay in her warmth of feelings."[6] Stone rarely displayed emotions.

Shoemaker was forty-eight years old the day Stone called her to his office. He was only three years older. No doubt small talk was held to a minimum and he got right to the point. Many universities were creating the position of dean of women, Stone said. He said he believed Purdue should do the same, and he asked Shoemaker to take on the additional responsibilities of the position. She hesitated. She said she was surprised and was not sure she was qualified. Stone's response has become among the most memorable quotes in Purdue history: "Be a man, Miss Shoemaker. Be a man. Do not let this or any other task worry you."[7] There is no doubt Stone believed acting like a man would make Shoemaker a better and

Carolyn Shoemaker, first (part-time) dean of women.

stronger person and leader at Purdue. But, in fact, he wasn't looking for someone who thought like a man for his dean of women. And in Shoemaker he would find a woman who was capable of any task placed before her.

She served in the position until her death in 1933. "Purdue was not a part of her life," said Stanley Coulter, dean of men from 1919 to 1926. "Purdue was her life."[8]

There were fewer than one hundred women on the Purdue campus when Shoemaker was named dean.

The right of women to vote—which would not be final until 1920—was being hotly debated throughout the country at the time, including at Purdue. The 1911 *Debris* called senior student Mary Catherine Kennedy "contrary Mary," who "got her start by instigating an unsettled argument on women's suffrage and she is willing to expound her theories at any time. She loves to defeat any man in an argument."[9]

The yearbooks also made it clear that what the future expected of Purdue female graduates was marriage and caring for home and family. The 1911 *Debris* said of senior Mary Keiffer, "She will probably teach china painting in some Southern seminary until one or the other succeeds in convincing her that he is of more importance than anything else in the world."[10]

In 1918, five years after Shoemaker assumed her position, there were 260 female students on campus. She reported in the *Debris:*

> Girls come to Purdue to get a technical, useful and practical education along with the cultural development. Here she has an opportunity to study scientific cookery. . . . Such girls as come to Purdue will give the world a message of simple and wholesome living. In the near future we expect to have a new woman's building and then with the increased number of girls that it will bring we have great ambitions. Girls will be president of the classes, will be elected to Prom Committee and will be admitted to Iron Key. Don't give up girls, there's a great day coming.[11]

Iron Key was an honorary organization for male students.

All of Shoemaker's predictions came true, although not for young women studying at Purdue in 1918 and not for their daughters. It came true for their granddaughters.

Part of the credit goes to federal legislation written by a Purdue graduate—and a man. (See part 3, chapter 30.)

14

The Purdue School of Medicine

1903 to 1908

No more important educational advance can be accomplished
in Indiana than to forget the rivalries and conflicts in
which our colleges have too often engaged.

—WINTHROP STONE

Early in the twentieth century Purdue and Indiana Universities became embroiled in a conflict like nothing seen before or since. Indiana University saw it as a fight for its very survival, and the controversy had the potential to severely damage Purdue.

It was a turning point in the histories of both universities and decided the future of higher education in Indiana. In a very real sense the cooperative relationship between the two major research and teaching universities emerged out of a disagreement that took them right to the precipice.

It was inevitable that as Purdue grew in stature, the missions of the two universities would collide. In 1906 and again in 1907 Purdue determined its Land-Grant Act mission included the teaching of medical sciences. It conferred MD degrees on 118 men and 4 women in 1906 and 68 men and 4 women in 1907. Among them was a man whose name would become among the best known in Indiana medical care: A. A. Arnett, co-founder of the Arnett Clinic in Lafayette.

As Purdue conferred degrees, Indiana University loudly proclaimed that it, and only it, could be the state's publicly supported medical school, asserting that Purdue's Land-Grant Act missions limited its focus to agriculture and engineering.

The disagreement lined up Purdue and IU graduates against one another at family dinner tables and in the Indiana General Assembly. Newspapers took sides. Neighbors argued and doctors disagreed. Everyone had an opinion. As the universities sought separate state approval for their dueling MD programs, they even started counting who had the most alumni in the legislature (it was IU).

Indiana University historian Thomas D. Clark wrote, "At no time in its whole history did Indiana University come nearer the brink of political chaos, if not actual self-destruction of the university itself, than when it began the serious business of organizing a medical school. In this adventure it found itself in a damaging public rivalry with Purdue University and a group of willful doctors in Indianapolis."[1]

Medical science had advanced rapidly in the late nineteenth century, but medical education lagged behind. Writing for the *Indiana Magazine of History* in 2002, Walter J. Daly said that during the 1880s medical schools in Indiana were similar to many others in the United States: "Largely free of state oversight . . . schools expected incoming students to be able to read—little else. Few U.S. medical schools required more. Nearly half required only a common school education. In fact, few universities viewed a medical school as little more than a trade school."[2]

Among the medical programs in Indiana were two private, proprietary medical schools in Indianapolis named the Central College of Physicians and Surgeons and the Medical College of Indiana. There was also a Fort Wayne College of Medicine. All three were owned and operated by doctors, and many people considered them inferior to publicly operated and regulated programs in some other states.

William Lowe Bryan became president of Indiana University in 1902 and quickly set it on a course of expanding professional and graduate programs. IU had to find itself in the new century.

Bryan told the IU trustees: "The University has expanded too much in the liberal arts and sciences in contrast to other universities—too little in professional fields. In my opinion, the University's future depends mainly on a successful change in this situation. The people will make a great university only upon the condition that the university does many kinds of important things that the people want done."[3]

What the people wanted done was the creation of a state-run professional education program for doctors, and almost as soon as he arrived at IU, Bryan worked to give them one.

In 1903 Bryan led the University in a plan creating two years of medical education at IU in Bloomington. After two years in Bloomington the plan was to send the students elsewhere to complete their degrees. To accomplish this, IU would merge with a private medical college in Indianapolis, or launch its own school in Indianapolis, or even send its students out of state for the final two years of study. It was essential that the final two years take place in a larger city, where students would have opportunities for clinical experience. Bloomington, as well as Lafayette, was too small to provide the clinical caseload these students needed.

On June 21, 1903, the *Indianapolis Star* reported that the Central College of Physicians and Surgeons had offered to merge with Indiana University, donating its facilities, provided they were used for the medical school.[4]

One month later the Medical College of Indiana offered to join the Central College of Physicians and Surgeons in seeking the merger with IU.[5]

Negotiations proved to be difficult. A plan to unite the two proprietary medical schools with IU was approved by faculty in the College of Physicians and Surgeons but rejected by the Medical College of Indiana faculty.[6] Still, the Medical College continued to negotiate with IU.

By the fall of 1904 negotiations were still ongoing, and if an agreement was accomplished, it was announced that the College of Physicians and Surgeons would join.[7] One sticking point in negotiations was Medical College dean Henry Jameson's insistence that a separate board, instead of the IU trustees, administer the merged medical school. A second alternative he proposed was a strong representation of the Medical College on the IU board. Bryan rejected both proposals.[8]

In addition, Jameson insisted that IU drop its two-year medical program in Bloomington and transfer all four years of study to Indianapolis. Bryan was also opposed to this. He believed the proprietary programs in Indianapolis were too strongly focused on clinical work with not enough time spent in classrooms and laboratories.[9]

Talks broke down, but Bryan assumed that at some point they would continue. Instead, Jameson turned to Purdue—without telling Bryan or making any public statements about what he was doing. Purdue president Winthrop Stone entered into negotiations eagerly, but only after being assured by Jameson that talks with IU had failed and ended.

On September 1, 1905, the Purdue University Board of Trustees approved a merger with Jameson's Medical College of Indiana, creating the Indiana Medical College, Purdue School of Medicine, in Indianapolis. Purdue had no plans for a two-year medical program on its campus similar to IU's in Bloomington, so that was not an issue. It was agreed that all four years of the Purdue School of Medicine would be in Indianapolis. Also, Jameson agreed that the Purdue School of Medicine would be under the control of the University's board of trustees.

It was further noted that legislation needed to be enacted authorizing Purdue to operate a medical school in Indianapolis. Therefore, the Purdue medical school would operate at first under a memorandum of understanding, pending authorization by the General Assembly.[10]

The board meeting where this agreement was approved took place in Indianapolis in the office of Governor Frank Hanly, who took part in the lengthy discussions, approved the agreement, and might have brought everyone together.[11]

The *Star* called the agreement "a wise step and one that is in line with the professional teaching of the day."[12] The *Indianapolis News* said as a result of the merger, "there will be increased strength for both institutions. . . . Some time ago the plan of uniting the medical school with the State University [IU] was considered and seemed likely to be adopted. But there was a failure to agree on details and now Purdue wins."[13]

With the Medical College of Indiana and Purdue successfully merged, the Central College of Physicians and Surgeons joined the agreement on September 25, 1905.

The Fort Wayne College of Medicine joined on October 9, 1905. The *Star* reported that 320 students from the new Purdue School of Medicine went to Union Station in Indianapolis to welcome students and faculty from Fort Wayne. They proceeded to march through the business sectors of the city and around the Monument Circle to the new home of the school, carrying a University banner.[14]

Bryan was not pleased. There were also angry protests from some medical students in Indianapolis who sided with IU. IU historian Clark wrote, "Purdue was accused of exceeding both its declared charter purpose and legal authority. . . . Partisans of Indiana contended that the Lafayette school was organized as an institution of mechanical arts and agriculture. The authority to 'embrace professions' was reserved to Indiana University by law."[15]

John Purdue's 1869 offer to the state of $150,000 and promise to find one hundred acres of land did require locating the University in Tippecanoe County. But Stone contended that the University would still be located there. It would just offer a branch in Indianapolis for the study of medicine. Purdue already was offering pre–medical school courses and it had a School of Pharmacy. On IU's side, an 1838 law had authorized it to form a medical school as soon as financially possible.[16]

In 1906 friends of IU purchased the Central College of Physicians and Surgeons Building in Indianapolis to be used by IU as a new medical school. The *Bloomington Telephone* summed up IU's concerns and its reasons for planning a medical school of its own: "If Indiana University's medical department is to be captured by Purdue and friends of the institution [IU] lay down, how many years will it be until some other department here goes the same way?"[17]

Some newspapers that sided with IU turned farmers against Purdue by incorrectly saying that the land-grant school was turning its interests away from agriculture to medicine.

The first doctors graduated from the Purdue School of Medicine in the spring of 1906 (they had started their studies with the proprietary schools). While classes were in Indianapolis, commencement was held with great fanfare in West Lafayette. Tippecanoe County historian Robert Kriebel wrote, "For Purdue's 1906 commencement . . . Stone invited members of the Tippecanoe County Medical Society. Some members marched to the Big Four Railroad Depot (in Lafayette) to escort the medical school faculty, the graduating class and friends from Indianapolis. Hundreds came along on special trains. There was a buffet luncheon, a procession from the library to Eliza Fowler Hall and a reception by the faculty and Medical Society in Agriculture Hall."[18]

In 1907 Purdue and IU both appealed to the General Assembly to settle the issue. The *Indianapolis Star* reported, "War is on now between the factions of the medicine men and scalpels are out. The person on the fence cheerfully remarks, 'Go it, Purdue! Go it, I.U. and may the best crowd win.'"[19]

The two sides met head-to-head at a meeting of the Education and Judiciary Committee of the Indiana Senate on January 22, 1907. Every corner of the meeting room was packed with people.

Stone told committee members, "The Medical School of Purdue University is now a self-supporting institution and ready to be given as a gift to the state of Indiana without condition." He said the Purdue School of Medicine in Indianapolis did not deprive IU of its rights. IU students could take their first two years of work in Bloomington and then transfer to the Purdue School in Indianapolis—just as Bryan had planned to do with one or more of the proprietary schools. Stone also pointed out that land-grant universities in other states had started medical schools.[20]

Bryan asked the legislators how far they wanted to go in using public funds to support duplications at IU and Purdue. "It has seemed to us that this might be settled in two ways," he said. "One would be to allow a large amount of duplication in the work of its state schools, or on the other the state might insist that there be as little duplication as possible in the state schools. If the state of Indiana is willing to pay for two state universities each doing what it pleases it is not for us to object. It has been our judgment, however, that the state would not approve such conditions."[21]

If Purdue had received approval for its medical school, IU would probably have responded by launching programs in engineering and agriculture in addition to medicine in order to compete.

The General Assembly adjourned without taking sides. On February 12, 1907, the *Bloomington Telephone* wrote, "doubtless [the universities] realize fully by this time that if they do not live together in unity they will die separately."[22] In fact, during the legislative session, Purdue's request for a budget increase was ignored.

Throughout the rest of 1907 the sides battled. By January 1908, Stone had experienced enough. He telephoned Bryan and asked for a meeting in Indianapolis. Governor Hanly, who initially supported Purdue's merger with the Medical College of Indiana, encouraged the sides to settle. They met at the Claypool Hotel. Negotiations were intense. Bryan and the IU Board of Trustees rejected several of Stone's proposals. But by spring they had an agreement. Purdue would drop its medical program in Indianapolis. IU would operate a four-year course of medicine in the city, and the program would be under the control of the IU trustees. IU historian Clark called the date of the agreement the University's "second foundation date."

"Had Indiana lost its battle to establish a medical school it would have dropped the key out of the arch of its university structure," he wrote. "No longer could it have claimed with much justification to be the state university, and no doubt it would have demoted to the role of a secondary college."[23]

Stone said:

Purdue's only purpose has been to advance medical education and serve the state. It has never been regarded as an essential part of the program of the institution to control medical instruction if it could be done equally well or better by any other institution. The failure of the legislature to lend some assistance to the plan to promote medical education as well as the attitude of Indiana University complicated the situation, but out of this has come a solution which seems to insure a cessation of strife and ultimate progress toward a medical school of which the state may be proud. What is of equal import is the putting aside of a course of contention between the two state universities which will enable them to apply their undiminished energies in their respective fields.[24]

In a letter to Stone, Bryan wrote, "The two state universities while organically distinct, should and do maintain a certain relation and interdependence, their curricula supplement each other in that they together comprise a general scope administered in certain universities in a single institution. There is no prospect of future consolidation. Therefore we should eliminate as many sources of conflict and friction as possible."[25]

In 1908 the Indiana General Assembly approved the proposal to allow IU to conduct a medical school in Indianapolis.

In a gesture of friendship, Stone was invited to speak on the Indiana University campus on June 5, 1908, during the dedication of a new library prior to commencement. "I trust that the time is forever past when the success or progress of any educational institution shall inspire any other than feelings of real satisfaction and approval among its sister institutions in the state," he said. "No more important educational advance can be accomplished in Indiana than to forget the rivalries and conflicts in which our colleges have too often engaged in the past and in their place to cultivate a broad spirit of cooperation in advancing the interests of higher education in Indiana."[26]

During the controversy, for the only time in their histories since the sport began, Purdue and IU did not schedule football games in 1906 and 1907. The schools did not play basketball in 1907. Purdue and IU did not compete one-on-one in track in 1906 and 1907, and they did not play baseball in 1907. Purdue had suspended the games, but the suspension wasn't initiated because of the medical school issue. Purdue believed Indiana was violating conference eligibility rules, that debate became bitter, and the athletic rivalry became dangerously intense.

On December 30, 1906, the *Indianapolis Star* published a letter from Dr. A. W. Bitting, the husband of Katherine Golden Bitting and the alumnus director of the Purdue Athletic Association. Bitting stated that Purdue had stopped athletic contests with IU "based upon the differences in the enforcement of the college conference rules relative to ineligibility of players and to differences in the methods of conducting business relating to contests."[27]

In announcing the resumption of games on January 14, 1908, the *Star* reported, "Intense rivalry has always existed between the two schools and the feeling grew into hatred following some unfortunate incidents at a baseball game in Lafayette in 1906. A riot almost resulted that day, but the ardor of both universities has since cooled down."[28]

While the medical school issue did not stop the games, Stone was not interested in negotiating a resumption while he was fighting with IU for rights to the medical school.

On January 10, 1907, as he prepared to make his case with the Indiana General Assembly, the *Purdue Exponent* reported on a speech Stone had given to the students. "He expressed the hope that Purdue will play ball with Indiana some time," the article said, "but expressed his opinion that for the present Purdue is quite right in not playing, since the relations of the large bodies of students are so strained."[29]

It was a difficult time for Stone. Following the horrible train wreck and loss of life, the stress on Stone from the medical school controversy and cessation of athletics with IU was intense. And there was more adding to his personal trauma.

His wife had abandoned him and their two sons.

15

A Man of Stone Quietly Faces a Personal Crisis

1907 to 1919

The daily papers make no mention of my
divorce. And I hope they will not.

—WINTHROP STONE

As Winthrop Stone was struggling over the medical school and unwanted disagreements with Indiana University, he was also going through enormous personal problems. They might even have been a factor in his ending what had become a fight with IU about the medical school. Perhaps he just had too much stress in his life.

In 1907 or 1908, at the height of the medical school controversy when he was spending a great deal of his time in Indianapolis, Stone's wife, Victoria, abandoned him and their sons.

By 1911 he had given up all hope that she would return and was making plans to move on with his life. To do that Stone had to divorce Victoria, the woman he had brought to the United States from her native Germany. His June 15, 1911, diary entry details the quick and very quiet divorce proceedings in the Tippecanoe County Courthouse. At that time his son David was a student at Harvard University. His younger son, Richard, lived at home. During a twenty-minute session in a Tippecanoe County court, Stone was granted a divorce and custody of the minor child.

In his diary Stone wrote, "The daily papers make no mention of my divorce. And I hope they will not."[1]

They did. On June 20, the *New York Times* carried a story that focused national attention on the situation with the scandalous headline "India Cult Causes Divorce, Dr. Stone of Purdue University Deserted By Wife, a Sun Worshipper." Stone confirmed to the newspaper that years earlier his wife

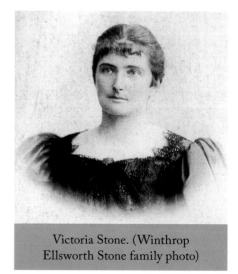

Victoria Stone. (Winthrop Ellsworth Stone family photo)

had withdrawn from the world, including separating from him and their children, to pursue the philosophy of yoga. "He and his two sons are heartbroken, and would eagerly have welcomed her back, but are unable to reach her," the article said.[2]

According to the *Times*, Victoria Stone had joined a Lafayette class in yoga philosophy in 1907–1908. "Many women and some men in the college town joined the class, which became the fad with certain people," the newspaper said. "It was taught that the complete power of Yoga involved withdrawal or separation from kindred and friends. The last heard from President Stone's wife she was in Germany but it is reported she has gone to Kabakon [Papua, New Guinea], a South Sea island to join a colony of the cult."[3]

Members of the cult, founded by German August Englehardt, lived on a diet of coconuts. Englehardt died in 1919. He was forty-three.

According to his diary, Stone told the Purdue University Board of Trustees on June 14, 1911, that he had decided to seek a divorce. He further noted that he had offered his resignation, but the board unanimously refused to accept it. No doubt he also told the board that he needed the divorce because he intended to marry Margaret Winter, a woman with whom he had fallen in love. He called her "M" in his diary.

Stone had just turned forty-nine in June 1911. Margaret Winter was twenty-seven, so there was a twenty-two-year age difference that might have raised some eyebrows on campus and in the community. But there was more to this story.

When Stone divorced Victoria, Margaret was living in Stone's home at 146 North Grant Street in West Lafayette, and she had been there since 1903, when Victoria brought her to West Lafayette.

Like Victoria, Margaret was born in Germany, probably in 1884, according to the 1910 U.S. Federal Census. The Census report said she immigrated to the United States as a child in 1888, so she must have come with parents or other family. Her parents died and Victoria took her from an orphanage in New Jersey and brought her to West Lafayette. The 1910 Census form said Margaret's last name in Germany was Binter and that she was a "roomer" with Stone in West Lafayette.

Stone's grandson was named for him. Born in 1928, the son of Richard Stone, he said the story of Margaret is a mystery in the family. "She was brought to the family by my grandmother [Victoria]," Stone said. "She helped out around the house and then when my grandmother left she stayed on."[4]

Stone waited one year after his divorce before proceeding with his marriage to Margaret. The pastor of his church, Dr. John Hale, performed the 3 p.m. marriage ceremony on July 10, 1912. Both the *Lafayette Daily Courier* and the *Lafayette Morning Journal* reported that the Stones were married "quietly" in their home. The *Journal* said there were about twenty guests and Margaret "wore a simple white wedding gown. After the congratulations the bridal couple left for northwest Canada where they will spend the summer climbing mountains. . . . Dr. Stone and Mrs.

Purdue president Winthrop Stone with his second wife, Margaret.

Stone will return to Lafayette early in the autumn."[5]

Stone's grandson said Margaret was always a part of the family from that time on. "We called her Aunt Margaret because my grandmother [Victoria] was still alive," he said.[6]

His father, Richard, was in Germany during the world war and remained there after the armistice. While there he found Victoria. For a time, before Richard Stone married, Victoria came back to the United States and kept house for him. When he married, she returned to Germany and lived with her brother, who was a professor of dentistry.[7]

"For a long time my dad [Richard] supported his mother," Stone said. "All during the 1930s he corresponded with her in Germany. Then during World War

II he lost touch with her. After the war he got the International Red Cross to help find her and she died shortly thereafter." Margaret never talked about her past. "She's sort of a mystery person in a way," Stone said. "And my father never talked about Victoria. Every once in awhile he'd let it slip and mention a cult. But I don't know. I've always wondered if she was really happy in this country."[8]

Near the end of Margaret's life, she asked the family to go through personal belongings she had in storage. "There were books, papers, dishes, stuff like that," Stone said. "And there were some letters I'm surprised Aunt Margaret kept. They were correspondence between my grandparents after Victoria had left the family but before my grandfather married Margaret. My grandmother was in Germany. She apologized to my grandfather and said she would like to come back. My grandfather said, no."[9]

Football Coaches Paid More Than College Presidents!

Stone's other major struggle during his presidency was with a group of alumni. Athletics and other extracurricular activities were at the heart of the disagreement, and the leader of this opposition was the University's most famous and popular graduate—George Ade. Under Stone the University took direct control over athletics. It had previously been run by an autonomous association that included alumni.

Stone believed in the importance of athletics for students. But he had concerns about football, and these even entered the speech he gave at the dedication of Eliza Fowler Hall in June 1904. Stone was also concerned about the number of extracurricular activities on campus.

In that Fowler Hall dedication talk he drifted onto the topic of college football and said it had led to "excesses and extravagance. . . . Under the head of athletics the student organizes his play into a business enterprise which, in many instances, exceeds in volume the entire income of many respectable colleges. He employs a teacher of football at a salary greater than that of a college president."[10]

He went further in other talks, including at a Purdue football banquet in the Lahr House in downtown Lafayette on December 18, 1905.

"The popular outcry against football is based upon the severe criticism of the dangers, the corrupt practices, and the exaggerated relations of the game," he said. "These criticisms are well founded. That it is a dangerous game the numerous recorded deaths and casualties testify; that the conduct of the game is attended by practices and methods which are unfair, unsportsmanlike and unworthy of college men and college spirit we are unfortunately convinced; and that the game occupies an undue prominence in college life is the consensus of opinion of those whose observation and experiences qualify them to speak with authority. These and other objectionable features of football have become so prominent as to obscure much that is commendable and meritorious in an attractive and important form of athletic recreation."

He wasn't alone in this thinking. The *Chicago Daily Tribune* reported that in 1904 alone there were eighteen football deaths and 159 serious injuries. Most of the injuries involved high school players. At the time, newspaper editorials across the country called for banning the game. The issue went all the way to President Theodore Roosevelt, who pushed for reforms.[11]

Stone's public statements about the sport were not popular with many alumni and fans, including George Ade, who in 1912 became the first alumnus elected a member of the board of trustees.

In 1916, Ade resigned from the board, citing disagreements with Stone's policies. He was further angered by the campus reception of a play he wrote for the annual Purdue Harlequin Club show that year. He had even prepared a speech to be delivered after the first act, but he wasn't given an opportunity. He left without seeing the end of the show, and his resignation was official the following day.[12]

In March 1917 Ade wrote a letter stating his unhappiness with the University. "I have not been at Purdue for about a year," he said. "I left there disgusted with Doc Stone and most of the faculty and thoroughly discouraged because of the lack of any real spirit or enthusiasm among the students. I have made no definite plans for returning at any time although I still have the kindliest feelings in the world for the Sigs [Sigma Chi fraternity]. As for Doc Stone, I wish him everything he wishes me and I could not say anything rougher than that."[13]

In 1918 Ade became editor of the *Alumnus*, and an open dispute with Stone began in its pages.

In a December 1919 article Stone wrote, "The constant multiplication of diversions and 'activities' among college students has reached the point where even the most liberal-minded of college and university authorities feel that great inroads are being made upon scholarship. Recognizing to the utmost the necessity for recreation and diversion among students one cannot help but feel that these things have gone to excess in our American colleges and that here at Purdue they are becoming one of the several factors which are operating against the realization of the best results in teaching."[14]

Ade shot back: "The student 'activities' for which the alumni have been pleading may carry with them some incidental 'recreation and diversion' but primarily they are intended to unite the student body in loyal support of their alma mater, strengthen the community spirit, cultivate initiative, teach the value of team-work, make the college four years a sentimental journey as well as a hill-climb, and give the young people that important training which comes only from many-sided contact with people of brains and ambition and sweet human qualities."[15]

In a May 20, 1919, letter to former Indiana governor Samuel Ralston, who had left office two years earlier, Ade said: "I am taking an awful chance in writing this letter because I don't know what you think about Doctor Stone of Purdue. What I think about him cannot be set forth in this letter as I do not wish to violate the federal statue against sending profane matter through the mail."[16]

Ade's campaign against Stone ended fourteen months later—but not in a way either man had ever expected.

16

A Bulwark of Defense in Time of War

1917 to 1918

Purdue was conceived and born in a spirit of service to the nation. There can be no doubt that the institutions thus created were intended to be bulwarks of defense in time of war.

—WINTHROP STONE

The book Winthrop Stone used to keep his diary during the year 1917 was a gift from Margaret, the woman he had married five years earlier. "May there be much happiness recorded for the year," she wrote on the cover.[1]

Less than four months later the United States entered a world war, and Stone found his days filled with more work, pressure, and concerns than he had ever known before.

The war on the other side of the Atlantic Ocean had started in the summer of 1914, and people in the United States hoped to stay out of it. Gradually events and sentiments shifted, and Congress approved a declaration of war on April 6, 1917.

On April 2 Stone wrote in his diary: "Congress assembles to a special call to hear the President's message on a state of war with Germany. Great excitement throughout the country. Special [Purdue] committee meeting for directing students in enlistment and a special faculty meeting this evening at the library. All present signed a statement of allegiance."[2]

On May 27 Stone gave a Memorial Day address in West Lafayette. His message was stark: "Our young men must lay down their peaceful pursuits and face death on sea and land," he said. "This is the lesson of the day, the necessity of sacrifice for the preservation of all that we value."[3]

Three days later he gave another Memorial Day address on campus: "To no class of people does this hour appeal so deeply as to us of the University," he said. "It is for us to set an example of practical patriotism not for any self-glorification, but as our plain duty as American citizens. . . . No other university has a higher responsibility to the nation than Purdue. Purdue, and others of like kind, was conceived and born in a spirit of service to the nation. . . . There can be no doubt that the institutions thus created were intended to be bulwarks of defense in time of war."[4]

Stone urged students not to enlist until they were called, continuing their studies as long as possible. "Do not desert your classes half trained," he said. "The war will last a long time and the need for trained men and women will steadily increase. . . . Chemists, physicists, biologists, nurses, pharmacists, engineers and farmers, in short exponents of every kind of teaching given at Purdue, will be greatly needed, but not in half developed stages."[5]

Two hundred eighty-five Purdue men did enlist by the end of summer, 1917. Twenty-four members of the faculty signed up. Those who remained on campus were left "to keep the home fires burning."[6] The School of Agriculture went immediately to work helping increase Indiana food production. Under Mary Matthews, Home Economics faculty stressed the importance of food conservation throughout Indiana. Purdue's Lella Gaddis was state leader of home demonstration, and she went to work hiring agents to help women in Indiana learn to conserve food.[7]

The opening of the school year in the fall of 1917 brought continued enthusiasm for the war effort. Enrollment had dropped from 2,226 in 1916 to 1,687. It was the lowest since 1904. By the fall of 1918 with the conscription age set at eighteen, the military draft threatened to take all the male students.

In 1918 the government created the Student Army Training Corps (SATC). Across the nation 157 colleges and universities participated, including Purdue. Under the program, universities pledged to train draftees in fields needed for the war effort. The SATC provided an opportunity for students to continue in college while also receiving military training.

Purdue pledged to house, feed, and train fifteen hundred SATC men during the 1918–1919 school year. The government paid the University one dollar per day for

A cornfield shared the location of these SATC barracks. Demolition of the barracks came on December 14, 1918. By mid-1919 new grass had almost covered the scars where the barracks and other camp buildings had stood.

each man plus twelve cents for tuition. Buildings were constructed to house the men, along with sheds to hold equipment.[8] Women also did their part.

In the 1918 *Debris,* part-time dean of women Carolyn Shoemaker said women at Purdue "have done their bit, if we measure by the number of knitted garments, of which ninety five have been made by the freshmen, one hundred five by sophomores, one hundred three by juniors and one hundred one by seniors; again if we measure by surgical dressings for which the girls have turned out regularly on Thursday nights."[9]

The war ended November 11, 1918. The *Debris* yearbook that came out in the spring of 1919 talked about the somber mood that had prevailed on campus at the beginning of the academic year that fall:

> A restless, nervous feeling of indifference pervaded the campus last September when school opened and studies once more were resumed. The cause of the abnormality was the new selective service legislation, which included all men from the ages of 18 to 45. As the law stood, it meant the doom of the university and colleges within a short period as it would eventually have drafted all physically fit students. The Student Army Training Corps, which had been introduced to offset the disadvantages offered by the new selective service act made a distinct appeal to the many thousand of students over the country. With only a month remaining before the S.A.T.C. became a reality the school activities were existing more or less in a state of coma and some activities were not resumed in the least at the opening of the school year due to the uncertain trend of affairs.

The *Exponent* did not publish. University societies and clubs did not meet. Debating clubs, literary societies, and dancing clubs disbanded until the war was over. The future of athletics, and especially the football program, was in doubt.[10]

The global influenza pandemic of 1918–1919, which infected about five hundred million people worldwide and killed at least fifty million, also impacted Purdue, where eleven people died. In a 1918 *Alumnus* editor George Ade reported: "Oct. 17—the whole state is still under lock and key on account of the flu and Purdue is running on one cylinder. Oct. 18—Further suspension of University classes announced by Dr. Stone (because of flu). Sad news received—Glossop, football captain in 1913, a fine athlete and popular fellow, dies at Camp Taylor of influenza."[11]

When the armistice came, Purdue students joined in celebrating with the people of Lafayette and West Lafayette. Parades of people marched up and down the streets all day and over to "Camp Purdue," as the *Lafayette Journal* called it. "It was a sight such as was never before seen in this fair Indiana city and the like of which undoubtedly will never be witnesses again. Thirty thousand people celebrated last night and it was not a peace celebration, but a victory celebration."[12]

Shops and stores closed. Offices, factories, and schools shut down and people paraded jubilantly through the streets to share their excitement and to mark a turning point in history and hope. The Purdue military band followed by University military detachments led the main parade. Purdue coeds paraded with the Victory Bell—the first time females had been given charge of the iconic bell.

"The city streets were thronged during the early morning but long before the time set for the start of the parade, the streets were impassable," the newspaper said. "The parade stood in line awaiting the soldiers and sailors from Camp Purdue. After a short wait the strains of the Camp Purdue military band's music floated from lower Main Street and the crowd cheered wildly."[13]

Behind the band marched fifteen hundred Purdue soldiers, three hundred naval reservists, and twenty army trucks. Soldiers, trucks, bells, bands, and joy stretched for three miles.

Four thousand Boilermakers served in the military, 16 percent overseas. Both of Stone's sons, Richard and David, served. Sixty-seven Purdue men died in service, seventeen in action. They were young men like Robert Earl Symmonds from the class of '15. Perhaps anticipating war, he left the University in 1914, entered the U.S. Military Academy, and graduated in 1917. By December he was heading overseas. On November 3, 1918, he reported to his commanding officer and asked to be assigned to a company in contact with the enemy. While leading his company in an attack that night he was wounded.

He died November 22. It was eleven days after the armistice.

17

Winthrop Stone Saw Nothing Higher

1921

How better could he have arranged the draperies of his soul for immortality than by an ascent into the mountain top where for a moment the horizon encompassed the transient glory of a world.

—Purdue trustee James Noel

On the evening of July 3, 1921, Winthrop Stone sat at a desk in his Grant Street home and entered a note in his diary. He said the weather was very hot and his wife, Margaret, was struggling with it. "M needs to go soon," he wrote. "Remained at home preparing camp equipment. All ready. Tomorrow a busy day."[1]

It was the last entry he made in his daily diary. An outdoor enthusiast, Stone was preparing for the great joy of his life: mountain climbing. He loved nature and was an experienced and even an expert climber. Margaret had learned to love it as well.

Two weeks after his last diary entry, Stone and Margaret were climbing Mt. Eon in southwestern Canada on the border of Alberta and British Columbia, west of Calgary. They were about to become the first people to scale the peak.

Stone was in the lead and called down to Margaret that he could see nothing higher. He had reached the summit. Just as he spoke the rock slab he was standing on gave way. Margaret watched in horror as her husband fell directly past her, still holding his pick axe. Stone had disconnected a rope that connected them shortly before the fall. He bounced off the rock wall of the mountain as he descended to his death in a crevice. Margaret later said he had fallen five thousand feet.[2]

She immediately began struggling down but became trapped on a ledge. She remained there without food and only a trickle of water from melting snow, hoping rescuers would find her.

At Purdue University, faculty, staff, and friends spending a quiet summer break from classes believed the Stones were having one of the most wonderful adventures of their lives, when the Associated Press broke the story on July 26. The news was quickly followed by a telegram received that afternoon by Helen Hand, Stone's secretary. The telegram read: "Dr. and Mrs. Stone missing several days. No word from search parties yet. Fear accident."[3]

President Winthrop Stone: "I can see nothing higher."

Because of the Stones' experience and careful natures, everyone hoped that they had run into trouble but would be found alive.

The Stones had set out from their base camp on Saturday, July 15, 1921, with enough food and water for four days, leaving word about when they expected to return. When they failed to come back, hundreds of men led by the "Father of the Rocky Mountain Climbers," Professor Charles Ray of Tufts University, set out to find them, following the route they believed the Stones had taken. As the rescuers progressed, they stopped and looked over the mountainside through binoculars. They saw nothing.

Suddenly one of the searchers stopped the others. He thought he had heard a woman's voice. They all stood silently and listened, and this time they all heard her voice. Searching the landscape with the binoculars, they saw a woman far below across the canyon. One hour later Margaret Stone was rescued—eight days after she became trapped on the ledge.[4] She was suffering from exposure and lack of food and water. They later recovered Stone's body.

A memorial service was held for Stone in Lafayette on October 12 after Margaret had recovered from her injuries. Among the speakers was James Noel of the Purdue University Board of Trustees. "His eyes were fixed upon the mountains and his spirit moved toward the heights," Noel said. "He ascended until he could see nothing higher and as the sun of a perfect day was lowering behind the mountains he passed into Eternal Life. How better could he have arranged the draperies of his soul for immortality than by an ascent into the mountain top where for a moment the horizon encompassed the transient glory of a world."[5]

In 1969 Margaret died in Jacksonville, Florida. She was eighty-five and had never remarried. Her remains were returned to Lafayette and Spring Vale Cemetery, where she was buried, without a headstone, beside her husband.

Upon the death of Stone, for the first time since 1900 and only the second time since 1883 the board of trustees was tasked with finding a new president. Board member Henry Marshall, publisher of the Lafayette *Journal and Courier*, was named acting president. The board would find their new leader. But he would not come from the east or from Indiana as in the past.

The next Purdue president would be the first to come from the west.

18

Edward C. Elliott
Wakes Up Purdue

1922 to 1945

It is my intention shortly to propose to the faculty . . .
that our curricula be reorganized to permit students
of ability and persistence to complete the work for
a degree within three years instead of four.

—Purdue president Edward C. Elliott

Searching for a new president to lead Purdue University, the board of trustees did not require that applicants know legends of the West such as William "Buffalo Bill" Cody and Annie Oakley. They didn't consider it a prerequisite to have a friendship with international heroes such as General John Pershing, commander of the American Expeditionary Forces on the western front in World War I.

But in Edward C. Elliott, that's exactly what they got—a man who grew up in Nebraska, who had served at the University of Wisconsin and the University of Montana, and who had never been closer to West Lafayette than Champaign, Illinois, where his bride found the weather far too hot.[1]

He had known some interesting people in his day, and instead of being a "granite from the east" like Winthrop Stone, in Elliott Purdue found its first president from

144

west of the Mississippi River. Not surprisingly, he was completely different from those who came before him.

Edward C. Elliott

Board members spent six months in a national search for Stone's successor. In April 1922 they found Elliott, forty-seven, the chancellor of the University of Montana, the son of English immigrants, whose father had been a blacksmith for a time. Born in Chicago in 1874, Elliott was the first Purdue president born after the U.S. Civil War. He lived for most of his childhood in North Platte, Nebraska, where he came to know fellow townsman "Buffalo Bill" Cody. In 1887 Elliott traveled with his family to England. Cody was there at the time, taking his Wild West Show on a European tour. The Elliotts were Cody's guests at the show, and he introduced them to the stars, including sharpshooter Annie Oakley.

In his book *Boilermaker Music Makers,* Joseph L. Bennett said Elliott's powerful demeanor led people to believe he was taller than he actually was. "Although he was five feet ten inches tall in his shoes, people thought of Elliott as a tall man," Bennett wrote. "He was ascetically thin with deep-set eyes, a nose large enough to look down, and a massive authoritarian chin. An excellent speaker as well as a good conversationalist, Elliott spoke in slow, deep, rolling tones and was fond of alliteration."[2]

Elliott graduated from the University of Nebraska in 1895 with a degree in chemistry, a field he held in common with Stone. He stayed at the University and in 1897 received a master of arts degree. It was at Nebraska that he met Pershing, who served as the University's military instructor. Pershing died in 1948 at the age of eighty-eight, and until the end Elliott remained in contact with him.

Purdue president Edward C. Elliott.

Elliott stood and sat rigidly straight—a military trait he said he learned from Pershing. Elliott's children remembered that at their family dinner table he told them to "sit up straight and get those shoulders back or I'll have to get a brace for you."[3]

A man who loved education, after receiving his master's degree in 1897 Elliott taught and became superintendent of schools in Leadville, Colorado, where he met his future wife, Elizabeth. They didn't marry until 1907.

Though they stayed in contact through letters, he left Elizabeth and Leadville in 1903 and accepted a fellowship at Teachers College, Columbia University in New York City. He received his PhD in 1905, took an assistant professorship in education at the University of Wisconsin, and in 1916 became chancellor of the University of Montana, his last position before coming to Purdue.

Elliott, who would serve Purdue until 1945, was second in a line of three consecutive Purdue presidents whose tenure (including interims) would stretch through most of the twentieth century—seventy-one years or nearly half the University's history at its sesquicentennial.

Elliott ran the University. There was no question who was in charge. Elliott led Purdue with his intellect and forceful personality through the Roaring Twenties, Prohibition, the Great Depression, and World War II.

"He demanded, fought, threatened, pacified, and encouraged, striving as if some barrier of ignorance or inertia were blocking his way and always looking for new means to increase the importance and prestige of his university," H. B. Knoll wrote in his book *The Story of Purdue Engineering*. "He had not only programs, but the unremitting drive to prod and promote, forcefully and dramatically, even melodramatically. Purdue was greatly changed under the impact of his personality."[4]

When Elliott arrived he thought Purdue was "sleepy and staid." But he thrived on challenges. He thrived on conflict. He scolded the faculty at his first meeting with them just days after he officially began his duties on September 1, 1922. He refused to let members of the board of trustees smoke at their meetings because students were not allowed to smoke on campus.[5] He played by rules. And he made everyone else follow them, too.

"There was an air of buoyancy and drama about him, even when he dealt with routine affairs," Knoll wrote. "When he received a faculty member in his office, he would be busy with papers at his desk, would thump his desk lightly, and would talk roughly, almost in a bark."[6]

He wrote memorandums for faculty, staff, and sometimes students, averaging about one a month. On February 6, 1924, seventeen months after he arrived, he issued Memorandum 22: "To: All members of the University staff. Reports have been received in this office that on a number of occasions electric lights not in use have been found burning and that water has been left running." He said the superintendent of the physical plant had been ordered to report all future such incidents directly to him.[7] Two months later Elliott reported a noticeable decrease in the use of electricity.

By May 10, 1928, in Memorandum 85, he was clearly fed up with the growing problems cause by automobiles on the campus.

> During recent weeks frequent complaints have been presented relative to the violations of motor traffic regulations on campus and also as to the fast and reckless driving of motor cars on the campus driveways. A preliminary examination has brought to light the regrettable fact that in not a few instances those responsible for the violation of parking and speed regulations are members of the University. . . . Instructions have been issued to the Superintendent of Buildings and Grounds to arrest and to prosecute in the courts all individuals guilty of violating either the parking or the speed regulations. I trust it will not be necessary to take extreme measures against any member of the University staff.[8]

In 1927 Elliott showed his softer side when Ladies Hall was taken down. He had the Old Pump, a symbol of romance outside the building that was part of the original campus, saved, reconditioned, and placed south of University Hall. In 1958 the Purdue Reamer Club Spring Pledge Class resurrected it at the southeast corner of what is now Stone Hall.[9]

Overall, the faculty liked and respected Elliott, and students were said to love him. Shortly after being named president, he showed up at a pep rally, took off his suit coat, rolled up his shirtsleeves, and gave an enthusiastic talk. The students were instantly sold on their new president. Stone had never done anything like that.[10]

A New Presidential Home

While the trustees had their man in April 1922, there were still details to be worked out. In May, Elliott presented a list of conditions to the board before he would accept the position, most of which were easily handled. He wanted his appointment to be unanimous among board members. It was. He asked for not less than $12,000 a year in pay. The board gave him $10,000, but Elliott agreed to that because the board agreed to a stipulation that he be given a house suitable for entertainment. All previous Purdue presidents owned their own homes. But times were changing. University presidents were being required to do more entertaining. Furthermore, Edward and Elizabeth Elliott had four children aged seven to fourteen and her mother lived with them. Seven people needed a large home, and they preferred not to be on campus.

Purdue president's home on Seventh Street in Lafayette,
first occupied by Edward C. Elliott.

While the board was deciding where the Elliotts would live, the large home of Dr. Guy Levering, at 515 South Seventh Street in Lafayette, came on the market.

Guy and Priscilla Levering had built the Seventh Street home four years earlier for about $70,000. In addition to being a physician and surgeon, Levering came from one of the most prominent and prosperous families in Lafayette. The couple had one daughter. In November 1922 Priscilla was granted a divorce from her husband. Eleven months later on September 8, 1923, Guy Levering married Clara Stitz. She was from Lafayette, but the marriage took place in Shreveport, Louisiana. The divorce probably led Levering to sell the house to the University quickly. He and his second wife moved to a new home nearby.

Elliott and his wife loved the Seventh Street home. The trustees bought it for $50,000 plus not more than $6,000 for furnishings. There would also need to be some remodeling. The Elliotts moved into Purdue's first presidential home in August 1923 and lived there until Elliott retired. In a warm and structured environment, the children enjoyed sitting at the dinner table in the large home and hearing their father tell stories—like the time he accidentally slipped his chair off a stage while awaiting his turn to speak at a sports banquet. He told them the fall was "the Notre Dame shift."

During his early tenure Elliott turned down several attempts by the board of trustees to increase his salary until April 1929, when the trustees raised his pay to $15,000 without giving him an opportunity to dissent. That same spring the alumni showed their affection for Elliott by presenting him with a new Packard sedan. It replaced his ten-year-old Hudson Superior Six touring car that had seen better days. It had been the first car he ever owned. The alumni presented him with a second Packard in the spring of 1937.

WBAA

Elliott's arrival on campus coincided with the launch of a radio station. WBAA received its license to operate in 1922. It is the oldest continually operating radio station in the state of Indiana. Early in its operation WBAA aired adult and distance education programs, preceding television and the Internet.

There were no textbooks telling them what to do when Purdue students began experimenting with radio, so electrical engineering students and faculty used their own knowledge and creativity. To improve their signal, they strung an antenna line between the old Powerhouse smokestack and the Electrical Engineering Building. They made their own condenser and used burlap to soundproof the studio. Students originally were only allowed to broadcast two to three hours on weekends, and they had to share the frequency with three other stations. They traded time slots with the other stations to air Purdue football games.[11]

In March 1929 fire destroyed the station, which was then located in the new Electrical Engineering Building, but it was rebuilt and back on the air by January 1930 with two eighty-five-foot towers located at the top of the building. The towers remained until the 1980s.

Big-name bands played for "mixer" dances on campus. In 1939 Tommy Dorsey was invited to play for the dedication of the north ballroom addition to the Memorial Union. He brought with him a young singer named Frank Sinatra and was disturbed when he learned the performance was being broadcast on a "teapot radio station."[12]

Waking Up a Campus with a Firm Foundation

If Elliott thought the University was sleepy and could do better, it was in sound condition with more than twenty-nine hundred students and a faculty that exceeded three hundred. The main campus in 1922 had thirty-one buildings, seventeen of them built in the previous nine years under Stone—and of those thirteen were in agriculture.[13]

Under Elliott's leadership graduate work was organized into a Graduate School overseen by a dean. In 1928 Maurice Zucrow was the first person in the modern era

Farm buildings west of Smith Hall about 1930.

to receive a Purdue PhD. Elliott issued invitations to Zucrow's dissertation defense and about four hundred people attended. After a career in industry, Zucrow returned to Purdue and started what became the world's largest academic propulsion lab.

David Ross and Virginia Meredith

Elliott arrived just as a man who would become one of Purdue's greatest benefactors was emerging into leadership—David Ross. An engineer and member of the Purdue Class of 1893, Ross grew up in the Lafayette area and became a very successful businessman. He started four companies and held eighty-eight patents. In 1920, he began volunteering for Purdue. He never stopped. In fact, he left his positions with the companies he founded and volunteered full time for the University.

His first Purdue assignment was to serve on an alumni committee raising funds for the Union. The Class of 1912 had started the drive to fund a Union Building. By the start of the Great War $17,800 had been raised.[14] In 1918 it was decided that the building should be a memorial to the 4,013 who had served in the war and the 67 who had died in service. The memorial tradition continued with the Union's Great Hall displaying the names of all Purdue students who died in service to the nation, including World War II, the Korean War, the Vietnam War, and conflicts since.

Ross started by making a contribution to the project and then helped lead fund-raising that reached nearly $1 million by 1924.[15] Construction actually began in 1922 and the building opened for partial use, uncompleted, in 1924. In 1929 the first

President Edward Elliott leans over bushes during a ceremony at John
Purdue's grave. Board president David Ross is third from right.

section was finally finished. That original first section included the south ballroom,
the Great Hall, lounges, a billiard room, a barbershop, a cafeteria, and meeting rooms.
The first section of the Union Club Hotel also opened in 1929. More improvements
came over the years, including the Union North Ballroom in 1939.[16]

Virginia Meredith played a key role in the Union design, construction, financing,
and management. She was a strong advocate for the building. On September 2, 1921,
Purdue trustees Meredith, Joseph Oliver, and Ross were named to the Memorial
Union Association Governing Board. Meredith served as president of the board
of governors and directed the financing of the building. Meredith also chaired the
groundbreaking ceremony where Ross turned the first spade of soil. At the groundbreaking she found a four-leaf clover that she took home and framed.[17]

In the fall of 1923 when pledges were slow to come in, Ross and fellow trustee
Henry Marshall borrowed money to put a roof on the building before winter arrived.[18]
Dean of Women Carolyn Shoemaker also played a significant role in seeking pledges
for the project.

In 1926 the Purdue Alumni Association established offices in the Union and
remained there until 2004, when it moved to the new Dauch Alumni Center on the
south campus. At the sesquicentennial it also had an office in the Union.

From his work raising funds for the Memorial Union to the day he died on June
28, 1943, Ross never stopped serving Purdue. He was named an alumni member of

Purdue campus, 1920s: Memorial Union, upper
right, and University Hall, center.

the board of trustees in 1921. In 1927 he was elected president of the board and served in that position until his death.

He had a special interest in research and let it be known that if any student had a good idea, he would pay to have it patented. In 1930 he and fellow trustee J. K. Lilly Sr., of Eli Lilly and Company, each gave $25,000 to start the Purdue Research Foundation (PRF). The Research Foundation was created "to promote educational purposes by encouraging, fostering, and conducting scientific investigations and industrial research; by training and developing persons for the conduct of such investigations and research."[19]

Ross introduced campus planning and purchased land for future University growth. For Ross, a lifelong bachelor, Purdue became his family, its students his progeny—just like John Purdue. During his life he gave Purdue in cash or the equivalent along with land about $900,000. In his will he left the greater part of his estate worth more than $1.7 million to the Purdue Research Foundation.[20] And that does not include an unknown amount of money he gave and loaned to students who were in need, especially during the difficult days of the Great Depression.

When he died he was buried at the highest point on campus, west of where Slayter Center would later be built, on land he had donated to the University. A granite slab covers his grave with the inscription: "David E. Ross, 1871–1943, dreamer, builder, faithful trustee, creator of opportunity for youth."

Robert Bruce Stewart

In 1925 Elliott made a significant hire—Robert Bruce (R. B.) Stewart. By the time Stewart retired as vice president and treasurer in 1961, his fingerprints were on everything at Purdue. He was the first vice president and treasurer Purdue ever had, and he left an indelible imprint nationally on the art and science of university administration. He played a key role in every new facility, devising innovative means for financing. He was the first secretary/treasurer of the Purdue Research Foundation, growing its assets from $50,000 to $22 million. He and his wife, Lillian, made personal loans and gifts to students, and they donated their home, Westwood, and land around it to Purdue. The R. B. Stewart Society for donors to the University is named in his honor.

Stewart's biographer, Ruth Freehafer, wrote, "In the opinion of some, Robert Bruce Stewart, as Purdue University's chief financial and business officer from 1925 to 1961 had a greater single impact on the institution's destiny than any other individual in the university's history."[21]

Complete a Degree in Less Than Four Years

Elliott loved to speak. He was funny, engaging, and had important, well-thought-out messages. He was asked to give addresses around the nation. He stood straight at

President Edward C. Elliott, hands in suit coat pockets during speech.

the podium and often jammed his hands into his suit coat pockets as he spoke. On May 3, 1924, he spoke at a University program marking fifty years since classes began.

One of his first messages to students, faculty, and staff that day concerned a theme that would be popular in the next century: how to reduce the cost of education. He said: "I am convinced that the present four-year program of training for all students is unwise, uneconomical and unjustified. It is my intention shortly to propose to the faculty of the University that our curricula be reorganized so as to permit students of ability and persistence to complete the work for a degree within three years instead of four."[22] Nothing

Gate over State Street marking Purdue's first fifty years of classes.

came of it, but more than ninety years later Purdue president Mitchell E. Daniels, Jr., made the same statement and programs were launched to accomplish it.

Student Residence Halls

Elliott, Stewart, and Meredith teamed up to create student housing on campus. By 1922 previous residence facilities had closed, and students either lived in fraternities or sororities, or they rented rooms in West Lafayette and Lafayette.

The situation was particularly difficult for women of that era. A small number of women were still housed in Ladies Hall. Most of the remaining hundreds of female students lived at home with their parents or in cooperative houses that the University rented. Some female students earned their room and board by working as home servants in Lafayette and West Lafayette.[23]

In 1927 when Ladies Hall, one of the original 1874 campus buildings, came down, displacing the fifty women who lived there, Meredith believed that surely the University's first residence hall for women would soon follow. Elliott named Meredith to chair a committee looking at residence halls on other campuses. Also on the committee were Dean of Women Carolyn Shoemaker, Dean of Home Economics Mary Matthews, State Leader of Home Demonstration Agents Lella Gaddis, Head of the Department of Physical Education for Women Gertrude Bilhuber, and Stewart.

In *The Story of Purdue Engineering* H. B. Knoll wrote:

Non-fraternity men ate in restaurants in the Village, or in boarding houses. One meal available was the "pint and four" which consisted of a pint of milk and four glazed buns costing ten cents. The boarding houses were in private homes in which the woman of the house—often called "Mom"—did the cooking for fifteen to twenty students. Students working for their board waited on tables and washed the dishes. West Lafayette was underbuilt . . . the housing problem, Elliott said, was so critical that the university could be said to have reached a limit in student enrollment.[24]

Elliott said the University needed to prepare for an influx of students. He believed that by 1974—fifty years in the future—enrollment would reach twelve to fifteen thousand. He was about half right.[25] In 1974 enrollment was 27,466.

Elliott wanted to build campus residence halls to meet the increasing demand for housing. The problem was financing. That was eased in 1927 when the Indiana General Assembly passed legislation allowing universities to sell bonds for the building and furnishing of residence halls.

In 1927, a Lafayette businessman came forth with an offer of $50,000 for a men's residence hall. Frank M. Cary had come to Lafayette from Detroit, Michigan, at the age of about twenty. In Lafayette he married Jessie Levering—Dr. Guy Levering's sister. Cary became president of Barbee Wire and Iron Works. Frank and Jessie had a son, Franklin Levering Cary, who died at the age of nineteen in 1912. It was a crushing tragedy for the couple. Frank Cary approached Elliott and Henry Marshall, who was then president of the board of trustees, about building a residence hall in memory of his son.[26]

Cary wanted the University to build a residence hall at the corner of State and Marsteller Streets, a location that would later be occupied by St. Thomas Aquinas Center. The location was very visible at the center of campus activity. Cary proposed a building for forty male students. The board of trustees and Elliott wanted a building for 100 to 150 students, and that required more land.[27]

Stewart talked with Cary and convinced him the two should take a trip to the University of Wisconsin to look at student housing there. Next, Stewart took Cary to land between Ross-Ade Stadium and Stadium Avenue. Professor George Spitzer had given it to the University. Cary didn't like it. According to Stewart biographer Ruth Freehafer, Stewart "must have driven Cary down Northwestern Avenue one hundred times to tell him that this was where Purdue planned to build dormitories."

Cary protested that the location was too remote; no one would ever see it. Stewart told him that if he gave the University $50,000, he would sell $100,000 in bonds and build a residence hall that would be the first thing everyone would see going to a football game.[28]

Franklin Levering Cary Memorial Hall for 157 men opened in 1928. The first of what would eventually be four buildings, it faced east.

Cary also wanted to memorialize Jessie, who had also died. In 1928 he offered Purdue $60,000 to build a residence hall for women named for his wife. The rest of the funds would again come from revenue bonds. The residence hall was planned north of State Street on Russell Street. The land was owned by Phillip Russell, whose parents had sold some of their land for $1 in exchange for the location of Purdue University. John Purdue, who negotiated the deal, argued long and hard to start the University adjacent to the Russell property. But he did not prevail. This time a Russell heir fought in court University attempts to buy the land. By 1930 Cary gave up and decided to memorialize his wife with the Jessie Levering Cary Home for Children on Eighteenth Street in Lafayette. He transferred his $60,000 Purdue gift to another men's residence hall near the first Cary Hall.[29]

Cary North opened in 1930. The remainder of the quadrangle was completed in 1938 and 1940.

But there was as yet no residence hall for women after nine years of effort by Meredith and others. Finally an Indiana Supreme Court decision gave Purdue title to 32.45 acres of Russell land at a cost of $32,002. Using funds from the federal Public Works Administration and other sources, Duhme residence hall for women was dedicated at homecoming, 1934. Duhme Hall was followed by Shealy in 1937, Wood in 1939, and Warren and Vawter in 1951. Collectively they became known as Windsor Halls for women.

The dedication of the residence hall for women was particularly gratifying for Meredith, who was eighty-five in the fall of 1934. She lived to take part in the dedication of the women's residence hall she worked so long to accomplish. And a year earlier in Chicago she had been honored at the World's Fair for her work forty years earlier on the Columbian Exposition. She died in 1936.

The Lafayette *Journal and Courier* said in an editorial: "Mrs. Meredith was remarkable in so many ways it is difficult to appraise her services without appearing to exaggerate . . . Keen of wit, zestful, humorous, understanding, sympathetic, widely read, quick at repartee, sure in decision as between right and wrong, courageous and frank in support of her convictions, Mrs. Meredith combined in one vibrant and friendly personality those myriad qualities that make for leadership."[30]

To Stewart, the buildings Purdue constructed for its students were not to be called "dorms." They were not just a place to sleep. They were homes for the students. "He termed them residence halls," Freehafer said. "And members of his staff who referred to them as dormitories were summarily informed of the difference in such a manner that they never made the mistake again."[31]

Stewart, with Elliott and Ross, was also instrumental in starting the Purdue Student Housing Corporation, which created the co-op housing system in the late 1930s, allowing students to cut living expenses by doing work themselves.[32]

Memorial Center Mural

Like Elliott, Stewart knew what he wanted and did not back away from confrontation. "He could engage in shouting matches with people," Freehafer wrote. "Then, when the air was cleared, go on as if there had been no disagreement. . . . One of his colleagues once said, 'He was an interesting person to work for, but it was much better if your ideas and his coincided.'"[33]

Mural, *The Spirit of the Land Grant College*. Left to right: R. B. Stewart, vice president and co-donor; J. H. Moriarty, director of libraries; Professor C. I. Calkin, head of art and design; Eugene Francis Savage, artist; Walter Scholer, architect and co-donor.

Among Stewart's crowning moments was the 1958 dedication of the Memorial Center, west of the Memorial Union. It included a four hundred–seat auditorium named Fowler Hall, since the previous Fowler Hall had been torn down to make way for the new building.

A $1 million gift to the Memorial Center from Lafayette retailer Bert Loeb and his wife, June, resulted in the center's 1,052-seat Loeb Playhouse. One of the most spectacular features of the center was a mural, *The Spirit of the Land Grant College*, painted by Eugene Francis Savage. Stewart told the trustees he would get the $25,000 needed for the mural if they would approve it. They did. Stewart left the meeting, met with architect Walter Scholer, who designed the building, and told him, "I just made a commitment to the trustees that you and I are going to provide the $25,000 for the mural." Scholer said "okay," but asked for time to talk with his wife first.[34]

At the mural dedication in the fall of 1961 Stewart said: "Without the land-grant colleges, I would be an uneducated man. . . . Land-grant is the priceless heritage of America."[35] In 1972 the Memorial Center, which had been titled during planning and construction an adult education student building and the Union–Hall of Music Annex, was renamed Stewart Center in honor of R. B. and Lillian.

In the last year of Stone's life he hired a new dean of engineering, who arrived on campus in September of 1921. The dean led Purdue engineering to international recognition. He was loved and respected by everyone.

But he remained silent about one important aspect of his life all the way to his grave.

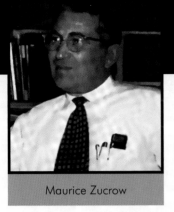

Maurice Zucrow

M aurice Zucrow was born before the Wright brothers' first flight. He lived to see a fellow Purdue University graduate walk on the moon, played an important role in aerospace research, and helped educate generations of engineers.

At his retirement from Purdue in 1966, then-President Frederick L. Hovde called Zucrow one of the nation's leading rocket engineers.[36]

Zucrow was born in Kiev, Russia, on December 15, 1899. His parents were Solomon, a Hebrew teacher, and Daba Zukroff. In 1900 Solomon and Daba, along with two daughters and their newborn son, fled Russian violent anti-Semitism for England, where a third daughter was born.

On November 6, 1915, Daba and the children sailed for New York on the RMS Saxonia, arriving November 16. Solomon remained in London, joining his family in December 1917. In the United States, the family changed their last name and Daba and all the children changed their first names. Moses Zukroff became Maurice Zucrow.[37] They became U.S. citizens in 1921.

The family settled in Dorchester, a neighborhood of Boston. Zucrow graduated from Boston English High School.

In the fall of 1918 he enrolled at Harvard on a scholarship and also enlisted in the U.S. Army. He was honorably discharged that December. Zucrow received Harvard's first engineering bachelor's degree magna cum laude in 1922.[38] He earned a master's degree at Harvard in 1923 and entered Purdue in 1924, where he taught, researched, and earned a PhD in 1928. He married and had a child in West Lafayette.

From 1929 to 1946 Zucrow worked in private industry. He helped develop the nation's first gas turbine in 1942 and a rocket used to assist seaplanes during takeoff.

He returned to Purdue to teach and research in 1946, wrote the first textbook on jet propulsion and gas turbines, and created what was later named Maurice J. Zucrow Laboratories, which became the largest academic propulsion lab in the world and a leader in combustion. Some of his research was used in the design of the space shuttle's main engine.[39]

Zucrow died in 1975.

His students called him "Doc." He called them his "boys." "Dear Doc," one of his students wrote shortly before he died. "I still have the first rough draft of my master's thesis you returned. The comments and corrections were nearly equal in length to the original manuscript. I learned more . . . from that one experience than from all my previous training. Thank you, Doc, for all you have done for 'your boys' and for so many others."[40]

19

The Dean: Andrey A. Potter

1882 to 1979

He knew all his men and all their work and had the capacity
to inspire them with his unfailing friendliness and good
cheer. Men liked him and liked to work with him.

—Purdue professor of engineering H. B. Knoll on A. A. Potter

Before officially accepting the position of president of Purdue University, Edward
C. Elliott presented the board of trustees with a list of stipulations. One of them
was that his appointment be unanimous. It was.

But that didn't mean he was the only person considered.

In fact, Purdue's dean of engineering, Andrey A. Potter, was the first person
approached about taking the position. Potter had been on campus less than one year,
but he was so highly regarded at Purdue and throughout the nation that the trustees
wondered if he would be interested in replacing Winthrop Stone, who had hired him.

Potter, his wife, Eva, and their two children arrived in West Lafayette in
September 1920. The family settled into a small house on Russell Street near campus,
later moving to a larger one just down the block.

Potter immediately impressed the trustees. After the mountain climbing tragedy in the summer of 1921, trustee James Noel brought Potter's name forward as
a person to be considered as Stone's successor. Trustee and acting president Henry

A. A. Potter, dean of engineering.

Marshall talked with Potter to find out if he was interested.

Robert Eckles, a Purdue history professor who wrote a biography of Potter, said Potter declined to be considered for the Purdue presidency: "In 1921 and 1922 the Ku Klux Klan was thriving in Indiana and Potter felt that it was not expedient for Purdue to have a president who was not a native and who spoke English with an accent," Eckles said.[1]

There was a lot more behind Potter's thinking. Potter preferred to forget his early years growing up in Russia.[2] He never discussed his heritage in the fifty-nine years he lived in West Lafayette.

Potter was Jewish.

While by the end of the decade it had declined rapidly, in 1922 Indiana had the largest Ku Klux Klan membership in the nation. Klan members were white supremacists, anti-Catholic, anti-Semitic, and anti-immigration. In addition to the Klan presence in Indiana, West Lafayette neighborhoods had written restrictions stating that "the ownership and occupancy of lots or buildings in this subdivision are forever restricted to members of the pure white Caucasian race." Jewish people understood this covenant was targeting them as well as African Americans and Asians. There was an early Jewish community in Tippecanoe County. Temple Israel was founded in 1849 and Sons of Abraham in 1903, both in Lafayette. But it was a small community, and although no one knew it, Potter was the first Jewish person hired on the Purdue faculty.

Harry H. Hirschl met Potter when he was a student at Purdue. Hirschl graduated in 1947 with a degree in mechanical engineering and worked in a family business until 1956, when he returned to the University as a staff member. He retired in 1986 as director of administrative computers. Hirschl said Potter never really hid his Jewish background. He just never told anyone about it and he was never asked. Many Jewish people, like Hirschl, understood Potter's background and his silence about it.

Why didn't he discuss it? "To me the answer is simple," Hirschl said. "Purdue, and the United States, were overtly anti-Semitic for many years, especially during his tenure."[3]

Rabbi Gedalyah Engel, who died in 2012, was director of Purdue's Hillel Center from 1955 to 1989. "Potter, as a child in Russia, had witnessed a pogrom." Engel said. "For the rest of his life his goal was to remove himself and his family from having to face such a man-made obstacle." Potter, Engel said, was "so afflicted by anti-Semitism in the land of his birth that he carried his Jewishness as a burden all the days of his life."[4]

Potter was born in 1882 in Vilnius, which was then under the brutal control of Russia. His family name was Podruch. At the turn of the twentieth century the city was about 40 percent Jewish. "In recalling the life he left, the dean has said that had he remained in Russia he undoubtedly would have joined an anti-czarist group, probably would have become an active revolutionary, and would have ended his life making the long trek to Siberia as a political prisoner," Eckles wrote.[5]

Potter grew up in a loving household with his mother, Rivza, who was Russian; his father, Gregor, who was Dutch; a grandfather, a brother, and a sister. His father, who was employed in the chemical industry, taught Potter to speak and read French, German, and Russian. When he was eight his father gave him a blue and white Dresden china ocarina, a wind instrument whose use dates back thousands of years. He kept it, cherished it, and played it all his life.[6]

Potter used the word "escape" to describe his departure from Vilnius as a teenager, first to England and then to Quebec, Canada, on a cattle ship, finally settling with an aunt, uncle, and four cousins in Boston. He arrived intending to become a rabbi, but his uncle convinced him to study for a different profession. Smart, studious, he learned English and at the age of seventeen was admitted to the Massachusetts Institute of Technology, graduating with a bachelor of science degree in 1903. He worked for General Electric Company in Schenectady, New York, before taking a position as an assistant professor of mechanical engineering at Kansas State.

In Manhattan, Kansas, he rented a room in the home of a Presbyterian minister. He developed a friendship with the minister and his family and attended church with them regularly. He first saw his future wife at a church social, and the minister later formally introduced them. She was a student at Kansas State and the daughter of a farmer.

They married in 1906, the same year Potter became a U.S. citizen. His career moved quickly. In 1913, Potter was named dean of engineering and director of the Engineering Experiment Station at Kansas State before being named dean of engineering at Purdue beginning in the fall of 1920.

Potter was active in the Lafayette community, joining Rotary. He and his wife attended Central Presbyterian Church in Lafayette, where he served as an elder.

He joined the Freemasonry and on September 26, 1946, was conferred with the thirty-third degree—the highest.

During a Purdue career that spanned from 1920 to 1953, Potter increased the number of professional engineering courses at Purdue. He oversaw the addition of four new engineering schools and the construction of three engineering buildings. Potter was enormously respected by faculty, staff, and students. He served on national committees and gained a reputation around the country as the "Dean of Deans of Engineering Universities."[7]

As a student, Richard Grace knew Potter. He later worked with him as a member of the engineering faculty and continued a friendship with him during the dean's retirement years. "Potter's circle of friends throughout America was enormous," Grace said. "He told me that when he would visit Thomas Edison's laboratory he would shout into Edison's hearing trumpet to no avail. Henry Ford once refused to pay Potter's consulting fee for a day's work in Detroit. Potter, who rarely spoke an unkind word, did call him a 'tight bastard.'"[8]

In his book, *The Story of Purdue Engineering*, H. B. Knoll wrote: "Among his staff he wanted above all a sense of cooperation and accomplishment. . . . He knew all his men and all their work and had the capacity to inspire them with his unfailing friendliness and good cheer. Men liked him and liked to work with him."[9]

Potter and Elliott respected each other, but they didn't always agree. Eckles wrote, "During an interview on the subject of Elliott's relations with the engineering schools, Potter was asked if he had to disagree or oppose the president very often. His answer was short and to the point. 'Every day,' he said."[10]

Just before Elliott came to the Purdue campus, he met with Potter in Chicago. They walked many miles together along the lakefront, talking about engineering at the University. Eckles wrote, "Later they might argue or even engage in heated discussions. [But] at this first meeting they came to some appreciation of each other's ideals and goals for Purdue. And there was never any real difference afterwards. . . . It is a tribute to both men that, although they possessed personalities and manners which were very dissimilar, they never allowed obstacles in the shape of temper, temperament, or concepts of leadership to stand in the way of accomplishing the best things for Purdue."[11]

Eckles called Elliott "abrasive, abrupt, eloquently powerful, with a strong rhetorical style. His executive orders, for example, were commands to hired help. The faculty and staff members coming to see him were given short shrift unless they answered in kind, then give-and-take would take place. He liked to see if a man would stand up to him. . . . Many recalled that a standard answer to a question was 'no!' However, if the questioner would persist, Elliott would change his manner and a sensible discussion would take place. He seemed to invite a rough-and-tumble verbal wrestling match. Those who would fight back were listened to and respected."[12]

Confrontation was not at all Potter's style. He was a gentleman and a gentle man. In *The Story of Purdue Engineering,* Knoll wrote:

The Potter-Elliott relationship, involving as it did two blazing stars, was a favorite topic of speculation, much of it erroneous. Elliott was sometimes rough with Potter in the executive committee and faculty meetings, as he was with other staff members, and there was a puzzlement that Potter did not fight back then and there. But a fight in public was not the Potter way. He would bide his time and, on occasion, he could talk turkey to Elliott as straight forward as anyone else. Elliott once wanted to cut the salaries of certain engineering professors for not attending his meetings, or at least said that he did, and Potter told him that he would have a better attendance if we would stage a better performance—the professors really could not be blamed for staying away.[13]

Knoll knew them both well and said that while their relationship was not close, they both got behind projects 100 percent once decisions were made. And both liked to "stir people to action. . . . The dean liked to praise and encourage; the president liked to topple a faculty member just to see how he would respond. They had their differences, and Elliott, as the chief administrator, must have looked on Potter as a tremendous force to be reckoned with. He must also have admired him, since he admired all men who stood up to him and got things done. . . . Neither man was known to carry a grudge. Some people thought that Elliott did, but they were mistaken."[14]

Potter was the great-uncle of Grace Hackel Lebow, now of Chevy Chase, Maryland. She loved Uncle Andrey, as they all called him. The relation is through her mother. She said:

He was very devoted to our family and came to visit us in Boston once a year or more. He was a very dear, generous, warm, loving gentleman and had a good sense of humor. He was so devoted to the family. At the very end we had to place my mother in nursing care and he would come and play his ocarina for her.

Uncle Andrey was really important to me. He helped support my family financially. He sent money to help me through college. There are a lot of stories about him in our family. When he came to visit when we children he would always bring little gifts, a harmonica or some toy that was clever. He got down on the ground and played with us.[15]

Elliott retired in 1945 and was named president emeritus. With Elliott's departure, Potter was again asked if he wanted to be president, and once again he declined. Potter retired in 1953 and was named dean emeritus of engineering. In 1957 Elliott suffered a stroke and never fully recovered. He was left partially paralyzed on his right side. His wife had died and he rarely left his home. But he had a caller every Saturday afternoon until he died in 1960 at the age of eighty-five.

The caller was Andrey Potter.

As Potter lay on his deathbed in late 1979, Rabbi Engel visited him. The dean died on November 5. At 10 a.m. on November 9 he was buried at Grand View Cemetery in West Lafayette with the minister of Central Presbyterian Church officiating. A memorial service was held that afternoon at the church.

The evening before, at the funeral home with only family members present, Potter's final request was fulfilled. According to Jewish custom, his body was clothed in a white shroud. Engel stood beside the closed casket and recited the Kaddish, a hymn of praise to God used in Jewish funeral services.

When he finished, Engel was in tears.

GIANT LEAPS IN LIFE

Orville Redenbacher

For millions of people the thought of popcorn conjures images of a man with white hair parted down the middle, wearing suspenders and a bow tie. His name was Orville Redenbacher and he was the Popcorn King. His fame came from television and print advertisements, where his down-home country style and old-fashioned, honest-to-goodness appearance and personality made him an instantly loved celebrity.

But his genius was not in marketing. It was in agricultural science. And he learned it at Purdue University.

Orville was born in Clay County, Indiana, to parents who farmed. When asked what his family did for fun when was growing up, Redenbacher replied, "We worked."[16] By the age of twelve, in addition to his chores, he was growing his own popping corn and selling it to nearby stores.

He started his education in a one-room schoolhouse, became the first person in his family to graduate high school, and turned down an appointment to the United States Military Academy at West Point. Instead, he enrolled at Purdue and studied agriculture, using savings from farm work and popcorn sales to help pay his way.[17]

Orville, who loved his studies and faculty, graduated in 1928. "If it hadn't been for a solid foundation of knowledge gained on the farm and later formalized at Purdue, my name wouldn't be a household word today," he said in 1991.[18]

He often said it was at Purdue that he learned to "toot his own horn." He played the helicon in the marching band.

After graduation he married his high school sweetheart and began a career that included work as a vocational agricultural teacher, a county 4-H agent, a county agricultural agent, and managing a twelve-thousand-acre farm. In 1951 he partnered with a friend, Charlie Bowman, to buy a hybrid seed company in Valparaiso, Indiana.

Throughout it all, Redenbacher never lost his interest in developing better popcorn. In 1965, after thirty-five years of work, he finally succeeded.

The popcorn was called Redbow—a combination of the names Redenbacher and Bowman. A Chicago advertising agency told them to change the name to Orville Redenbacher Popcorn and to use Orville's image to promote it.[19] By the mid-1970s they had one-third of the nation's unpopped popcorn business. The company was sold in 1976 to Hunt-Wesson.

In 1988 Orville received an honorary PhD from Purdue.

He died in 1995 at the age of eighty-eight with the right to officially call himself Dr. Popcorn.

20

An Era of Change

1922 to 1937

We had a talk one day about a year ago and confided to each other
that we had an ardent longing to see that stadium before we died.

—George Ade

I t was their first meeting. George Ade and David Ross were successful, wealthy
bachelors, Purdue graduates, and active alumni. They had a lot in common and
they became fast friends. They would soon have even more in common than Ade
ever anticipated.

Ade and Ross first met in the downtown Lafayette office of Tippecanoe Superior
Court judge Henry Vinton. Ross had requested the introduction, and Vinton was
surprised the two had not gotten together years before. It was early in 1922, and after
some small talk in Vinton's office, Ross asked Ade to join him for a drive. There was
something he wanted to show the famed writer. They drove over the Wabash River
to West Lafayette, around the Purdue campus so familiar to both of them. Ross
took Ade north on Northwestern Avenue and suddenly pulled the car to the side of
the road. Ross got out of the car and Ade followed him across the street and onto a
field north of campus. It was filled with weeds, many more than two feet high. Both
men wore suits, and they had to crawl through barbed wire, being careful not to tear
their clothing.

"I'm a little curious to know what you're up to," Ade finally said. Ross let the comment pass but in time he stopped, looked over the landscape that dropped in front of them, and said, "Here it is."

"Here is what?" Ade asked indignantly.

Ross told him "here" was where the University was going to build its new football stadium and recreational fields. "Fine," Ade replied. And then he uttered what might be the most clueless words he ever spoke: "Just how does all this concern me?"

"The fact is," Ross said, "I've been wondering if you'd be willing to join me to help finance the whole thing. How would you feel about that?"[1]

Ade didn't need any time to think before agreeing. He was passionate about Purdue football and loved to attend the games. He also had a passion for Purdue and believed the University needed better athletic facilities.

While it was easy for Ross and Ade to reach an agreement about the location of the new stadium, there were details that took time to iron out, including land sales and swaps. The field that Ross and Ade walked on that day had been acquired by the Shook Agency, a Lafayette real estate company established in 1915 by Robert H. Shook and his son Charles W. Shook. On the west edge of campus along University, Waldron, and Russell Streets, the faculty had been building homes, providing them with a short walk to campus. The area was filling up, and the Shook Agency had acquired land north of campus for a new housing development.

So money from Ross and Ade bought land on the other side of Northwestern Avenue, including the Frank Tilt Dairy Farm, and swapped that with the Shooks for the stadium site. The Shook Agency developed the land they received into the Hills and Dales subdivision. It became home for West Lafayette High School and a model for west side growth and development.

The University created the Ross-Ade Foundation to receive the gift from the two alumni. Through the foundation the University could do things it otherwise by law could not. State law did not allow Purdue to sell bonds to finance a football stadium. But the foundation, as a separate corporation, could sell bonds. Work on the financing began.

In the fall of 1923 Ross and Ade wanted to speed the project along. So they each offered an additional $10,000 if alumni and friends bought life stadium seats for $200 each. In one evening 173 seats were purchased, and the number eventually reached 201.[2]

The original 13,500-seat stadium with standing room for another 5,000 people was far from the biggest in the Big Ten. Eight of the Big Ten schools had stadiums under construction at that time or had recently completed them. The Ohio State stadium

Purdue president Edward C. Elliott with writer and alumnus George Ade.

had a capacity of 60,000, Minnesota 55,000, Michigan 42,500, Wisconsin 40,000, and IU 25,000.[3]

The Boilermakers loved their new home. The Purdue stadium was dedicated during a football game on Saturday, November 22, 1924. The evening before, Ade spoke at pre-dedication festivities: "When Dave Ross and I acquired that tract two years ago, we believed that some day there would be a stadium up in that hollow. . . . We had a talk one day about a year ago and confided to each other that we had an ardent longing to see that stadium before we died."[4]

Ade joked that he didn't like the name of the stadium. He said "Ross-Ade" sounded like a drink such as lemonade.

At the dedication Ross said that it was his hope as well as Elliott's that "no professional player ever would enter [onto the playing field]. This place is for students," he said. "It is for our kids."[5]

The football opponent on dedication day was Indiana University. The stadium was packed with cheering Boilermakers who loved the new facility, and they also loved the outcome of the game. Purdue won 26–7. The world's biggest drum was part of the day. It had been introduced as part of the band in 1921.

Ross-Ade Stadium shortly after completion.

The Old Oaken Bucket

The following year, 1925, Purdue and Indiana University alumni introduced the Old Oaken Bucket trophy. Chicago alumni from both universities had met earlier in the year to discuss a trophy for the rivalry. Many of them could remember from their rural nineteenth-century youths drinking cold, clear well water from an old oaken bucket on hot summer days. That, they believed, would be the ideal symbol for the game. They found an oaken bucket in southern Indiana between Kent and Hanover on "the old Bruner farm." According to local lore, Confederate general John Morgan and his men drank from it during the Civil War on a raid into the north in 1863. More importantly, Harvey Wiley—the "Father" of the Pure Food and Drug Act and a frequently mentioned candidate for president of Purdue—said he had taken drinks from the old oaken bucket on the Bruner farm when he was a boy.

The first year of the trophy, Purdue and IU battled to a 0–0 tie, so an "I" and a "P" were added to the bucket. Each year thereafter the winning team added its letter to the chain. The letters in the chain are bronze, except for the 1929 "P," which is made of gold. That signified an undefeated season for the Boilers. A "P" with a diamond on it represented Purdue's Rose Bowl victory in 1967. IU has a gold "I" with a diamond representing its trip to the Rose Bowl in 1968.[6]

David Ross and Andrey Potter

Ross was filled with ideas for Purdue. Elliott, who walked to work almost every day from his home on South Seventh Street in Lafayette, often stopped on his way to the University to talk with Ross at his Main Street office on the north side of Courthouse Square. Ross initiated a master plan for the University. He believed that Purdue needed room to grow, and he purchased land and donated it for the future campus.

Ross and Andrey Potter became great friends. Both had a sense of humor, said Maurice Knoy, who graduated from Purdue in mechanical engineering in 1934 and became president of the board of trustees. Ross "loved Potter," Knoy said in a 1972 interview: "One of his favorite stories, which he must have told me three or four times, concerned Potter." Ross had a farm in west Tippecanoe County. He had a beautiful country home there and liked to take people out for weekend visits. During one of those, Potter was in Ross's car and when they reached the farm, Potter jumped out to open the gate across the private drive. But he couldn't get it open. He struggled and struggled and fumed while Ross sat in the car, laughing. Finally Ross said, "Dean, the gate opens from the other side." The sight of perhaps the most prominent dean of engineering in the nation trying to open the wrong side of a gate gave Ross years of storytelling and laughter.[7]

The Nation's First University Airport

Ross was competitive and he wanted Purdue to be the best. By 1930 the field of aeronautics was developing quickly, and wind tunnels had become an essential at engineering schools. "Every engineering college was trying to build a bigger wind tunnel than anyone else," Potter remembered in a 1969 interview. "MIT built a certain size wind tunnel and Michigan built a bigger one and California a bigger one still. . . . Ross talked to me about it and wanted to know if we should get a real big wind tunnel. I told him you can't always learn from size, that the type of wind tunnels we have at Purdue were made by our own staff and students to illustrate principles. I said what I would like to see is a flying field within walking distance of the campus where tests on real planes could be carried on."[8]

Within a few days Ross stopped at Potter's office and took him for a ride, just as he had done years earlier with Ade. It was a short ride because what Ross wanted Potter to see was a field of land adjacent to the south campus. He had an option to buy it and wanted to know if Potter thought it would make a good airfield. Potter thought it would be a splendid airfield.

It was an idea that had been in the minds of Ross, Elliott, and others for several years. So in 1930 when Potter approved the site for an airport, Ross wasted no time. The director of the Purdue Research Foundation, G. Stanley Meikle, did most of the planning. On November 1, 1930, officials from the U.S. Coast and Geodetic Survey and the U.S. Department of Commerce inspected the location—a windsock was hung from a dead tree and the landing area was marked by colored limestone. The site was officially declared an airport—the first on any university campus in the nation.[9] It would be another four years before the University started building its first hangar.

Amelia Earhart

Among the first world-famous people to use the Purdue Airport was a woman whose contributions to the history of flight were equaled by her determination to show that women could do any job that men could do.

Amelia Earhart had emerged seemingly out of nowhere to become an overnight international celebrity in June 1928, when she was the first woman to fly in an airplane with two male pilots, nonstop across the Atlantic—one year after Charles Lindbergh's solo flight. She didn't fly the plane. She said she was nothing more than "a sack of potatoes in the cockpit." Still, the world was captivated that a woman had taken the risk and joined the expedition.

In 1932 she became the second person after Lindbergh, and the first woman, to fly solo, nonstop across the Atlantic, and her fame grew ever greater. She set flight records, gave speeches, and wrote books along with newspaper and magazine articles. And her interests went even higher than flight. She had lofty dreams for women.

Earhart was at the peak of her career when Elliott first met her at the Waldorf Astoria Hotel in New York City on September 26–28, 1934. The two were taking part in the fourth annual Women's Conference on Current Problems, sponsored by the *Herald Tribune* newspaper. New York mayor Fiorello LaGuardia and First Lady Eleanor Roosevelt were among the speakers. Also speaking was the U.S. secretary of labor and the U.S. attorney general. The evening session on the opening day focused on new opportunities for youth and included several university presidents in addition to Elliott.

The nation was deep into the Depression. Elliott spoke on "New Frontiers for Youth." Earhart focused on women in aviation. Elliott said that in the previous five years, ten million young people had passed the age of sixteen and ten million more would do so in the five years ahead. These twenty million Americans, he said, were "the working capital for the maintenance of the future national virility and virtue." He quoted a study of Purdue alumni showing that in spite of the Depression, 90 percent were employed and two-thirds of them were working in their chosen professional fields. "This study . . . warrants for me as an American and as a public steward not less but more faith in the fundamental business of training [educating] youth of the country,"[10] he said, standing at the podium with his hands in his suit jacket pockets.

Elliott did more than talk at the conference. He listened—especially to Earhart's presentation about opportunities for women. Elliott began to consider the possibility of connecting the aviation legend, who advocated careers for women, with Purdue. He had become concerned with what he termed "the problem" of education for women. Almost all women at U.S. universities in the 1930s were studying home economics, education, or nursing (Purdue did not yet have a nursing program). Most of these young women did not go on to have careers, but instead worked for a short time, or not at all, and then married and became mothers and homemakers. Elliott was dedicated to helping women achieve. He wondered—why was our country educating women if not for careers? He had built a women's residence hall. In 1939 he launched a liberal science program for women, and later men. And a few women were being placed in leadership positions at the University.

Class of 1924 home economics students.

In 1933 Carolyn Shoemaker, Purdue's first part-time dean of women, died. She had been called "the mother of all the Purdue girls."[11]

Elliott hired Dorothy Stratton to replace Shoemaker, and this time as Purdue's first full-time dean of women. Stratton had a PhD from Teachers College of Columbia University, where Elliott had also received his doctorate. When he contacted her about the Purdue job, she was dean and vice principal at Sturges Senior High School in San Bernardino, California. Stratton was described as "outdoorsy, easygoing, and attractive, with short, dark, wavy hair, thin lips that smiled pleasantly and confidently, and almond-shaped blue eyes."[12] She was also very intelligent and a natural leader.

Stratton enjoyed sports, and in her free time she shot baskets in the women's gym where she met Helen Schleman, who worked part-time in the Purdue Department of Physical Education. Schleman had an undergraduate degree from Northwestern University and a master's from Wellesley. The two women formed a bond that would last their lifetimes. Stratton hired Schleman in the Office of the Dean of Women. Schleman became director of the new women's residence hall, and the two would go on to impact Purdue and the lives of female students. (More about Stratton and Schleman in coming chapters.)

In New York, Elliott had lunch with Earhart and her husband/promoter, George Palmer Putnam. Earhart told him her primary interest in life was not aviation. Instead, she said her main interest was "the problem of careers for women."[13] In less than a month Elliott had Earhart speaking at Purdue. She talked to female students and faculty about "Opportunities for Women in Aviation," very likely the same talk she gave in New York.

She liked Purdue. Purdue liked Earhart. Elliott offered her a position. She would be an adviser to Purdue aeronautics. But more importantly, she was named a career counselor for female students. Earhart claimed she was the first person to hold that title at any university in the nation. From 1935 through 1936 she spent several weeks at a time on the Purdue campus, living in the recently completely Duhme women's residence hall, meeting with students and faculty, and encouraging women to be what they wanted to be. She was paid about $2,000 a year.

Another speaker at the September 1934 conference in New York City was Lillian Gilbreth, an engineer, who spoke on the home as a major industry. She was among the first female engineers with a PhD, and she was considered the first industrial/organizational psychologist. Perhaps Elliott also talked with her at the conference, because he also hired Gilbreth as a visiting member of the faculty and the first female professor of engineering. She also lived in the women's residence hall and had the title of consultant on careers for women. It might have been Potter who encouraged Elliott to bring Gilbreth to campus, because he had a long friendship with her and her late husband, Frank. Two of the Gilbreths' twelve children wrote books about the family titled *Cheaper by the Dozen* and *Belles on Their Toes* that were made into movies with the same names. Lillian Gilbreth was a pioneer, applying scientific management to household tasks. She created the "work triangle" in household kitchens, came up with the idea to add shelves inside refrigerator doors, and invented a waste container that could be opened with a foot. In addition, she interviewed thousands of women to come up with a proper standard height for kitchen counters and appliances.

Lillian Gilbreth.

Potter had great respect for Gilbreth, but not for Earhart. To the end of his life he said that Earhart was courageous, but not well educated. She did not have a college degree in any field, and he believed she was not helpful in advancing aeronautical engineering education for his students.[14]

Most people at Purdue did not agree with Potter on this point. While Earhart was not an engineer, what Potter somehow missed was that because Earhart flew daring missions in the airplanes his young male students wanted to design, she inspired them. And often before you can teach students, you have to first inspire them.

In addition to working with students and advising the aeronautics program, Earhart analyzed the Purdue curriculum. "Lowering the fortress walls between schools would serve to eliminate some of the condescending attitudes on the part of men students toward girls," she wrote in her report. "Today it is almost as if the subjects themselves had sex, so firm is the line drawn between what girls and boys should study."[15]

She urged the female students to become mothers and housewives if that's what they wanted to be. But she also told them they could study to become medical doctors or engineers, or to work in the sciences. "All women are not suited to aviation any more than all women are suited to be lawyers or even housewives," she said. "Someday people will be judged by their individual attitude to do a thing and [society] will stop blocking off certain things as suitable to men and suitable to women."[16]

She talked to students in classrooms. In the residence hall in the evenings she sat with them in circles on the floor. She told them their careers didn't have to stop with marriage. She told them they could keep their own last name after marriage. She talked about careers in radio.

Earhart loved buttermilk. When she first arrived on campus, she ordered buttermilk in the dining room. It wasn't long before lots of students followed her lead and ordered buttermilk, which was readily available since Purdue had its own dairy and creamery. Earhart wore slacks on campus—outside the residence halls—which was against the rules for females in the 1930s and for many decades after. When the female students asked Dean Stratton why they could not wear slacks on campus, she told them that when they flew an airplane solo across the Atlantic Ocean, they could wear slacks at Purdue.[17]

Earhart and Stratton, who were only one year apart in age, and Schleman became close friends. Stratton didn't cook, but she would invite Earhart to her home for Sunday dinner—which was waffles. Earhart loved waffles. Her recipe was printed in *This Week*, a national magazine.[18]

During one campus visit, Elliott asked if Earhart had plans for any more of her "adventures." She said she was interested in a flying laboratory to do aviation research, flying around the globe roughly at the equator. Elliott was intrigued, and he talked with people who might help finance her adventure. Ross and fellow Purdue trustee J. K. Lilly Sr., of Eli Lilly and Company in Indianapolis, each offered $20,000. The money was given to the Purdue Research Foundation and made available to Earhart.

On March 20, 1936, Earhart purchased an Electra 10-E airplane from Lockheed for $42,010. According to the purchase order, that price included deductions totaling

Smith Hall, Purdue Creamery about 1930.

$22,400 for "equipment to be furnished by the purchaser" and a "rebate." Earhart put down $10,000 on the purchase and owed $32,010.[19] Earhart signed the purchase order herself. It was her airplane.

"That plane is a dream come true," she said of her Electra. "Thanks to the cooperation of Purdue University I have an airplane equipped with every tried appliance of modern flying. . . . I hope the accomplishments [of the flight] may measure up to the opportunities both for the advancement of science and in the interest of women."[20]

According to a report from her husband, Putnam, to Meikle, of PRF, dated July 27, 1936, in addition to the $40,000 donated by Ross and Lilly, the foundation provided additional gifts to Earhart totaling $7,000 from the Bendix Corporation, Standard Oil, and William Horlick, whose company bought and refurbished surplus

President Edward Elliott with Amelia Earhart outside the presidential home on Seventh Street in Lafayette.

military equipment. Putnam reported they had received $2,500 from the *New York Herald Tribune* as an advance on articles about the flight she would write for the newspaper. Total income and donations were $49,500. Putnam reported expenditures of $46,110.55, including $44,000 to Lockheed for the airplane ($42,000 for the original contract and $2,000 in extras), leaving $3,389.45 on hand. Another $2,500 was due from the *Herald Tribune* and $1,500 from Standard Oil.[21]

Amelia Earhart initially planned to fly west and began in March 1937, but damaged her plane in Hawaii attempting a takeoff on the second leg of her flight. The plane was sent by ship back to Lockheed for repairs. On May 18, 1937, Lilly wrote a letter to Elliott saying: "Yessir! I am quite willing to give $2,500 to the Research Foundation for the specific purpose of aiding Miss Earhart in her work."[22] The letter indicated that Elliott and the Research Foundation were helping to find donors to repair the plane. Earhart had planned to write a book about her flight titled *World Flight*. It was posthumously published as *Last Flight*.

By the time the plane was ready again, weather conditions were not ideal for a flight going west. So Earhart decided to fly east, leaving from Miami on June 1, 1937. At 7:40 in the morning in the middle of the Pacific Ocean on July 2, 1937, the voice of Amelia Earhart burst into the radio room of the U.S. Coast Guard ship *Itasca*. "We must be on you, but cannot see you," Earhart radioed from her airplane. "But gas is running low. Have been unable to reach you by radio. We are flying at 1,000 feet." They were among the last words she is known to have spoken.

Attempting an around-the-world flight roughly at the equator, Earhart and her navigator, Fred Noonan, disappeared that July morning, launching first a rescue and then a recovery search that continued into the twenty-first century. She was on the last leg of the flight from New Guinea to Howland Island when she and Noonan disappeared.

She had emerged suddenly before the public in 1928 just before the start of the Depression, and during the hard years that followed, she was a beacon of light for people who struggled daily and hoped for a better tomorrow. She had burst into the public consciousness like a shooting star. She blazed across the sky of the 1930s and suddenly, as quickly as she appeared, her star burned out. And she was gone.

Putnam gave many of Earhart's papers to Purdue, but kept some in his possession. In 2002 his granddaughter, Sally Putnam Chapman, donated five hundred more items. Purdue's Virginia Kelly Karnes Archives and Special Collections Research Center includes the largest archive of Amelia Earhart papers in the world.

People continue to focus on what happened to Earhart on that last flight. Perhaps they always will. Perhaps someday the mystery will be unraveled, perhaps not. But amidst all the unknown, one thing is certain.

Television host Ed Sullivan with Al Stewart and the Purdue Glee Club. The Glee Club performed on the Ed Sullivan Show in New York City in 1955.

Amelia Earhart would not want people to focus on how she died. She would want them to focus on how she lived and what she said.

No Money for Music at Purdue

It was January 1931 when twenty-three-year-old Al Stewart walked to the office of Purdue University president Edward Elliott. Stewart had a wonderful idea. Of course, it would cost a little money, but not much. Stewart was the unsalaried director of the Women's Glee Club at the time, and he wanted to start a male and female University Choir.

When he walked into Elliott's office to request funding, the president continued to read papers at his desk. As Stewart began to speak, Elliott gradually looked up.

When he finished, Elliott stood, looked Stewart square in the eye, and remained silent for a moment. Finally he spoke, slapping his hand on his big wooden desk. "Never!" Elliott shouted. "Never, as long as I am president, will this university spend one damn penny for music on this campus, young man. Get that through your head!"[23]

In a short time Stewart got his University Choir from Elliott. He also became director of the Men's Varsity Glee Club, taking over from Paul Smith, who died, and Smith's wife, Helen. Helen Smith was the only female director of the Glee Club. Stewart ultimately became the iconic head of Purdue Musical Organizations.

And he got a whole lot more from Elliott than one penny for music at Purdue.

Francis Beck
(Photo courtesy of Sonny Beck)

Beck's Hybrids is the largest family-owned retail seed company in the United States, serving farmers in eleven states. It all started in 1937 on six acres in Hamilton County, Indiana, planted with hybrid seed corn from the Purdue University Botany Department.

Francis Beck and his father, Lawrence, started the company with teamwork, integrity, innovation, adaptability, commitment, and passion—qualities that continue to define the company today.

The Beck family came from Germany to the United States in the 1830s and settled in Tipton County, Indiana. It was family tradition that when each child married they were given eighty acres of land. But when Lawrence's time came there were only forty acres left. He sold it and in 1901 bought an eighty-acre farm in Hamilton County. It's the home of Beck's Hybrids today.

Francis was born in 1908 and raised with the work ethics and values of his parents. In 1929 and 1930 he attended the eight-week winter course in agriculture at Purdue, providing participants with the latest information on modern agriculture.

"He profited from what he learned at Purdue," said Sonny Beck, Francis's son and now CEO of the company. "Purdue was teaching new practices and that put him ahead."[24]

In 1937 Purdue had higher yielding hybrid seed corn available and invited all farmers in Indiana to apply for a three-acre allotment. With his education at Purdue, Francis knew this was the way of the future. He and his father planted six acres with a horse-drawn two-row planter. They harvested by hand.

"My father's connection and education at Purdue allowed him to take that small thing Purdue gave him, grow seed corn, sell it, buy more land, and grow more seed corn," Sonny said.[25] Today the company has thousands of acres of seed planted around the original eighty acres Lawrence purchased at the start of the twentieth century.

Lawrence died in 1938 and Francis carried on the business himself with the personal commitment "helping farmers succeed," including a 100 percent free replant policy. He continued applying the best scientific practices to his farming, created equipment for detasseling, and slept in the seed house next to his coal furnace to keep it going all night during drying. The family connection to Purdue grew stronger with Sonny and more Beck generations becoming Boilermakers.

Francis worked until the end of his life in 1999. In 1972 he had been named an Indiana Prairie Farmer Master Farmer "for a job well done and life well lived."

21

A Hall of Music in the Great Depression

1929 to 1940

The air we breathe is charged with the spirit of rejoicing.

—Edward C. Elliott

The day was May 3, 1940, and Edward Elliott was feeling particularly proud. With his hands thrust into the pockets of his dark suit, he addressed students, faculty, staff, and community leaders at the dedication of the newest building on the Purdue University campus.

"Here and now," Elliott said, "Purdue celebrates another triumphant success; the product of the united effort of the University, the state, and the nation. From now on each of you students and each of your successors through the years to come possess new opportunities to be a vital unit of the living and serving university. And so the air we breathe is charged with the spirit of rejoicing."[1]

The University president who nine years earlier had stood from his office chair, slapped his hand on his desk, and declared, "Never, as long as I am president, will this university spend one damn penny for music on this campus!" was dedicating the crown jewel of his administration: the Hall of Music.

Adding to the irony, in 1958, long after his retirement, the Purdue University Board of Trustees would honor him by forever attaching his name to the building, and therefore making a music hall the single accomplishment for which he is most remembered.

The Hall of Music was Elliott's project. He had help from University vice president and treasurer R. B. Stewart, architect Walter Scholer, and others. But it was Elliott who first saw the opportunity to build the Hall of Music and who guided it each step of the way. It cost $1.205 million, which is a lot of pennies, and it had been accomplished at a time when the nation was still in the grips of the Great Depression.

The Great Depression

The University had struggled with tight finances since the day Elliott arrived in 1922. The Depression began seven years later, and Elliott persevered, trying to do more at Purdue with less.

The impact of the Depression was not immediate, but it was felt soon enough. In his biography of Elliott (which Elliott refused to call a biography but instead called a "dissertation"), Frank Burrin wrote, "From 1932 to 1937 virtually every problem of the university was related to finance and there were no ready solutions to most of them."[2]

Working with Stewart, Elliott was particularly adept at making use of federal funds for support of education, including grants from federal agencies such as

Board president David Ross, trustee J. K. Lilly Sr., and Dean Stanley Coulter, May 1940.

the Public Works Administration, the Works Progress Administration, and the National Youth Administration. From 1933 to 1940 federal grants to Purdue totaled $3.75 million.[3]

In 1932 the Indiana General Assembly responded to declining state revenue by taking steps that cut state funds for its public universities. Purdue's state appropriation was reduced by nearly 30 percent in 1932–1933. In response, the Purdue University Board of Trustees reduced salaries 10 to 15 percent—including Elliott's. The state also mandated that public universities could not increase salaries or add staff without approval from the State Budget Committee.[4] There would be more cuts in the years ahead.

In September 1932 Elliott met with staff and representatives of student organizations. He told them, "The country is facing a crisis today such as it has not known since the Civil War."[5]

Enrollment at Purdue dropped in 1932 and again in 1933, moving from 4,655 to 3,699—a 20 percent decrease. But in 1934 it started climbing back up again, and by 1939 it reached an all-time high of 7,119. Why did enrollment increase through most of the Great Depression? Elliott was concerned that students were attending Purdue simply because they couldn't find a job, and the additional enrollment was causing a financial strain on the University.

Although times were hard, students on campus still had fun. There were many student dances throughout the Depression. The big fall dance was the Military Ball. And in the spring the Junior Prom was white tie and very formal. Students danced to dreamy music of the 1930s: "The Way You Look Tonight," "Begin the Beguine," "They Can't Take That Away from Me," "Mood Indigo," "Cheek to Cheek," and jazzier numbers such as "Sing, Sing, Sing."

Students at the 1934 Junior Prom in the Memorial Union.

The *Debris* of the 1930s featured whole sections of "beauties." Up to a thousand students attended events at which dozens of young women paraded before judges, hoping to be named a campus beauty and win a photo spread in the yearbook.

Harry's Chocolate Shop and Prohibition

The *Debris* yearbooks through the Depression years featured no stories about what people did without and how they suffered. The yearbooks focused on the love of campus life and the joy of being young. Intercollegiate athletics and men's and women's intramural sports programs continued. And students went to Harry's Chocolate Shoppe in West Lafayette for ice cream and a soda. It was never a speakeasy as lore suggests.

Harry Marack Sr. and Horace Reisner founded the Chocolate Shoppe at 329 State Street in 1919. Reisner owned a book and student supply store next door. The business was advertised in the 1923 *Debris* as "the home of malted milks and fine candies."[6]

In 1927 the business faced competition when the Sweet Shop opened in the Union, selling ice cream from Purdue's creamery. Frank "Pappy" Fox worked there for more than thirty years. It was later renamed Pappy's Sweet Shop.

Prohibition ended in 1933, the partnership between Marack and Reisner dissolved, and Marack added another business, a bar named Harry's, selling beer at Lindberg Road and Salisbury Street—the first alcohol license in West Lafayette. In 1934 Marack combined his two businesses and Harry's Chocolate Shop on State Street became a bar. It was licensed to also sell wine in the 1950s, and later liquor.

Students in the Memorial Union Sweet Shop, mid-1950s.

In 1960 Harry Marack Jr. became a partner. Marack Jr.'s wife, Helen Posthauer Marack, said that for many years Harry's closed at 10 p.m. When a woman entered, Harry warned men to watch their language. The bar closed during home football games so the Maracks could attend. Its popularity was due to the students' love for Harry. "It wasn't so much about the place as it was about the man who ran it," Helen said. Harry's was sold out of the family in 1971 and continues with the same name. It's a campus tradition that has served Boilermakers for two thirds of the University's first 150 years.

Hall of Music

In March 1933 Franklin D. Roosevelt became president of the United States, and in June the National Industrial Recovery Act created the Public Works Administration (PWA) to spur the economy and increase employment through large-scale projects including building schools.

Lafayette architect Scholer, who designed the campus master plan and many Purdue buildings, said at one point under instructions from Elliott and the board of trustees, he, along with University vice president and treasurer R. B. Stewart and Allison Stuart, attorney for the board, filed applications for six Purdue projects at the same time with the federal PWA regional office in Chicago. "The general idea was if you filed six you might get a third of them, or a fourth of them, or one, so we better file plenty," Scholer said in a 1971 interview. "All of these projects had been discussed and were needed."[7] The projects were the Fieldhouse, a portion of Cary Hall, a women's residence hall, additions to the Memorial Union, the Chemical Engineering Building, and the Hall of Music. It was an aggressive application.

Scholer said: "One day Dr. Elliott called me and said that he had just got word that all six of the projects had been approved. He said, 'Weren't you planning on going to Europe?' I said, 'Yes, we're leaving next Wednesday. . . . My wife and two boys are going.' He said, 'Well, I think maybe you had better stay here.' I said, 'I'm sure I should stay here.'" Work began immediately. One of his toughest jobs was designing cantilevered balconies. All the math and engineering was correct. But as they were filling with people on dedication day, he kept looking back to make sure they would hold.[8]

There were other PWA projects on campus. One of them was the Executive Building. On October 4, 1935, the Lafayette *Journal and Courier* reported that Purdue had received a $151,875 PWA grant for the Executive Building that also received state funds and money from other sources. "The granting of these federal funds will make possible the early completion of another link in the extensive building program of the University," the newspaper reported. It said the location just west of the Oval off Northwestern Avenue would become the center of campus.[9] With six, three story classic columns and steep front steps, it was designed as the most impressive building on campus and the focal point of activities.

The full name of the facility was the Student Service and Administration Building. In 1975 it would be renamed Frederick L. Hovde Hall. By the end of 1936 and continuing into early 1937, offices were moved into the new facility. But even before it was finished, the *Journal and Courier* reported that it would "eventually serve as the foyer for a general assembly hall."[10]

Hall of Music under construction behind the Executive Building, 1939.

Elliott first began discussing that assembly hall in 1934, when he said the University needed an auditorium that would seat five thousand people. That would be huge and big enough to seat all the nearly four thousand students enrolled at Purdue that year.

Financing for the hall ultimately included $542,000 from the PWA, $300,000 from the General Assembly, $300,000 from revenue bonds, and $62,750 from various other sources. By a 1929 state law, Purdue could sell revenue bonds specifically for a Hall of Music. So it was planned as a Hall of Music since that was to be one of its main purposes, in addition to hosting lectures and religious programs, commencements, and campus assemblies. At the same time the PWA and the state provided money for the Purdue Hall of Music, funds were also awarded for an auditorium at Indiana University. Additional money went to Ball State and Indiana State.

The Hall of Music was constructed directly behind the Executive Building and connected by a bridge. Campus lore said Elliott planned the hall as an annex to the Executive Building in order to receive funds. But the connection was not needed for the state funding or the bonds, and there is no evidence that the PWA required it. Why, then, are the buildings so close together and connected? One explanation is exactly as the *Journal and Courier* explained in 1936: the Executive Building would be the foyer for the Hall of Music.

The Executive Building and the Hall of Music were both constructed with main floors on the second level, above the ground floors. The second level is the location of the bridge connecting the two buildings. The foyer outside the theater in the Hall of Music is small for the size of the facility, whereas there is a foyer in the

The cantilevered first balcony in the Hall of Music was properly designed, but it worried architect Walter Scholer as people filled it during dedication, May 1940. He said that if it collapsed he wanted to be underneath so he would be the first to go.

Executive Building, which was located on Oval Drive where there was parking in front of the building. Off-campus patrons were to park on Oval, walk up the steps of the Executive Building to the main floor, and walk across the bridge to the Hall of Music. In March 1939 J. Andre Fouilhoux, the New York consulting engineer for the project, complemented plans that called for access to the Hall through the Executive Building.[11]

On July 18, 1938, the *Indianapolis Star* wrote: "The campus is arranged for an auditorium which would be built next to the Executive Building, with the entrance to that building serving for both." On August 31, 1939, the Lafayette *Journal and Courier* said, "The Executive Building serves as a foyer for the music hall." Elliott put all the money available to him into maximizing the number of seats using existing space in the Executive Building for the foyer and entrance.

The seating capacity in the art deco theater changed slightly over the years from the original 6,146 to 6,005. It is one of the largest proscenium theaters in the world, larger than Radio City Music Hall. It was designed so a curtain could be dropped from the first balcony, closing off the back of the theater for small events.

Elliott was frugal. But he provided flourishes for the Hall of Music, including exterior sculptures done by John Johnson, a Frankfort, Indiana, artist.[12]

There were two days of dedication ceremonies, Friday and Saturday, May 3 and 4, 1940. At the Saturday event Elliott said men, women, and children for many years into the future would come to the Hall "to see, to listen, to think and thus be made over and higher, little by little, through their seeing, hearing and thinking. . . . Over the entrances to this building there should be carved 'that men may better know the worth of the harmony of sight and of sound and of feeling and of reason.'"[13]

Life at Purdue resumed quietly in the fall of 1940, through the spring of 1941, and again the next September, although storm clouds were already moving through much of the world.

On December 7, 1941, the Japanese attacked Pearl Harbor. Everything changed—for the nation, for Purdue, for its students and faculty, and for Elliott, who was only days away from his sixty-seventh birthday.

Nothing would ever be the same.

GIANT LEAPS IN LIFE

Charles A. Ellis

Charles A. Ellis kept a photograph of the Golden Gate Bridge above his desk at Purdue University, where he was a professor of structural engineering from 1934 to 1946.

If anyone asked him about it, he said, "I designed every stick of steel on that bridge."[14]

Purdue always celebrated Ellis as designer of the bridge the American Society of Civil Engineers calls one of the Seven Wonders of the Modern World. But, for eighty years his name was erased from all involvement with the iconic structure.

Joseph Strauss, who did not have an engineering degree, was the force behind the Golden Gate Bridge. He campaigned for it and overcame opposition. His Chicago company had built four hundred bridges, but the design he personally submitted for the Golden Gate was rejected. So in 1929 he turned the design work over to Ellis, who was his company vice president.

Ellis worked twelve- to fourteen-hour days and filled eleven volumes of notes with his calculations made with a circular slide rule and hand crank adding machine.[15] But there was tension between Strauss and Ellis. Just before Christmas, 1931, Strauss fired Ellis, believing he was too slow with his calculations. He also believed Ellis was creating problems between Strauss and the bridge governing board. When the project was completed, Strauss would not allow Ellis to receive public credit for designing one of the world's most famous bridges.

Strauss died in 1938, one year after the bridge opened. Ellis died in 1949 having never officially received the recognition he deserved. But some people knew. In his obituary the Associated Press credited Ellis with designing the bridge.[16]

In 1986 John van der Zee wrote *The Gate: The True Story of the Design and Construction of the Golden Gate Bridge*. In it he credited Ellis as the designer. After the book was published, Lewis McCammon, who studied under Ellis at Purdue, came forward with his professor's papers, including letters exchanged with Strauss and other documents proving beyond doubt that Ellis had designed the bridge. The papers are in the Purdue University Virginia Kelly Karnes Archives and Special Collections Research Center.

After its own investigation in 1992, the Golden Gate Bridge District concluded Ellis was the designer but declined to place any marker acknowledging his work.

In May of 2012, on the seventy-fifth anniversary of the bridge, the American Society of Civil Engineers finally placed a plaque at the Golden Gate.

It names Charles Ellis as the designer.

22

Preparing Students for War

1941 to 1945

Today all the people are being matriculated
in the College for War.

—Edward C. Elliott

On December 15, 1941, eight days after the Japanese attack on Pearl Harbor, the Purdue University community gathered in the Hall of Music. Most had been present nineteen months earlier for the dedication of the building and could still recall the words of President Edward C. Elliott that day: "The air we breathe is charged with the spirit of rejoicing."

In mid-December 1941 the holiday joy that usually filled the campus was subdued. This gathering had been called to hear the first of what would be three war messages from the president of the University. It was broadcast live by WBAA radio, located in the lower level of the Hall of Music.

"This is a solemn and significant hour for all of us," Elliott said. "Many trying and uncertain days are ahead. But the fogs of time and events do not conceal the grim fact that the tragic happenings of the moment are certain to produce sweeping changes in the life plans of all of us."[1]

Like President Winthrop Stone during World War I, Elliott encouraged the male students to continue their studies as long as they could before being drafted. "The nation must plan for a prolonged conflict," Elliott said. "This conflict is a contest of the strength of the machines and of the technical knowledge of men who make and operate those machines. The final test of your allegiance to the country's cause is to be found in the quality of your determination to get that training, here available to each of you as one of the fortunate few. Modern warfare requires a complicated organization of men and machines. The effectiveness of this organization is dependent upon the skillful fitting of the right men to the right jobs. They also serve who only stay and study."[2]

He presented a "Bill of Responsibilities" for Purdue students:

1. Apply yourself to your work as students;
2. Reduce to a minimum your recreation and social activities that make demands upon your money, your time, your concentration upon your serious tasks. Avoid waste;
3. You have a responsibility for rigid economy to your personal affairs;
4. Help the University to be economical;
5. Stay informed, make yourself rumor proof;
6. Maintain your health and physical fitness.[3]

Within three weeks Purdue entered an accelerated teaching schedule to move the students through to their degrees and into the military. Three sixteen-week terms were created. Final exams and summer vacations were eliminated. Students graduated in three years, and the men moved on to active duty, the women to service on the home front.[4] In early 1942 Elliott told a student convocation, "The primary business [of Purdue] is to prepare you for some meaningful job in the war."[5]

Elliott Goes to Washington, D.C.

Elliott himself was quickly called to national service. In January 1942 he was named to the Committee on War-Time Requirements for Specialized Personnel. He turned down another appointment as a consultant to the Office of Production Management. Later in 1942 the board of trustees granted him a leave of absence to become chief of the Division of Professional and Technical Employment and Training, part of the War Manpower Commission. He left campus June 22, 1942. Administration of the University was left to a committee: Vice President and Treasurer R. B. Stewart, Dean of Engineering A. A. Potter, Dean of Agriculture H. J. Reed, and Vice President Frank Hockema.

David Ross, Purdue trustee from 1921 until his death in 1943, president of the board the final sixteen years.

Only a few weeks after Elliott's departure, board president David Ross had a massive stroke. He never fully recovered and died on June 28, 1943. George Ade died less than a year later, on May 16, 1944.

The campus began to look like a military camp. In a sense, it was. About 450 staff members entered the military services and others went into civilian work.

Dorothy Stratton and Helen Schleman Serve the Nation

The day Elliott left Purdue for Washington, D.C., he told the faculty, "Each and every one of you should be prepared to give not one hundred percent but 110 percent of your effort for the success of the war program."[6] That's exactly what Dean of Women Dorothy Stratton and her assistant, Helen Schleman, did. But they didn't need Elliott to tell them. They took their advice from visiting faculty member Lillian Gilbreth, who was advising many people and organizations. Once the war started and factories began preparing to hire women to replace men who had enlisted, companies were confused about how work would change with a female workforce. Gilbreth was asked what industry needed to do. "Build separate restrooms," she answered.[7] That was the only change they would need to make.

At Gilbreth's urging, Stratton enlisted in the Navy Women's Reserve, the WAVES. She was called to duty on August 28, 1942, with the rank of senior lieutenant. She was soon ordered to a meeting in Washington, D.C., where "a room full of admirals" questioned her and said they were looking for a woman to head the Women's Reserve of the U.S. Coast Guard. They offered her the position almost immediately, and she began the task of forming and naming the organization. She selected the name SPAR based on the Coast Guard motto, "Semper Paratus, Always Ready."[8]

Congress approved legislation creating the SPARs on November 23, 1942, and Stratton was sworn in as lieutenant commander. Her first job was to hire an executive officer. It was an easy decision, and she contacted Helen Schleman at Purdue. On December 14, 1942, Schleman was sworn in as the second officer to join the SPARs. Her rank was lieutenant, senior. Stratton and Schleman would lead

President Edward C. Elliott with Helen Schleman (right) and Dorothy Stratton during World War II. Stratton was the first director of the SPARs, the U.S. Coast Guard Women's Reserve. Schleman attained the rank of captain in the SPARs.

the SPARs through the war.[9] Assistant Dean of Women Clare E. Coolidge served as interim dean of women after Stratton's departure.

Stratton was promoted to captain in 1944 and was awarded the Legion of Merit medal in 1946. At the end of the war she resigned from Purdue. As the first full-time dean of women at Purdue in 1933, she played a major role in the growth of female enrollment from five hundred to more than fourteen hundred. She created studies for women and an employment placement center, and she oversaw construction of women's residence halls. A scholarship named in Stratton's honor supports women's participation in Purdue's Naval ROTC program. But there is no building on campus to honor her memory and contributions.

Stratton served as the first director of personnel at the International Monetary Fund from 1947 to 1950. Next she served for ten years as executive director of the Girl Scouts of the USA. She was also the United Nations representative of the International Federation of University Women and chair of the women's committee within the President's Commission on Employment of the Handicapped. She retired to West Lafayette and died in 2006 at the age of 107. In 2010, First Lady Michelle Obama christened the Coast Guard cutter *Dorothy C. Stratton* in her honor.

Schleman also attained the rank of captain in the SPARs. She returned to Purdue in 1947 as dean of women. She was about to witness a sea change in the role of women, and in fact, she helped lead it.

Navy V-12 Program

Beginning in 1943, the army and navy launched training programs at Purdue. The army program trained 1,348 men. About 2,730 men served in the navy V-12 program during the war, and 400 of them graduated with bachelor's degrees in engineering. They were allowed to participate in student activities, including football. With Purdue players plus naval personnel relocated to Purdue from other universities, the Boilermakers had their first perfect football season since 1929.

In his book *The Story of Purdue Engineering*, H. B. Knoll said: "A Naval Electrical Training School for more than three years provided the most distinct military feature on campus. Its headquarters were in one of the men's residence halls and the men, unlike those in other groups, marched from class to class, their white uniforms making the campus look like a port. They carried a heavy schedule and were trained

Some of the 123 members of the Class of 1942 being sworn in
as ensigns in the navy by Captain J. C. Fremont.

Navy shipmen march to class on Purdue campus during World War II.

to be electrician mates. The normal complement of men was 800 at a time and the total trained over the years was 6,000."[10]

The low point of student enrollment during the war was 3,762 students in 1944.

Manhattan Project

More than a hundred people at Purdue worked for the Manhattan Project, helping to develop the nuclear bomb. There was other war-related research on campus. "Chemistry had all the war research it could handle," Knoll wrote. The faculty worked with government agencies and industries in work such as developing synthetic rubber, antimalarials, and explosives.[11] Part of the physics faculty moved to MIT and worked with researchers there developing radar.

Every school at Purdue was involved in the war effort in some way. "Agriculture, through the School, the Experimental Station and Extension, probably more than ever before spread its influence throughout agriculture in Indiana," Knoll wrote. "To cite just a sampling of its work, it developed special training courses; kept the fertilizers on the market which were best suited to crop production and yet were conservative of raw materials; improved the palatability, nutritive value and keeping qualities of dried eggs and dehydrated vegetables; and helped in the placement of almost 7,000 farm workers annually during three of the critical war years."[12]

In *A Century and Beyond: The History of Purdue University*, Robert Topping said the most dramatic research was done in physics on the properties of germanium crystals for microwave radar. Under the direction of Professor Karl Lark-Horovitz, it helped lead to the development of the transistor.[13]

Penicillin

John A. Leighty received his PhD in chemistry from Purdue in 1936 under the direction of Ralph Corley. Leighty went on to a distinguished career with Eli Lilly and Company in Indianapolis. Lilly was among the first companies to develop a method to mass-produce penicillin-G, the world's first widely available antibiotic, marking the beginning of a sustained effort to fight infectious diseases. Leighty was part of the team that accomplished this. Penicillin saved thousands of lives during World War II. Leighty was also involved in developing, producing, and supervising other pharmaceuticals including erythromycin, vancomycin, streptomycin, and Darvon. He served as Lilly's executive director of scientific research, leading 650 scientists in seven research divisions.

Curtiss-Wright and RCA Cadets

As men joined the military a huge need arose for women in business and industry. American industrial production of ships, airplanes, tanks, guns, and all manner of equipment needed for war played a major role in the Allied victory. Women "manning" the home front did the work. Two groups of female students at Purdue during the war were the Curtiss-Wright Cadettes and the Radio Corporation of America (RCA) Cadets.

The Curtiss-Wright Company had a contract to produce dive-bombers for the war effort, and they were in danger of defaulting on their agreement. In 1942, seven U.S. universities contracted with the company to train qualified women in an accelerated engineering program. More than nine hundred women were trained at Purdue and other universities, and upon completion of the program they were immediately put to work around the country.[14] The first one hundred cadets arrived at Purdue on February 12, 1943. More came in the years ahead.

The Advanced Development Group of Radio Corporation of America sent RCA Cadets to Purdue in the spring of 1943 to study math, drafting, shop, electric circuit theory, electronics, radio theory, and much more. Purdue was the only university chosen for the program. The women graduated after intensive studies and went to work at RCA plants around the nation.[15]

Women also came to Purdue for technical studies from Wright Field in Dayton, Ohio.

At the end of the war when the men returned, women were sent back to being housewives. Their contributions to victory have been too often neglected and even forgotten.

Purduettes

By the fall of 1942 the enrollment of male, nonmilitary students on campus had declined. Many of the males studying on campus were in the military and had restrictions on their travel. Al Stewart's all-male Varsity Glee Club could not move around the state and nation representing Purdue.

But Stewart had an idea. He started an all-female group, called them the Purduettes, and they became the premier vocal group at the University for the duration of the war. There were thirteen female students in the first group and audiences loved them. When the war ended the Varsity Glee Club once again became the top vocal group at Purdue. But the Purduettes also continued to perform, giving young women an opportunity to enjoy their talents and love of music. In the twenty-first century the Purduettes had become a much larger group and remained more popular than ever.

Entertainment

The campus was not all studies and war. Men and women romanced, fell in love, and married, sometimes before the soldiers and sailors were shipped overseas. They danced and listened to music like "I'll Be Seeing You," "I'll Be Home for Christmas (If Only in My Dreams)," and upbeat recordings like the Andrews Sisters' "Boogie Woogie Bugle Boy." The Hall of Music became a place for major entertainment. In 1943, Al Stewart, of Purdue Musical Organizations, came up with the idea of a series called Victory Varieties, bringing big-name entertainment to campus. R. B. Stewart gave him $1,500 to see what he could do, thinking he'd lose it. On the first evening, October 23, 1943, more than two thousand people arrived to see Milt Britton and Bonnie Baker.[16]

The programs continued after the war, usually focused on a hoped-for Saturday afternoon football victory to add to the merriment of the evening programs. Many of the most popular performers of the day came to the Hall of Music, which was large enough to attract them as they toured from Chicago to Indianapolis, Cincinnati, Louisville, and other cities.

Elliott Returns to Campus and Retires

In April 1943 several board members wanted Elliott back on the campus full-time. He resigned his Washington, D.C., duties and returned to Purdue. In April 1944, the Purdue University Board of Trustees informed him that he needed to step down as president at the end of June 1945. University regulations required that he step down at the end of the fiscal year after his seventieth birthday. Elliott had helped establish the retirement rule, but he didn't think it applied to him. He really wasn't ready to retire. But the board was firm.[17]

Elliott's service had been remarkable. But times had changed dramatically since 1922, when he arrived. A new, younger leader was needed. Elliott's last day as president was June 30, 1945.

Victory

Tuesday, May 8, 1945, was Victory in Europe Day. The Germans surrendered and that part of the war ended. It was different in the Lafayette area than at the end of World War I. Celebrations were actually subdued. U.S. flags fluttered in the spring breeze. Fire stations sounded their sirens. Some church bells rang. Schools were let out and taverns were closed. People had known the end was at hand for several weeks, so no one was surprised. And there was apprehension about what would come next in the final battle with Japan.

It was a different scene when President Harry Truman announced Japan's surrender on the evening of August 14, 1945. Police estimated that through the evening thirty to forty thousand people packed downtown Lafayette. Stores closed and did not plan to reopen for two days. Schools were closed, Purdue was closed, there were special church services. Church bells, factory whistles, and car horns added to the noise of people celebrating in the streets. There were snake dances and impromptu parades. Factories were closed, and sailors from Purdue got out the University's Victory Bell and paraded it downtown. Celebrations went on deep into the night.[18]

During World War II about 17,500 Purdue men and women were in the armed forces. More than 500 of them died. They were people like Harry J. Michael. He majored in animal husbandry at Purdue, was commissioned a second lieutenant in December 1944, and arrived in Europe in February 1945. On March 13, while leading his men in battle, he became a hero, single-handedly taking out enemy positions and capturing German soldiers. His leadership and bravery inspired his men, and doubt-less saved many of their lives. For all of this he was awarded the Congressional Medal of Honor, the military's highest award for valor. It was presented posthumously. He was killed in action on March 14.

His death came eight weeks before the end of the war in Europe, and the day after his twenty-third birthday.

Frank P. Thomas Jr.

Frank P. Thomas Jr. graduated from Purdue University in 1941 with a degree in mechanical engineering and, along with most young men his age, enlisted in the military, serving as an army air force major in World War II.

When he came home he went to work with a small Indiana company his father had started. Thomas went on to touch the lives of millions of people by creating the nation's first patented soft serve ice cream machine, developing a system to flame broil hamburgers, and creating what at one time was the fastest growing fast food restaurant in the United States and the second in number of franchises after McDonald's.[19]

Founded in 1958 and based in Indianapolis, the restaurant, Burger Chef, had twelve hundred locations nationwide by 1971.

Thomas's father, who had just a fourth-grade education, was a carpenter and inventor.[20] During the 1930s he founded the General Equipment Company and manufactured frozen custard machines and amusement park rides. Frank Thomas Jr. joined the company and became president when his father retired. He improved his father's design and patented the Sani-Serv soft ice cream machine and the Sani-Shake to make milkshakes. In 1956, at the request of another restaurant owner, he developed the Sani-Broiler for flame-broiled food.[21] It could turn out a thousand hamburgers an hour, and later models doubled that.

Since they were already making all the equipment they needed to start a restaurant, Thomas and his brother, Donald (Purdue 1950), with two other partners started Burger Chef. It featured colorful buildings and advertising with a cartoon character named Burger Chef and a boy named Jeff—"Burger Chef and Jeff." They promoted "incredi-burgible" sandwiches.

In addition to competing with McDonald's, Burger Chef also was challenged by Burger King. In the late 1960s, Pillsbury acquired Burger King and Thomas decided he could not compete against the food giant. In 1968 Burger Chef was sold to General Foods for $15 million.[22] In 1981, General Foods sold the restaurants to Hardee's.

The last known Burger Chef closed its doors in Cookville, Tennessee, in 1996.

After selling the restaurant, Thomas and his wife, Jill, traveled extensively in their twin engine Cessna airplane. They also traveled internationally, studying urban renewal.[23]

Thomas died in Taos, New Mexico, on June 16, 2008, at the age of eighty-nine, having left a legacy in some of America's favorite foods: soft serve ice cream, shakes, hamburgers, and fries.

James S. Peters II was born in 1917 on his grandmother's farm in segregated Arkansas and grew up in Monroe, Louisiana. Near the end of his life he was in the first class of inductees into the Connecticut Veterans Hall of Fame, where he was cited for helping to bring about integration in the U.S. Navy.

James S. Peters II
(Courtesy of James S. Peters III)

Among nine others inducted with Peters was U.S. President George Herbert Walker Bush.

Peters entered Ph.D. studies in the Purdue University Division of Education and Applied Psychology in 1953, finishing in 1955. He went on to a career in education that included serving twenty-five years as associate commissioner and director of vocational rehabilitation for the Connecticut Department of Education, twenty-six books, and a lifetime championing the rights of physically and mentally disadvantaged people.[24]

He did research on segregation while serving at the U.S. Naval Training Center at Great Lakes, Illinois, during World War II. The navy selected Great Lakes for its first training of African American sailors. The training programs were segregated, but that policy began to end in 1944, and by 1945 all training at Great Lakes was integrated.

Peters was head of remedial education at the Naval Training Center and a specialist teacher/first-class psychologist doing research for the Special Training Unit and the Neuropsychiatric Unit. "The Bureau of Navy Personnel received monthly reports from both units and passed the results to the Secretary of the Navy and Department of Defense," Peters said.[25]

"My research showed that African-American sailors, if given the opportunity, could do the same things that white sailors did," Peter said in a 2008 interview.[26]

He did undergraduate work at historically black Southern University and A&M College and graduate work in psychology at the Hartford Seminary Foundation in Connecticut. After the war he worked for the Veterans Administration in Chicago, continuing graduate work in psychology at the University of Chicago and Illinois Institute of Technology on the GI Bill. When the Veterans Administration decided its psychologists needed a PhD, Peters enrolled at Purdue. His wife, Marie, also enrolled and received a master's degree in sociology. She became the first African American tenured professor at the University of Connecticut.

Although Peters said that in the mid-1950s "Indiana was a segregated state," he called his experience at Purdue "glorious," and supported the University throughout this life.[27]

He retired as Connecticut deputy commissioner of education in 1982 but continued in private practice, taught at universities, served on boards, and wrote. He died in December 2008.

23

Redbrick Buildings and Quonset Huts

1945 to 1950

We had old throw rugs with fringe on the edges and in the
winter the fringe would sometimes stand straight up from
the cold wind blowing in. But we were happy there.

—Purdue graduate student John Hicks

A sudden downpour drenched Washington, D.C., on a hot, humid day, April 25, 1945. Thirty-seven-year-old Frederick L. Hovde stood before the woman who had summoned him as rain dripped from his clothes to the office floor.[1]

His meeting was with Kathryn McHale, a national advocate for women's rights and general director of the American Association of University Women. She was also a trustee of Purdue University, and she wasted no time getting to her reason for the meeting. McHale asked Hovde if he was interested in becoming president of Purdue. Hovde responded without hesitation. He was very much interested.

That brief and soggy meeting led to the longest presidency in the history of the University.

With the retirement of President Edward Elliott at the end of June 1945, the dean of engineering, A. A. Potter, was once again approached about his interest in being considered for president. He said he was not and was named acting president until a new leader could be found.

Elliott moved out of the president's home on Seventh Street and became the first leader in Purdue history to remain in the community during retirement. The previous five presidents either died in office or moved from the area after leaving the University. As president emeritus, also the first person to receive that title, Elliott played an active role in selecting the man who would replace him.

On July 9 the board members were in agreement. Hovde was told he had their unanimous support. The official vote was taken at a meeting on July 24.

Young, energetic, and very bright, Hovde was a perfect fit in 1946 to lead Purdue into the future. He was very different from Elliott. Elliott loved to speak and was inspirational before an audience. Hovde was not. He often did not use a script and sometimes stopped mid-speech to look up and gather his thoughts before going on. He was neither confrontational nor overly friendly with faculty and staff. Purdue historian Robert Kriebel described him as "regal."[2] He was an amazing leader.

Hovde was born in Minneapolis, and his family moved to North Dakota when he was five. He was a good student and an outstanding athlete, playing football, basketball, and track. He led his high school football team to the state championship.

Hovde returned to Minneapolis in 1925 to enter the University of Minnesota, graduating in 1929 with a degree in chemical engineering. He was not a big man, average in height, weighing about 150 pounds. But he played on the Minnesota basketball team and in 1928 played a game against Purdue in the Memorial Gymnasium. Purdue won, 45–17. Hovde was a star on the football team, winning all Big Ten honors, and was named to the second-string All-America team his senior year. After graduation he went to Oxford University in England as a Rhodes Scholar. In addition to his studies at Oxford, Hovde joined the rugby team and became only the third American to win a "blue" at the school—the highest athletic honor given.

Returning to Minneapolis, he took a teaching job at the University of Minnesota that soon led to an administrative position. In 1936 Alan Valentine, the president of the University of Rochester in Rochester, New York, selected Hovde as an assistant. Hovde and his wife, Priscilla, called "Pris," headed east.

In 1940, before U.S. involvement in World War II, Hovde took a job in Washington, D.C., with the National Defense Research Committee. The war work took him to London for seventeen months. After being named president of Purdue in 1945, he closed out his work with the government as the war in Japan ended and prepared to move to West Lafayette with Priscilla and their children, Boyd, Jane, and Linda.

Hovde and Priscilla arrived at Purdue the first week of January 1946. They checked into the Memorial Union Hotel while their future home on Seventh Street in Lafayette was undergoing remodeling, and Hovde immediately went to work with no fanfare or celebration. He knew what was coming—the largest influx of students to university campuses in the history of the nation.

GI Bill

President Franklin Roosevelt signed the Servicemen's Readjustment Act on June 22, 1944. It was for veterans who had served in the military more than ninety days after September 16, 1940. It offered federal loans, hospitalization, and other benefits—including up to $500 for tuition and a cost of living stipend while attending college. With no jobs at home as the United States converted from a war to a peacetime economy, millions of veterans took advantage of the bill to become the first person in their family to attend college. The Land-Grant Act of 1862 and the Servicemen's Readjustment Act of 1944 are considered the two greatest pieces of educational legislation in the history of the nation, providing the opportunities of a university education to millions of people.

Purdue vice president and treasurer R. B. Stewart was chairman of the Administration of Veteran Affairs Advisory Committee on Vocational Rehabilitation, Education and Training. He played a leading role in implementing the GI Bill.

In September 1945, with the war against Japan just ended the month before, Purdue enrollment was 5,628, many of those in military service, By September 1946 it had more than doubled in one year, reaching 11,462—the largest one-year increase in Purdue history. In 1947 it shot up to 14,060 and in 1948, the peak postwar year, enrollment hit 14,674. It was part of a national trend. By 1947, 49 percent of all college students in the United States were military veterans.

The question was, where to put all these students. Barracks were taken down at nearby Bunker Hill military base and rebuilt on the Purdue campus to house the students. Quonset huts and "temporary" buildings went up, and some remained in use "temporarily" for the next sixty years. Retired faculty members were called back to the classroom. Faculty wives with degrees and expertise were recruited to teach.

There were other challenges. The students were vastly different from those of previous years. Some of them were eighteen, nineteen, and twenty years old and had never been out of their Indiana home counties. Others were in their mid- to late twenties or older and were veterans of a brutal war. In addition, a number of the veterans came with wives and children, presenting a housing problem the University had never faced before. In 1946 Purdue needed two hundred new units for married students in addition to more housing for those who were single.[3] Married student housing quickly became known as "fertility acres" as former servicemen and their brides hurried to start families.

In Cary Quadrangle the normal capacity of 936 was increased to 2,050 by doubling and tripling single rooms and putting bunk beds in the attics. One hundred ten bunk beds were placed in the attic of the Agricultural Engineering Building. The second floor of the airport terminal building housed students.[4] Attics in other buildings were put to use. Men living in the attics showered in the Fieldhouse. Three hundred students were housed in an unused factory building in Lafayette.

President Frederick L. Hovde walks on campus with a staff member and a student.

In 1947, Richard Grace arrived at Purdue from Chicago as a seventeen-year-old freshman. "I was a little overwhelmed," he said. "I was living in Bunker Hill Dormitory. It was located right where Tarkington Hall is. We had four large rooms. One was a study room and three were dormitories. We had approximately eighty or ninety guys living together in [each] of the dormitories. We had bunk beds. We had everybody sleeping head to toe. It was a wonderful existence." His mother didn't think so. She cried after dropping him off. "I was thrust into classes with captains and colonels from the army who were twenty-five and twenty-eight years old," Grace said. "They weren't kids like me. These were GIs and they were determined to get an education. It was an awakening. It was highly competitive. These guys were there for a reason, to get grades, to get married, and to get on with their lives. They were very serious." They also partied. "We learned from them to drink beer a little earlier than we might have," Grace said. "It was crowded. But the mood on campus was happy. It was an energetic time. Part of the pulse of the campus was the desire of these veterans to get their degree."[5]

Grace graduated in 1951 with a degree in metallurgical engineering. He completed his PhD at Carnegie Mellon and in 1954 joined the Purdue engineering faculty. He retired in 2000 as vice president for student services and founding head of the Undergraduate Studies Program (later renamed Exploratory Studies).

Not only was Purdue short of facilities, the town of West Lafayette had no ability to house the flood of students, faculty, and staff. One hundred fifty new homes for faculty and staff were built in an area along State Street, later replaced by married students housing. Sections of the prefabricated homes, which cost $2,006 each, were assembled in the Purdue Armory. They were called "Black and Whites" because those were the colors.

The number of women enrolling at Purdue increased in the postwar period—but not enough as far as the men were concerned. In 1940 there were 1,405 females on campus. By 1946 the number shot up to 2,023. The majority still studied home economics and only worked a year or two after graduation before marrying and starting their families. Many underclass women dated senior men and withdrew from the University before graduation to marry. Men outnumbered women 6–1. It was called *the ratio* and everyone knew what it was. A photo in the 1947 *Debris* yearbook shows an exasperated young man with a telephone to his ear. "A three months' wait to get a date," the caption read.[6]

These students had grown up in the Great Depression and had known hardships all their lives. They had lived through the war, so the peace years that followed were happy and exciting. Many of the men didn't have much in the way of civilian clothing, so they wore half military uniforms and half slacks and shirts that still fit from before the war. Some of the men wore military pants called "pinks" that were tan with a pink tinge. Others wore their military flight jackets, army boots, and navy peacoats. The men carried slide rules in holsters attached to their belts. Women wore sweaters, skirts, bobby socks, and black-and-white saddle shoes. Women still could not wear slacks outside the residence halls. There were dances, including some that started at 11 p.m. and went all night. Big-name entertainers such as Bob Hope came to the Hall of Music.

In a 1995 interview Jim Blakesley, Purdue '50, remembered married student housing with his wife Rosemary and their first two daughters in the postwar period:

> These housing units were wooden structures with a roof. They did have inside plumbing and the floor had nice openings in it. If there was dirt, you just swept it in the openings. They had iceboxes—real, old-fashioned iceboxes. If you put a lot of caulk on the windows you could keep warm. You could hear what was going on next door. I was always helping to take women to the local hospitals to have babies while their husbands were away [at classes]. I remember the woman next door knocking on the wall and saying "it's time to go!" In one short period with my help four wives were admitted to two different hospitals. The nurses must have wondered about my involvement.[7]

John Hicks arrived at Purdue with his wife, "Swiftie," in 1947. He received a PhD and in 1950 became a professor of agricultural economics. "You can't believe how difficult it was to find a place to live in Lafayette or West Lafayette in those days," Hicks said.

Black and Whites: prefabricated campus housing.

When my wife and I first got to town we rented a room in a private home. The homeowners sent their kids to live with grandparents so they'd have the space to rent to couples like us. Later Swiftie and I moved into an area that was formerly military barracks. The buildings were wooden frames covered with tarpaper. The apartment had a little kitchen with an icebox—no refrigerator—a living room with a natural gas stove to keep the whole place warm, a small bedroom, a bathroom and that was about it. We had some old throw rugs on the floor with fringe on the edges and in the winter the fringe on those rugs would sometimes stand straight up from the cold wind blowing in. But we were happy there.[8]

Hicks would later become a special assistant to Hovde and an acting president of Purdue.

With the veterans returning, many athletes from different high school graduating classes arrived at Purdue at the same time. In the fall of 1946, 350 men tried out for the basketball team.

Kris Kreisle Harder arrived at Purdue in 1946 to get her "MRS" degree. She was looking for a husband. "My four years at Purdue were among the best of my life," she said in a 1995 interview.

We were young and carefree. I met my husband, Frank, at Purdue during my sophomore year at a broom dance in the union. They guys took brooms to the dance. When a guy cut in on a couple, he handed the broom to the guy he was cutting in on. Then that guy had to dance with the broom until he cut in on someone else. There were pinning serenades [at the sorority] after hours. [When a man gave a woman his fraternity pin, it was a prelude to engagement.] It was so much fun. We would be getting ready for bed, the fraternity would come to the house and we'd all go out on the porch. The girl who had just been pinned and was being serenaded would stand together with her pin man while the fraternity guys sang to them. Then the sorority would sing.[9]

Harder studied home economics like almost all of the women. Some females enrolled in engineering, and some faced hostile faculty who told them they were using a spot in the program that a man needed so he could become an engineer and support a family. In 1950, Sarah Margaret Claypool Willoughby became the first woman to receive a PhD in engineering at Purdue. During her professional career she went to conferences where advance organizers didn't realize she was female and assigned her male roommates. "If I'm going to have to room with a man, I would at least like to review the applicants," she responded.[10]

Some women in engineering were accepted by the men. Patricia Bagley Ross grew up in the Chicago suburbs and graduated high school in 1946. She wanted to study engineering, but said all the openings were held for returning veterans. "The University wasn't taking women in engineering unless they had pull," she said in a 1995 interview. She studied at the University of Illinois, Chicago, and in 1948 when she got some "pull" she entered Purdue engineering. "While I was at Purdue I don't think there were twenty-five women in all the engineering branches. There was never another woman in any of the engineering classes I was in. We were treated beautifully. Everyone thought we might be discriminated against, but we weren't."[11]

Ross had hoped to work in engineering after graduation in 1950. But she married and had a baby, and then another, and another, and another. When her children were older, she did go to work. She became part of a national wave of women who had graduated in the 1950s and 1960s who went to work after their children entered high school or college and when technology had shortened the time needed for work around the house.

Given an opportunity, women had remarkable careers. One of them was Ruth Siems, who graduated from Purdue in 1953 in home economics and was a key inventor of a product that has been in virtually every kitchen in America.

The postwar college students had started learning science, engineering, and technology in the military and used the GI Bill to advance their education in college in those fields.

"World War II was the scientist's war," Hovde said in a 1972 interview. "The tremendous advances in electronics meant that the whole electronic world would be revolutionized. Another great revolutionary instrument that came out of World

Students in home economics class, 1953.

War II [was] the computer."[12] In 1962 Purdue created the nation's first department of computer science, headed by Samuel D. Conte.

In the immediate postwar era at Purdue liberal arts courses were in the School of Science. Art and design were in the School of Home Economics. There were no courses in philosophy when Hovde arrived, but he started a program in 1950.

To provide additional opportunities for women, Purdue continued to offer a program of study that emphasized the humanities called Liberal Sciences.

Among those crowding into Purdue during the postwar period was a student who would become one of the University's most distinguished alumni.

Birch Bayh had skipped a grade in elementary school and was only seventeen when he entered Purdue in 1945 at the end of World War II. As soon as he turned eighteen he left college, enlisted in the military, and joined Allied Occupation forces as a military policeman. He returned to campus in 1948, played shortstop and third base on the varsity baseball team, and in 1949 won the Purdue Golden Gloves Light-Heavyweight Novice Championship. He was active in Alpha Tau Omega fraternity, serving as president, and was also president of his Purdue senior class.

Bayh graduated in 1951 and returned to the farm, but soon was lured into politics and state government. A Democrat, he was elected to the Indiana House of Representatives in 1955, and in 1959 at the age of thirty he became the youngest

Speaker of the House in the state's history. He studied law at Indiana University in Bloomington, even as he served as Speaker. In 1962 Bayh ran for the U.S. Senate against the popular, eighteen-year veteran Republican Homer Capehart. No one thought Bayh had a chance, but he won and served three terms.

As chair of the Senate Constitutional Amendments Subcommittee Bayh authored and sponsored two additions to the U.S. Constitution—perhaps more by any one person since the Founding Fathers. The first, the Twenty-Fifth Amendment, dealt with the succession to the presidency and vice presidency in the event of death and procedures to respond to presidential disabilities. The second, the Twenty-Sixth Amendment, lowered the voting age from twenty-one to eighteen.

Bayh planned to seek the nomination as the Democratic candidate for president in 1972, but decided against running when his first wife, Marvella, was diagnosed with cancer. She died in 1979.

While at Purdue Bayh observed that there were few opportunities for women to participate in athletics, and there were only two women in his agriculture classes. He would one day lead efforts to change that and had a profound impact on the lives of women and equality on college campuses throughout the nation. (See part 3, chapter 30.)

The great GI Bill bulge of students who were World War II veterans was past its peak by 1950. Purdue enrollment dropped below ten thousand in 1951, 1952, and 1953. But by 1960 it exceeded fifteen thousand. And the baby boomers, the children of the "Greatest Generation" from World War II, caused universities' enrollment growth to continue.

The Class of 1950—that first big group to attend Purdue for four full years after the war—graduated on Sunday, June 18. Before the commencement program the *Exponent* reported that it was the largest graduating class in Purdue history. Two thousand two hundred ninety-seven people received degrees. "There are a few requests to seniors during the procession," the newspaper said. "Act dignified and try to keep from smoking and waving the commencement program around. If possible wear dark clothing and avoid wearing white buck shoes."[13]

One week later, on Sunday, June 25, on the other side of what had become a very small world, the Korean War began.

GIANT LEAPS IN RESEARCH

Gertrude Sunderlin

In the aftermath of World War II, Purdue Foods and Nutrition professor Gertrude Sunderlin and her students created something that touched the lives of everyone: Master Mix, a mix for baking biscuits, muffins, pancakes, chocolate cake, and much, much more.

Master Mix provided bakers with a precise list of dry ingredients and fats that could be prepared in large quantities and kept unrefrigerated in a household kitchen for six weeks. Then when people wanted to bake muffins, or something else, all they had to do was add sugar, eggs, and water to the correct of amount of Master Mix and put it in the oven.

Master Mix eliminated the need to prepare every baked food from scratch, cutting the time needed to blend ingredients by 75 percent.[14]

Commercial baking mixes in the United States began to appear in the 1930s. But after World War II they became more prevalent, and the Purdue Master Mix played a role in the development of convenience foods that required less time for preparation.

Sunderlin's research was a giant leap in the lives of homemakers. Commercial mixes were expensive and could only be used for one specific baking purpose. "Why wouldn't it be possible for the homemaker to prepare a mix in her own kitchen from which she could prepare several baked products," Purdue student Jean Margaret Billings wrote in a 1947 paper.[15]

Purdue students, including Billings and Lucy Goetz, helped develop and test the mix. It was announced nationally in the January 1947 issue of *Better Homes and Gardens*. From that one announcement, 150,000 copies of Master Mix and recipes for using it to bake many foods were sent throughout the United States. Sunderlin and her students also developed special mixes for baking a wide variety of cookies and cakes. By the time Sunderlin retired in 1954 after twenty-three years with Purdue, Master Mix and the cookie and cake mixes, along with recipes, had been distributed to several hundred thousand U.S. homes.[16]

Sunderlin had already broken barriers before arriving at Purdue. Born in 1894, when professional opportunities for women were few, she studied at Iowa State University, becoming the first woman to be granted a PhD from its home economics program. At Purdue she also perfected a method for freezing jelly and jam.

She lived until 1990 and saw the full impact of twentieth-century research in foods and nutrition on the American home and lifestyle.

Ruth Siems

Turkey is the star of every U.S. Thanksgiving dinner. But stuffing is a close runner-up. And since stuffing is traditionally cooked inside the turkey, the two go hand-in-hand in American culinary history.

So a Purdue University graduate brought major change to kitchens when she discovered an easy way to cook stuffing without the turkey. Ruth Siems made stuffing possible on family dinner tables any day of the year. If few people remember her name, almost everyone knows her product: Stove Top stuffing. In 1975, U.S. Patent 3,870,803 was awarded to General Foods for the stuffing, and Siems is the first person listed as an inventor. Her partners were Anthony Capossela Jr., John Halligan, and C. Robert Wyss.

The New York Times called the stuffing "an enduring emblem of postwar convenience culture. Its early advertising tag line, 'stuffing instead of potatoes' remains in the collective consciousness."[17]

"Stove Top made it possible to have the stuffing without the turkey, probably something no cook would ever have dreamed of but people eating Thanksgiving dinner might well have thought of: 'Take away everything else; just leave me here with the stuffing,'" said food history author Laura Shapiro. "It's kind of like eating the chocolate chips without the cookies."[18]

Shapiro called Stove Top stuffing, "one of those little milestones in the long road of the history of convenience foods."[19] It needs only water and butter and can be prepared in five minutes. It is now owned by Kraft.

Siems was born in Evansville, Indiana, and studied home economics at Purdue, working with Professor Gertrude Sunderlin on recipes and promotion of the baking Master Mix.

After graduation Siems worked for General Foods in Evansville, later moving to a company plant in Tarrytown, New York.

The idea for the stuffing came from the company's marketing department, but researchers had to find a way to create it. Siems's contribution was key. She determined the exact size for the breadcrumbs. If the crumbs were too small or too large, when water was added they turned either soggy or like "gravel."[20]

Siems retired in 1985 and returned to southern Indiana, living quietly with the knowledge of her contribution to American cooking. She died of a heart attack at the age of seventy-four. The New York Times announced her passing ten days later, on November 23, 2005.

It was the day before Thanksgiving.

24

Basketball Bleacher Collapse

February 1947

All the seats got pushed forward and it was
like riding a roller coaster down.

—Purdue student Bill Creson

I t happened in an instant.

The evening of February 24, 1947, was the date of the greatest Purdue tragedy since the train wreck of 1903. And as with the train wreck forty-four years earlier, this tragedy was wrapped in the excitement of an athletic event.

People packed into the Purdue Fieldhouse that Monday evening for a basketball game between the Boilermakers and the league-leading Wisconsin Badgers. Thousands of temporary bleacher seats were set up on the east side of the court to accommodate the rapidly increasing Purdue student enrollment. It was a back-and-forth contest. Purdue led 34–33 at halftime, and the Boilermakers stood to cheer as the players walked off the floor toward the locker rooms.

Just at the start of halftime, somewhere, deep under a section of the temporary bleachers that held nearly four thousand students, a support snapped. In an accordion-like motion, the bleachers began collapsing forward.

WBAA radio sportscaster John DeCamp was talking on the air with his assistant, Ward Carlson. "Oh, God, John!" Carlson exclaimed, and they both fell silent. "The

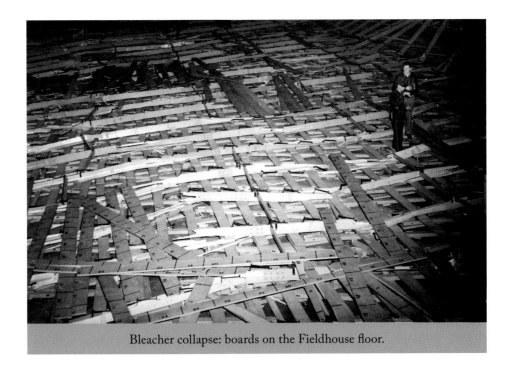
Bleacher collapse: boards on the Fieldhouse floor.

whole east end fell," DeCamp remembered years later. "They just surged forward and fell. I kept on broadcasting. I said there had been an accident at the Fieldhouse and we needed doctors and ambulances."[1]

Ambulances, hearses, buses, bread trucks—anything that could carry the injured—hurried to the scene. Doctors, nurses, and people who just wanted to help rushed from all corners of the county as DeCamp described the carnage. There was no panic among the injured and the people who witnessed it. Veterans who had experience treating wounded combat buddies helped those lying on the gym floor. An empty Greyhound bus arrived and took forty-five injured students to a hospital, among them Jim Hitch, '47, who majored in science.

Hitch said the temporary bleachers where students were sitting were "rather a dinky operation and rickety. We had actually talked about what would happen if the bleachers crashed. We decided we would raise our legs above the seats and ride down on top of it." His friends were able to do that, but Hitch caught his leg on a bleacher support and was injured from his calf muscle to his ankle. He was treated at the hospital and later released.[2]

"I was in the fifteenth row, near the top," Bill Creson, who was a student that evening, remembered many years later. "Everyone stood up all at once and gave a cheer. Then everyone started to sit down. At that instant the bleachers collapsed. It was like riding a roller coaster down."[3]

Students lay all over the basketball court. Stretchers were made out of broken bleacher boards. The next day the Lafayette *Journal and Courier* reported: "The injured themselves were stoical and calm, permitting themselves to be attended without murmur or outcry even though many of them must have been in great pain. One physician said a pretty girl lying on the basketball floor had a compound fracture of one leg, yet she lay quietly smoking a cigarette. Many suffered cuts and other injuries which caused them to bleed profusely. Many had blood streaming down their faces. Others with lesser injuries were hobbling around on one foot or nursing bruised arms."[4] Students had cuts on their faces caused by broken, flying splinters of wood from the bleacher boards.

Parents of students rushed to the Fieldhouse to look for their children among the injured—and the dead. Roger Gelhausen, twenty-two, of Garrett, Indiana, died almost instantly when a splintered board pierced his chest. He was a navy veteran and a freshman. William Feldman, twenty, of East Chicago, died at 10 p.m. at St. Elizabeth Hospital. He was a sophomore. Ted Nordquist, twenty-five, of Gary, died at 11 p.m. the following evening. An army veteran, he survived sixty combat missions in Europe during the war and came home to die watching a basketball game at his college.

More than 250 people, most of them students, were injured. One hundred ten people were hospitalized.

Purdue vice president and treasurer R. B. Stewart saw to it that the University paid burial and other expenses. When Hella Feldman, mother of William Feldman, said she had purchased an expensive headstone monument that cost $700, Stewart told her Purdue would pay for it.

Nordquist's wife, Maxine, had given up her studies at the University of Chicago when they married. Stewart personally arranged for the University of Chicago to provide her with a scholarship to finish her degree.

The second half of the Purdue–Wisconsin game was played three weeks later at Evanston High School in Evanston, Illinois, a Chicago suburb. Wisconsin won 72–60 and would go on to place third in the NCAA tournament that year.

Few Boilermakers would remember the final outcome of the game. But those on campus the evening of February 24, 1947, never forgot the bleacher collapse.

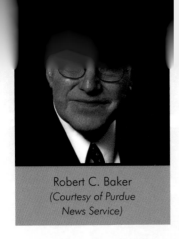

Robert C. Baker
(Courtesy of Purdue
News Service)

He was called the "Thomas Edison of the poultry industry"[5] and the "George Washington Carver of poultry."[6] He created nearly sixty food products, including chicken patties and chicken and turkey hotdogs. His chicken barbecue sauce is famed throughout his home sate of New York.

But Robert C. Baker is most acclaimed for creating a process for binding chicken parts together and coating it with batter—a precursor to what became one of the most popular American food favorites—chicken nuggets.

When he died in March of 2006 at the age of eighty-four, the Associated Press said he had "changed American cuisine with his poultry innovations."[7]

Baker was a long-time professor of food science at Cornell University. He did his undergraduate work at Cornell, received a master's degree from Penn State, and in 1957 a PhD in food science from Purdue. The three land-grant institutions set him on a career using science and research to help poultry farmers.

He grew up during the depression in a rural community just west of Syracuse, where his parents were struggling farmers. He understood firsthand the problems facing people in agriculture.

In post–World War II America, poultry farmers were struggling. Chickens were sold whole, too much for one person and often too little for large families. Meanwhile, beef and pork were being sold in a variety of forms, and chicken fell behind them in the eyes of consumers.[8]

At Cornell, the dean of agriculture challenged Baker with finding ways to help poultry farmers compete. Baker had already created a popular chicken barbecue sauce. At Cornell he promoted it even more and became known as "Barbecue Bob."[9] It became among the most popular foods each summer at the New York State Fair.

But Baker believed what poultry farmers really needed was more ways to sell their product, and that's when he came up with chicken and turkey sausages, patties, and more. They caught on with the public.

In 1963, working with a student, Baker discovered a way to bind ground chicken meat together and cover it with batter that would stick. They packaged and sold it frozen in grocery stores. Within six weeks they were selling two hundred boxes a day.[10]

Baker never patented any of his products. He gave away his processes and recipes to the food industry. Today, various recipes for chicken nuggets are used in sales at restaurants and supermarkets around the world. McDonald's introduced its Chicken McNuggets in 1981.

In 2004 Baker was elected to the American Poultry Hall of Fame.

Joel Spira
*(Photo courtesy of
the Spira family)*

The incandescent light bulb brought previously unimaginable illumination to homes throughout the world, making twentieth-century living incredibly bright. Purdue University graduate Joel Spira gave us the ability to dim it in homes and businesses globally.

Spira invented and patented the first electronic solid-state light dimmer to fit in a standard wall box, offering homeowners the possibility of smooth continuous dimming. To produce and sell his dimmer, he started Lutron Electronics—a Pennsylvania-based company that today offers more than fifteen thousand lighting and shading control products for residential and commercial applications.

Lutron systems are installed at some of the world's most iconic locations, including the *New York Times*, the Statue of Liberty, and the Shanghai Tower.

In addition to making lighted rooms more relaxing and even romantic, Lutron technologies save $1 billion in energy costs per year, help reduce carbon emissions, and contribute to more comfortable, beautiful places to live, work, and play.[11]

The company has about three thousand patents, and individually Spira is credited with more than three hundred design and utility patents.

In 2010, Spira's commitment to innovation was celebrated by the Smithsonian's National Museum of American History. The museum took possession of ten Lutron artifacts for its Electricity Collection, including Spira's first inventor's notebook and an early dimmer model in original packaging.

Spira was born March 1, 1927, in Brooklyn, New York. He was an inquisitive boy and spent time in the library studying airplanes, with plans to become an aeronautical engineer. He graduated high school at age sixteen and enrolled at Purdue to study physics.[12]

Before completing his degree, Spira joined the U.S. Navy at the end of World War II. He spent two years in military service, after which he returned to Purdue and graduated in 1948. He worked for several companies doing defense work and got his idea to work on a dimmer switch while working in the defense industry.

Spira invented the electronic light dimmer in 1959, working in a spare bedroom in his Manhattan apartment. Light dimmers were available for commercial use, but they were too large for homes. Spira's wife, Ruth, was very involved in many aspects of the business, including testing his designs. They founded Lutron in 1961. The couple donated generously to Purdue, including endowing the Ruth and Joel Spira Outstanding Teacher Award in electrical and computer engineering.

Spira died in 2015 at the age of eighty-eight.

"Someday I'll be gone," he said in a 2012 interview. "But dimmers will be with us forever."[13]

25

African Americans Apply to Purdue Residence Halls

1944 to 1947

It was an injustice for our own state college to discriminate against the citizens and taxpayers of Indiana, regardless of their color.

—Frederick Parker

I n the summer of 1946 Winifred and Frieda Parker, two sisters from Indianapolis, prepared for their freshman year at Purdue University. They were excited and nervous like all freshmen.

Excellent students, after being admitted to Purdue they applied for a room in a residence hall where all freshmen women were required to live. But Winifred and Frieda were denied.

They were African Americans.

During World War II, by federal mandate, African American male military personnel had been housed in Cary Hall—something that had not occurred throughout all the previous years of the University's existence.

But when the war ended and military personnel left, Purdue continued its practice of refusing African American students housing in residence halls. Some men were placed in a converted home that was occupied by international students. The rest, including women, had to find housing with families across the river in a segregated Lafayette neighborhood around the Lincoln School at Fourteenth and Salem Streets. African American children attended that segregated school through eighth grade and then went to the integrated Jefferson High School. All Lafayette school grade levels were not integrated until 1951. In West Lafayette there were written restrictions in neighborhoods against renting or selling property to people who were not "pure white."[1]

One African American student denied access to a Purdue residence hall went on to become chief justice of the U.S. Court of Appeals for the Third Circuit in Philadelphia—one step below the U.S. Supreme Court. During his long, successful career, Judge A. Leon Higginbotham Jr. recounted his experiences at Purdue and credited them with convincing him to become an attorney and civil rights advocate.

"In a career of energetic accomplishments and unambiguous liberalism, Judge Higginbotham received much recognition as a legal scholar and civil rights advocate, including the nation's highest civilian honor, the Presidential Medal of Freedom," the New York Times said in his December 1998 obituary. "Some historians say Judge Higginbotham was one of a handful of black jurists President Lyndon B. Johnson considered as candidates to integrate the Supreme Court before he named Thurgood Marshall in 1967."[2]

Higginbotham enrolled at Purdue to study engineering and was placed in International House, where students slept together in a top-floor room with open windows. The room was very cold in winter. It was believed that open-air circulation helped stop the spread of viruses, and cold dorms were common in fraternities and sororities. Some students liked them. Others, including Higginbotham, did not.

In 1944 Edward Elliott was president of Purdue, and his office was always open to students. Elliott had met Higginbotham previously. Higginbotham and another student protested that the Union barbershop would not cut the hair of African American students. Elliott told them he would buy clippers to be kept in the barbershop so they could cut each other's hair.

Now Higginbotham was entering Elliott's office with another request. He asked to be placed in a Purdue residence hall where students slept in warm rooms.

What happened next was retold many times by Higginbotham but was never mentioned by Elliott. "In emotional speeches, the judge . . . described how a white college president at Purdue University had flatly told him when he was a college freshman in 1944 that the school was not required under law to provide black students with heated dormitories and, therefore, never would," the New York Times stated.[3]

According to Higginbotham, Elliott told him he would never be allowed to live in the residence hall, and if he didn't like it, he should leave Purdue. Higginbotham did leave. In the fall of 1945 he entered Antioch College in Ohio and in 1949 was admitted to Yale Law School.

It's not known how many African Americans were enrolled at Purdue in the mid-1940s, because the University did not then keep records of students by race. African Americans who attended Purdue at the time estimated there were not more than twenty-five.

In a paper titled "Evolution of the Black Presence at Purdue University," Alexandra Cornelius said, "Between 1940 and 1951, African Americans began attending Purdue in increasing numbers. Despite their proven abilities, African American students in every college in the state were discriminated against both on and off campus. . . . West Lafayette restaurants did not admit African Americans. Blacks were only allowed to eat in the Union Building or the cafeteria during lunch time."[4]

Nationally, *The Negro Motorist Green Book,* published from about 1936 to 1966, was a guide to African Americans as they traveled through the country. Jim Crow laws that enforced segregation were in effect not only in the South but in many parts of the country. In their hometowns, minorities knew where they were welcome and unwelcome. But they did not know this when they were on the road. So the *Green Book* was their guide, listing hotels, restaurants, taverns, gas stations, and all manner of businesses where they would be welcome and safe. The 1949 guide lists only one business open to African Americans in Lafayette—the Pekin Café at Seventeenth and Hartford Streets. That was in the heart of the segregated African American neighborhood. There were other establishments in Lafayette open to minorities that didn't make the *Green Book.* But discrimination was the norm of the day.[5]

A Social Action Committee at the Purdue Methodist Church Wesley Foundation was opposed to segregation and sent African Americans to various businesses in Lafayette and West Lafayette to see what would happen. In February 1947 the committee reported that of nineteen restaurants investigated, five refused to serve African Americans, six did serve them but were rude and hesitant about it, and some of the employees who did serve blacks were not aware they were breaking management policy. Three movie theaters would only permit African Americans to sit in the balconies. No barbershop in Lafayette, West Lafayette, or Purdue would cut the hair of African Americans.[6]

Winifred, the younger of the two Parker sisters, had graduated number one in her high school class in May 1946. Frieda was second in her class and had graduated five months earlier in January. She completed a semester at West Virginia State College but decided to transfer. In early July 1946 both young women were admitted to Purdue. Their father, Frederick Parker, was a graduate of Amherst College in Massachusetts who had done graduate work at Harvard and Indiana Universities.

He was the former head of mathematics at the African American Crispus Attucks High School in Indianapolis and had moved on to become a consultant to the Indianapolis Public Schools system. Their mother, Frieda Parker, was a graduate of Butler University in Indianapolis.

Immediately upon their being accepted, Mrs. Parker took her daughters to Purdue to meet with administrators. According to a statement written and signed by the Parkers in September 1946, they met with the registrar, who told them that while admissions had been closed prior to July, Winifred and Frieda were accepted "as a special case" because of their outstanding personal and academic records.[7]

The Parkers next went to Clare Coolidge, acting dean of women, to obtain student housing in a residence hall. Coolidge gave them residence hall applications and told them to complete and mail them as soon as possible. The applications from the Parkers were rejected with a statement that the number of students seeking housing exceeded the spaces available.

The following day the Parkers received a letter from Coolidge. Parker said Coolidge was "gracious" to the family throughout the experience, but her letter was blunt. It stated: "I was informed by the [Purdue] President's Office yesterday that at its last meeting the Board of Trustees of the University had discussed the question of housing negro students in the residence halls and felt it was not feasible for the present. Had I known of this action, I would, of course, have saved you and the girls the embarrassment of making an application which can only be refused."[8]

The most recent meeting of the board of trustees had been on June 22, 1946. The written minutes of that meeting make no mention of a discussion concerning African American students in the residence halls. However, at a meeting of the board on April 17 and 18, 1946, the topic did come up, without any official decision being recorded. Minutes of the meeting state: "[Board] Secretary [and University vice president Frank] Hockema presented an application for admission to Men's Residence Halls from a negro student in the University. After discussion and by consent, the foregoing report was received as information."[9] A newspaper reporter was at the meeting but he did not write about the discussion.

Parker contacted Faburn DeFrantz, director of the African American Senate Avenue YMCA in Indianapolis. DeFrantz was a civil rights leader in the state who had worked on integrating housing at Indiana University. DeFrantz told the Parkers they should all meet with Hovde. "Our hope was that we could show President Hovde that it was an injustice for our own state college to discriminate against its own supporters, the citizens and taxpayers of Indiana, regardless of their color or their religion," Parker said. "We had heard that this new president was fair and just and therein lay our hope and prospects."[10]

They were unable to arrange an appointment with Hovde and instead met with Hockema, who told them the residence halls were full. Asked if the University had a policy that prohibited African Americans in the residence halls, Hockema said no. But "he admitted that such might be the case from 'custom and usage.'"[11]

Taking part in the conversation was one man who was Caucasian. After the meeting with Hockema he went to the office of the director of residence halls and said his two nieces were enrolling as freshmen in the fall. He was told there were several rooms available for them.[12]

Parker and DeFrantz returned to Hockema. "We pleaded and we begged," Parker wrote. "We insisted a man in so esteemed a position must be able to make the decision even if he had no precedent."[13]

Hockema said he could not go against the position of the trustees, who had told him to hold off on admitting African Americans to the residence halls.[14]

Following their meetings at Purdue, Parker and DeFrantz contacted Indiana governor Ralph Gates, who was "sympathetic" to their position.[15] Gates contacted one or more of the trustees and probably Hovde.[16]

On December 16, 1946, Parker received a five-page letter from Hovde, who went into great detail, stating that he opposed discrimination in any form, but societal change would take many years and "will not be completed in your lifetime or mine."[17]

The University had no written policy excluding African Americans from the residence halls, Hovde said. But he also noted that over the years three or four African American women had applied to be admitted and had been rejected. "These have been refused," he wrote, "not because we wished to discriminate, but because the administration did not think it wise to jeopardize the successful operation of the halls, jeopardize our incoming number of women enrollees in the halls; because administrative officers and faculty felt that neither our women students nor their parents were ready yet for true social democracy in living. The situation, therefore, exists, not because we of Purdue believe in discrimination of any kind, but because the decision was thought to be in the best interests of the University and its welfare."[18] They were afraid students would refuse to live in a residence hall with African Americans, placing a financial strain on the University's ability to retire construction bonds.

In his letter to Parker, Hovde concluded, "You realize and understand, I am sure, the difficulties I have in persuading and influencing others to my point of view in these matters. . . . I must use my own method of accomplishing my aims. I am personally sure they are right and will produce results more quickly than other methods. Otherwise I would act differently."[19]

Frustrated, Frieda and Winifred attended classes at Purdue in the fall of 1946 and lived in a home in the segregated Lafayette community. They took buses to and from campus. But change came. Hovde ended discrimination at the Union barbershop, and in January 1947 the Parkers were admitted to Bunker Hill Residence Hall.

Soon after the Parkers moved in, the students voted on leadership for their residence hall.

Winifred Parker was elected president of Bunker Hill.

Frank Brown Jr.
*(Photo by John Underwood/
Purdue University)*

The parents of Frank Brown Jr. couldn't read or write, but his mother encouraged him toward a career that included a PhD from Purdue University and a position as chief manufacturing scientist at Eli Lilly and Company in Clinton, Indiana.

"My mother said not all people are treated equally in this society and to prepare yourself you have to get a good education," Brown said. "A good education is the one thing that will stick with you and take you far in your life."[20]

Brown was born in 1942 in the inner city of Detroit. His father worked at Ford Motor Company "and had the lowest job," cleaning oil off the factory floors.[21] His mother took a keen interest in his schoolwork and understood the difference between the letter grades A and B and knew that 100 percent was a perfect score.

"I didn't realize we were living in a ghetto," Brown said. "We were too young to understand that. We were kids playing ball in the street. My mother told me there was a good university called Wayne State near where I lived, and there was an even bigger university called Michigan State. She said, 'Maybe you could even go to State!'"[22]

When Brown was nine years old his father was injured; he took early retirement and the family moved to Louisiana. His mother died two years later. When Brown turned thirteen his father wanted him to quit school and get a job. Instead, he left home, walked sixteen miles to live with his grandmother, and continued his education.[23]

He attended Southern University and A&M College, a public, historically black college in Baton Rouge. Other public universities in southern Louisiana did not admit African Americans.

An engineer who graduated from Purdue befriended him and counseled him to leave the South to continue his studies. "I had little prior knowledge of Purdue," Brown said. "He told me Purdue was very white, but he knew that they would treat me well."[24]

Brown arrived in 1964, received his PhD in chemistry in 1969, and accepted employment with Lilly Research Laboratories. After retiring from Lilly in 2000, he became director of the pharmaceutical sciences program at Purdue and adjunct professor in industrial and physical pharmacy and chemistry. In 2016 he established the Frank Brown Distinguished Professorship of Chemistry.

"There were many challenging problems at Lilly that I had to tackle, working with other scientists," Brown said. "But Purdue taught me what excellence is and to always seek the root cause."[25]

GIANT LEAPS IN LIFE

Forest Farmer
(Photo courtesy of Purdue
Intercollegiate Athletics)

F orest Farmer was born in the industrial town of Zanesville, Ohio, in 1941 with an ambition to do something different. He did. He helped to rebuild and lead a major American automaker.

Farmer's father worked for Armco Steel and his mother did domestic work, creating a middle class living for the family of four.

Farmer was a good student and an exceptional athlete, all-state football and basketball. In 1959 he was recruited for football by most of the Big Ten schools. One of his first visits was with Ohio State's Woody Hayes. Hayes treated him "okay," but when Farmer told the coach he wanted to study business administration, Hayes told him as an African American he'd never find a professional job in industry.[26]

Purdue coach Jack Mollenkopf told Farmer he could study whatever he wanted and the University helped get his original partial scholarship raised, keeping his annual expenses to $44.

Freshmen were ineligible for varsity, but beginning his sophomore year Farmer started at end on offense and defense.

Drafted by the Denver Broncos after his senior year, he injured his Achilles tendon during a preseason practice and was unable to play. He returned to Purdue, worked as assistant freshman football coach, and received his degree in science in 1965.

Farmer taught school in Indianapolis for several years and in 1968 was scheduled to be the head football coach at Shortridge High School. But the position fell through.

"So I decided I was going to look around and see what else was out there," he said.[27] He got in his car and drove around Indianapolis, eventually passing a Chrysler plant. He went inside and twenty minutes later he was a foreman trainee. His Chrysler career lasted until he retied in 1993. He was one of twenty-seven people chosen to reorganize the company after it neared bankruptcy in the late 1970s. Farmer worked as an industrial engineer, plant manager, director for advanced manufacturing planning, and president of the Chrysler subsidiary Acustar.

"Purdue had a lot of do with what happened in my life." Farmer said. "Purdue was good to me, and I've given back to it. When you walk into a company looking for a job and your resume says Purdue University—that means something. Purdue has been a big influence in my life.[28]

"Woody Hayes told me I couldn't get a job in industry," he said. "But I ended up doing what I wanted to do."[29]

26

New Schools and New Traditions: Everything's Coming Up Roses

1940s, 1950s, 1960s

This is the greatest day of my life.

—FOOTBALL COACH JACK MOLLENKOPF

I n 1962, the Krannert Graduate School of Industrial Administration became the first named school at Purdue University. But it wasn't named for a Purdue alumnus. It was named for a graduate of the University of Illinois.

And the naming emerged from Purdue's School of Agriculture treating sick cows.

Considered among the top programs in the nation, what became the Purdue Krannert School of Management had its origins in 1958. At that time one of Purdue's most distinguished faculty members, Professor Emanuel Weiler, was heading two programs in two different schools—the Department of Industrial Management and Transportation in the School of Engineering and the Department of Economics in the School of Science. This took some juggling and necessitated dealing with two different deans. So when Weiler was offered the position of dean of the Wharton

School of Business at the University of Pennsylvania, he was very tempted to take it. But first he decided to talk with President Frederick Hovde.

Weiler told Hovde about his job offer. And then he said something that intrigued Hovde. Weiler said that rather than take over leadership of an established program such as Wharton, he would really like to build something new from scratch. If Hovde named him dean of a new School of Industrial Management at Purdue, Weiler said he would stay. Hovde didn't do things on the spur of the moment. He took time to think about the offer. After consideration, he agreed with the proposal and a new school was born. It wasn't yet Krannert. But it was a school.

Calling the school "management" was a careful decision, Weiler said in a 1970 interview. "President Hovde wanted us to develop a program in business, but defined in such a way as to make it uniquely Purdue's program," he said. "He did not want Indiana University to object to a new school of business on this campus."[1] IU had a School of Business and the new Purdue program was of concern in Bloomington.[2]

Hovde saw Purdue's School of Management as something different from the IU School of Business. His school would focus on developing engineers into managers in industry. It would be focused on the "production" side of the economy. Students in the undergraduate program were required to basically have two majors—one in management and another in a technical area such as engineering, computer science, physics, or chemistry. An advanced degree called the master of science in industrial administration, or MSIA, was soon developed. In its first years it was limited to those with a bachelor's degree in engineering and was intended to help them move quickly into management. Everything went well with the new school. It was about to get even better.

Normandy Farm

Herman Krannert was a businessman and philanthropist in Indianapolis. In 1925 he had founded Inland Container Corporation with six employees. By the 1970s it was the nation's second largest manufacturer of corrugated shipping containers.

Krannert was a serious, hardworking man who rarely socialized. His wife, Ellnora, often worked beside him late into the evening. They had no children and were generous with their wealth, especially at the University of Illinois—Herman Krannert's alma mater.

In addition to the company, the Krannerts had an eighty-five-acre farm at 7043 West Seventy-Ninth Street north of Indianapolis. Called Normandy Farm, it was where Mrs. Krannert kept her world champion Guernsey cows. In 1959, the cows were showing signs of illness. Milk production had dropped.

She called the Purdue School of Agriculture to ask for help. Fred Andrews, a professor of animal science and head of the Department of Animal Sciences, was dispatched to meet with Mrs. Krannert.[3]

She liked Andrews, and Purdue developed a plan to manage Normandy Farm as a research project with the Krannerts providing the funds. Among their first discoveries was that calves were in buildings that had formerly been used by poultry. They found a chicken parasite that was harmful to calves and solved the Krannerts' problem.[4]

In 1960 Herman Krannert was so impressed with the work Purdue had done on the farm that he wondered if the University could help with his company. He asked if Purdue faculty would run a program for his managers. Hovde decided Emanuel Weiler, dean of Purdue's School of Industrial Management, was the perfect man for the job.

Weiler traveled to Indianapolis, met with Krannert, learned about his company, and bluntly told him he needed to adopt some new management methods. "He was intrigued by someone who would be that blunt with him," Weiler said.[5] Krannert agreed to do it. The University developed a management training program specifically for and financed by Inland Container, and it was very successful.

In 1961, Weiler proposed to Krannert the idea of a named program at Purdue—the Krannert Graduate School of Industrial Administration. The school was created in 1962 with a $2.73 million endowment from the Krannerts. They provided additional funding for a new building.

Walter Scholer and Associates designed the building. It became one of the few on the Purdue campus not made of red brick. It was made of limestone. Why?

Because that's what Mrs. Krannert wanted.

School of Veterinary Medicine

Starting a School of Veterinary Medicine was not as easy. Purdue had support in the effort from the Indiana Farm Bureau, but it took many years of work to overcome objections.[6]

The Purdue trustees started talks to launch the school as far back as the late 1940s. The need seemed obvious to Hovde. Before World War II there had been only ten veterinary medicine schools in the entire country. After the war, seven new schools were launched, including two in the Big Ten—Illinois and Michigan. Because there were so few programs, it was very difficult for a nonresident student to gain admission to them.[7]

A committee of Purdue trustees and faculty issued a report in 1948 that said the state had 526 veterinarians and their average age was fifty. Since a number of them were nearing retirement, more would be needed to replace them. This was a critical issue to Indiana farmers.[8]

"There was general support for the establishment of the School of Veterinary Medicine, even through the profession was somewhat divided," Hovde said in a 1972 interview. "A good many members of the practicing profession were, of course, fearful that if a school were established the incoming new professionals would be a threat to their livelihood."[9]

The Purdue school was introduced to the Indiana General Assembly in every biennial session beginning in the 1950s, but opposition continued from some members of the Indiana Veterinary Medical Association.

The issue was finally settled about 2:30 one morning in 1957 as the General Assembly faced adjournment. Indiana governor Harold Handley received Farm Bureau support for his $15 million state office building in return for a $2 million appropriation for the veterinary school.[10]

The School of Veterinary Science and Medicine, as it was initially called, enrolled its first class of fifty students in 1959. The new building that would house the school was nearly complete, but the students had to crowd into the existing Veterinary Pathology Building for that first year, until the new structure was finished. At the dedication ceremony in 1960, it was announced that the new building would be named after Purdue benefactor and longtime trustee Charles J. Lynn, who was instrumental in establishing the school. Before the first class graduated in 1963, the school received accreditation from the American Veterinary Medical Association.

Carol Van Paemel Ecker applied to the Purdue School of Veterinary Science and Medicine in 1958. While she had been accepted to two out-of-state schools, she was told the Purdue program would not admit women. She took the rejection to the dean of women, Helen Schleman, who went straight to Hovde. "I'll never forget the image of Helen standing toe to toe with President Hovde," Ecker recalled. Schleman said, "Why can't women become veterinarians at Purdue?" She said it was unacceptable and Hovde agreed.[11]

Ecker was one of two women in the school's second graduating class. She later became the first female president of the Indiana Veterinary Medical Association and served on the Purdue University Board of Trustees from 1988 to 1997. At Purdue's sesquicentennial, about 80 percent of its veterinary medicine students were women.

Regional Campuses

Purdue is a regional university with campuses in Westville, Calumet, and Fort Wayne, and a shared presence with IU in Indianapolis in addition to West Lafayette. These regional sites were not created overnight; it was a gradual process and they came about at least partially in response to community colleges. Part of Purdue's regional campus system developed during and immediately after World War II, beginning with extension centers launched in communities in response to specific needs.

Purdue Calumet got its start when the University offered technical courses to train defense plant workers. The courses were offered in cooperation with the federal government. When the war ended, Purdue remained in Calumet and offered credit courses at various locations beginning in 1946. It moved into a permanent home in 1951.

In Fort Wayne, Indiana University began courses in 1917. A Purdue University Extension Center was located in the city in 1941. The Purdue center was focused on providing a place for students to begin their undergraduate studies before transferring to West Lafayette.

What became Purdue North Central got its start after World War II with extension centers offering technical courses in La Porte and Michigan City. In addition to technical training, the centers offered Purdue's complete freshman engineering program.

After the war Purdue's extension campuses helped to meet the needs of huge numbers of World War II veterans going to college on the GI Bill.

During the 1950s discussion heated up in Indiana concerning the development of community colleges. First, many of the state's major urban areas did not have four-year colleges in their communities. Second, the baby boom generation had appeared, and it was clear that beginning in the 1960s and 1970s there would be large numbers of students applying for admission to college, requiring more facilities and faculty.

Purdue and IU were opposed to community colleges. Hovde and IU president Herman Wells believed their schools could provide better educational opportunities through regional campus systems. They were also concerned that once two-year community colleges were established, they would soon evolve to four-year programs and compete with IU and Purdue for students and funds from the state.

In 1964 Purdue and IU opened a combined campus in Fort Wayne. Formal merger of the IPFW (Indiana University–Purdue University Fort Wayne) administration managed by Purdue came in 1975. Purdue's programs in Calumet were redefined as a regional campus and renamed Purdue Calumet in 1962. Also in 1962 Purdue purchased land in La Porte County and created Purdue North Central.

Purdue and IU had separate programs in Indianapolis when in 1968 then mayor Richard Lugar called for an independent state university for the city. He also supported a board of regents to control all of Indiana higher education. To fend off the plan, the universities merged into IUPUI (Indiana University–Purdue University Indianapolis) in 1969. Indiana University was charged with its governance.

In 2014, Purdue announced a plan to unify the administrations of Purdue Calumet and Purdue North Central. It was completed in 2016 and the campuses became Purdue Northwest.

Purdue was charged with governance of the combined IU and Purdue Fort Wayne campus—IPFW. But in 2016–2017 the universities approved realignment into separate and different programs named Purdue University Fort Wayne and Indiana University Fort Wayne.

In addition to regional campuses, the Purdue Polytechnic Institute offers degree programs in ten Indiana communities.

In 1963 Ivy Tech was founded as Indiana's vocational technical college to provide technical and vocational education. In 2005 it became Ivy Tech Community College with campuses throughout the state.

School of Humanities, Social Sciences, and Education

Disagreements had festered for years about the role of the liberal arts in a land-grant university focused on engineering and agriculture.

Initially the liberal arts were part of the School of Science. Hovde started changes that would culminate in the liberal arts having a school of their own. In 1953 the name of the School of Science was changed to Science, Education, and the Humanities. In 1959 Hovde and the board of trustees approved a bachelor of arts degree, and they began movement toward a master of arts degree.

Helen Schleman, who had long fought for the degree along with Dorothy Stratton and others, said the liberal arts degree had not been offered years earlier because of an "agreement" with Indiana University. "David Ross, president of the Board of Trustees, told me at least fifty times that there was a gentleman's agreement that Purdue would never give the B.A. degree," she said. "When I tried to find out who the gentlemen were who made that decision, I never could find out. But, of course, gentlemen stick by their decisions."[12]

In 1962, Hovde and the trustees approved a reorganization plan that created a School of Science and a new School of Humanities, Social Sciences, and Education.

It was called HSSE, pronounced *hissy.* More changes in the liberal arts were coming in the years ahead.

School of Technology

The Purdue School of Technology got its start in 1962. Among the first departments in the school was aviation technology, headed by James Maris, who had been teaching flight at Purdue since 1955. Purdue became the first university in the nation to offer a four-year bachelor of science degree in flight.

Maris had plenty of time to lobby for Purdue's program when he flew Hovde around the state. "At that time when President Hovde had to go to Fort Wayne, he didn't have any transportation except to drive," Maris said. "So I would fly him. We were making these trips with President Hovde all over the state and we decided maybe he should learn how to fly. Whoever was flying with him would teach him the fundamentals of how to take off, land and navigate. We prepared him to qualify for his license, but he never had time to take the FAA tests. One time he was coming back from Fort Wayne and he went off the runway and bent the propeller. So we put it on a mahogany base and give it to him. He kept it on display in his office."[13]

Nursing

Nursing began at Purdue in the fall of 1963 with thirty students in a two-year associate degree program. Helen R. Johnson was the first director of the program. A four-year baccalaureate program began in 1970.

Johnson was successful in obtaining a federal grant leading to construction of a $1.6 million Nursing and Allied Health Sciences Building on campus. In 1977 it was renamed Johnson Hall of Nursing in her honor. In 1979 the board of trustees created the School of Nursing and placed it in an administrative unit named the Schools of Pharmacy, Nursing, and Health Sciences. In the twenty-first century it became part of the College of Health and Human Sciences.

U.S. Secretary of Agriculture

Earl Butz studied and taught at Purdue and was head of Agricultural Economics from 1946 to 1954. He was one of the major leaders of his time in Purdue and U.S. agriculture. In 1957 he was named U.S. assistant secretary of agriculture and also was appointed chairman of the U.S. delegation to the Food and Agriculture Organization of the United Nations. In 1957 he resigned both offices to become the dean of agriculture at Purdue. In 1968 he was promoted to dean of education and vice president of the Purdue Research Foundation. President Richard Nixon appointed Butz secretary of agriculture in 1971, a position he also held under President Gerald Ford. During his career he was a tireless advocate for agriculture, and at Purdue he helped bring University programs into national prominence. He retired from Purdue in 1972. When he died at the age of ninety-eight in 2008, the *New York Times* noted: "Mr. Butz was a forceful, sharp-tongued figure who engineered legislation sharply reducing federal subsidies for farmers. Mr. Butz maintained that a free-market policy, encouraging farmers to produce more and to sell their surplus overseas, could bring them higher prices."[14] Farm income did rise, along with the cost of food to consumers.

In 1974 it was that sharp tongue that caused him problems. He privately told a joke that was racist, obscene, and sexist, and when it became public Butz was forced to resign as secretary of agriculture. In 1981 he pleaded guilty to federal tax evasion charges and was sentenced to five years in prison with all but thirty days suspended.

Enormous Social Change Marked the Hovde Era

Women expanded more and more out of their traditional studies during the Hovde years. Rita Rossi came to Purdue as a freshman in September 1952. The daughter of immigrants from Italy, she lived in Beverly Cove, Massachusetts. An excellent student, Rossi applied to schools in the east such as Radcliffe and Tufts. But she didn't have enough money. Her sister, who taught at Purdue, suggested she apply; she did and she received scholarships. Enrolling at Purdue "turned out to be one of those wonderful, fortuitous outcomes because had I gone to Radcliffe I very likely would have ended up doing women's studies—literature or the humanities," she said. "By attending Purdue, I was exposed to engineering and to science and found that that's what I really liked."[15]

From 1998 to 2004 Rita Rossi Colwell served as the director of the National Science Foundation, supporting research and education in all nonmedical areas of science and engineering. She was the first woman to hold the position.

International Students

In 1951 Hovde hired Arthur "Art" Tichenor from the University of California, Berkeley where he had worked in foreign student programs. He became head of the international students programs at Purdue and an adviser to international students until his retirement in 1982. When Tichenor arrived at Purdue there were 148 international students on the campus. When he retired the number had reached fifteen hundred.

From 1953 to 1960 Purdue helped revitalize an existing university in Tainan, Taiwan. Money for the program came from the U.S. government. Purdue professor Norris Shreve led the project, and a number of faculty from West Lafayette lived and worked at Tainan during the effort. Among them was Lillian Gilbreth, who had retired as a visiting faculty member in 1948 but continued her involvement with Purdue. The school became National Cheng Kung University.

Al Wright, Director of Bands

Among other changes at Purdue in the 1950s was the "All-American" Marching Band. By 1954 there had been seven Purdue presidents, but only one person had served as full-time director of bands—Paul Spotts Emrick. He retired and in 1954 a thirty-eight-year-old man arrived in town from Miami, Florida. Everything changed.

Al Wright introduced the Golden Girl, the Girl in Black, the Silver Twins, and flag carriers. The band grew in size under Wright, reaching more than four hundred members. He changed the marching and performance style. He had help from his wife, Gladys, who was also a band director.

Al Wright, director of Purdue Bands.

In 1966 Wright started one of his most popular traditions at the urging of Lafayette *Journal and Courier* publisher Jack Scott. There was political unrest on college campuses across the nation, and Scott, a World War II marine, wanted Wright to show the patriotism at Purdue. Wright wrote and the band performed "I Am an American," which is performed at every Purdue home football game: "I am an American. That's the way most of us put it, just matter of factly. They are plain words, those four: you could write them on your thumbnail, or sweep them across this bright autumn sky. But remember too, that they are more than just words. They are a way of life. So whenever you speak them, speak them firmly, speak them proudly, speak them gratefully. I am an American."

Purdue Pete

The University officially adopted Purdue Pete in 1944. "Red" Sammons and "Doc" Epple, founders of University Book Store, created Pete in 1940 and used him to advertise their business. When the 1944 *Debris* used a drawing of Pete, they asked Epple the name of his character. He blurted out "Pete," because that's what came into his head. Purdue Pete was born and he's been a "student" at the University ever since. His looks have changed several times over the years to reflect different eras. His facial expression has changed from happy and innocent, to determined and a little scary, to a more friendly and neutral look.[16] But he's still Pete, attending athletic games and University events. In 1958 he attended a new event, the Purdue Grand Prix go-kart race. It has been held on campus ever since.

Purdue Pete, 1950s.

The Boilermaker Special

While popular, Pete is not the official mascot of the University. That title belongs to the Boilermaker Special—a train. In 1939 students decided the mascot for their athletic teams would be a train engine. Fund-raising began and on September 11, 1940, Boilermaker Special 1 was dedicated. New updated models have since been introduced.[17]

A smaller Xtra Special debuted in 1979. It is four feet wide, twelve feet long, and seven feet high, small enough to go places the Purdue Special cannot.[18]

The 1967 Rose Bowl

The 1966 Boilermaker football season was among the greatest in Purdue's history, led by senior quarterback Bob Griese and defensive back Leroy Keyes. Expectations for the team were big that season, and they were met.

The Boilers were working their way toward a Rose Bowl bid when the team ran into trouble in a game against the University of Illinois. Griese threw five interceptions, but

Purdue quarterback Bob Griese.

his comeback passing led his team to an ultimate victory. When they won their next two games against Wisconsin and Minnesota, the Boilermakers were virtually assured of a trip to the Rose Bowl. Michigan State would win the Conference that fall, but a Big Ten rule stated that a school could not travel to the West Coast game in consecutive years. Runner-up Purdue went. "This is the greatest day of my life," Coach Jack Mollenkopf said at the end of the Wisconsin game, after being carried off the field by his players.[19]

In Pasadena on January 2, 1967, the Boilers defeated Southern California 14–13 when the Trojans missed a two-point conversion after a touchdown late in the game.

In his three years of varsity football with Purdue (freshmen were not eligible), Griese threw for 4,402 yards and twenty-eight touchdowns. He earned All-Big Ten and All-American honors his junior and senior years and was runner-up to Steve Spurrier of the University of Florida for the 1966 Heisman Trophy. He was inducted into the College Football Hall of Fame in 1984. In the NFL he played fourteen seasons with the Miami Dolphins, leading them to three consecutive Super Bowls and winning two of them. He was inducted into the Pro Football Hall of Fame in 1990.

Keyes played defense during his sophomore year, when the Boilermakers went to the Rose Bowl. During the regular season that year he returned a fumble 95 yards for a touchdown against Notre Dame. His junior and senior years he was a running back and fans shouted, "Give the ball to Leroy!" He set school records with thirty-seven touchdowns, 222 points, and 3,757 all-purpose yards. After his last game against Notre Dame, Fighting Irish coach Ara Parseghian told him, "I'm glad you're graduating." Keyes finished third in Heisman Trophy voting his junior year and second his senior year. O. J. Simpson from Southern California finished

Football coach Jack Mollenkopf considered going to the Rose Bowl the greatest day of his life. (Photo by Bob Mitchell)

second and first those two years. In 1987 Keyes was voted the all-time greatest Boilermaker football player. He was inducted into the College Football Hall of Fame in 1990. He played five years in the NFL.

Mackey Arena

Also in 1967 one of Purdue's most iconic facilities was dedicated—the new basketball arena. In 1972 it was named for Guy "Red" Mackey, a 1929 Purdue graduate. He was a Boilermaker football assistant coach and in 1942 was named athletic director until his death in 1971.

The dedication game for the arena was December 2, 1967, and it returned to Purdue one of its favorite alumni, John Wooden, coaching his UCLA team that was on a thirty-four-game winning streak. Rick Mount, who had been the first high school basketball player ever to be featured on the cover of *Sports Illustrated* magazine, led the Boilermakers. Lew Alcindor, later named Kareem Abdul-Jabbar, a center who towered more than seven feet tall, led UCLA. Mount, a sophomore, scored a game-leading twenty-eight points, but UCLA won 73–71 on a last-second shot.

The big winner that day was the arena. People loved it and have ever since.

Brian Lamb
*(Photo by Mark Simons/
Purdue University)*

Brian Lamb started his media career working as a student disc jockey for his high school radio station. He retired as CEO of C-SPAN, the cable television company he founded that brings history—live sessions of the U.S. House and Senate and other programming—into one hundred million American homes daily.

Lamb was born and grew up in Lafayette, Indiana, and he never forgot his roots and work for his hometown radio and TV stations that helped pay his way through Purdue University.

The University played a significant role in his future. "More than anything, Purdue lets you do things—it lets you experience things," Lamb said. His classes in political science, history, and philosophy opened him up to new ideas and thought. He even got a chance to produce and narrate a documentary about the 1960 Purdue Mock Political Convention (Mock P), at which students picked John Kennedy and Lyndon Johnson for president and vice president months before the political conventions and election.[20]

In 1963 Lamb graduated from Purdue, where he majored in speech, and entered the U.S. Navy. He served two years on an amphibious attack ship before going to the Pentagon in public affairs. He became a social aide to President Johnson and walked Lady Bird Johnson down the aisle at their daughter's White House wedding. He later worked in the White House Office of Telecommunications Policy and served on a secret committee that helped prepare Gerald Ford for the presidency.

It was his work for *Cablevision* magazine that led to C-SPAN. In the mid-1970s the cable TV industry was growing rapidly. As the U.S. House debated televising its daily sessions, Lamb put together a plan for a cable industry–financed nonprofit network to provide the service. In 1979 it was launched with four employees and live sessions of the House. The Senate was added in 1986, and more programming followed. The history C-SPAN captures daily is archived in the Purdue Research Park.

In addition to management, Lamb had on-air responsibilities and distinguished himself for impartial journalism. Among the many honors he received are the Presidential Medal of Freedom, one of two highest civilian awards in the nation, and the National Humanities Medal. In 1993 he was inducted into the Indiana Journalism Hall of Fame.

He received an honorary PhD from Purdue in 1986, and in 2012 the University's School of Communications was named in his honor.

And he still loves to talk about working as a high school radio disc jockey.

Delon Hampton
(Photo courtesy of Delon Hampton)

Delon Hampton, who received master's and PhD degrees from Purdue University and founded one of the nation's leading engineering firms, credits his success to two people: the mother who died shortly after he was born and the aunt who raised him on the South Side of Chicago, where drugs, gangs, and guns were a constant threat.

"Prior to her passing my mother—at age 25—found the strength and courage to make my father promise to send me to live with my aunt and uncle, Elizabeth and Uless Hampton, on the south side of Chicago," Hampton said.[21] He further stated that his life was guided by wisdom and foresight from the mother he never knew and the love and devotion of the woman who raised him.

As a boy on the streets of Chicago, Hampton, along with his friends, had to run from gangs who shot at them. There were fights. Some of his friends became involved with drugs. Elizabeth, who became a single parent, had only an eighth grade education, but Hampton did well in school and graduated from the University of Illinois before receiving his Purdue civil engineering degrees in 1958 and 1961. At Purdue, Hampton was influenced by Professor Eldon Yoder, who encouraged him to complete his doctorate. He received an honorary PhD from Purdue in 1994.

In 1973 he founded Delon Hampton and Associates

The big break for his company came early when it won a $20 million contact to work on the Washington, D.C., metro system. "From there we marched around the country designing heavy rail facilities," Hampton said.[22]

Headquartered in Washington, D.C., Delon Hampton and Associates provides civil, structural, and environmental engineering and construction management services. It is one of the nation's leading civil engineering companies, with award-winning projects such as Baltimore's Fort McHenry Tunnel, Washington's Dulles and Ronald Reagan airports, Los Angeles International Airport, and the Atlanta, Chicago, and Los Angeles metro systems.[23]

In 1991, Chicago opened a new baseball stadium to replace Comiskey Park. It was built next to the old one in Hampton's boyhood neighborhood where he once had to flee from gangs. His company was part of a team that helped to design and build it.

In September of 2012 Purdue named its Civil Engineering building in honor of Hampton and the Aunt Elizabeth who raised him.

It followed a gift of $7.5 million from Hampton to honor the aunt who raised him and who played a major role in his life and his success.

Jules Janick
*(Photo by Tom Campbell/
Purdue University)*

Jules Janick was born in 1931, the stepson of a superintendent at an Upper Manhattan apartment building and son to a mother who had aspirations that he would pursue a professional career.

"With a Jewish mother a child gets asked a lot, 'What do you want to be when you grow up?'" Janick said. "The correct answer was 'a doctor or a lawyer.' I didn't want that. I wanted to be in agriculture."[24]

From the steel skyscrapers and crowded concrete streets of New York City, Janick pursued a career that led him to the American Society for Horticultural Science Hall of Fame and a designation by the Cornell College of Agriculture and Life Sciences as "one of the world's most highly respected and best known horticulturists."[25]

The Purdue University Horticulture Building has been his work home since 1951, although he spent two years on a Purdue project in Brazil. It is believed he's been at Purdue longer than any other member of the faculty.

The James Troop Distinguished Professor of Horticulture, Janick helped create more than twenty cultivars (varieties) of apples and pears. He has edited and authored more than 150 volumes and founded the annual review series *Horticultural Reviews* and *Plant Breeding Reviews*.

Growing apples resistant to disease, he created cultivars with names such as Pixie Crunch, CrimsonCrisp, and GoldRush, the official fruit of Illinois. Another is named Pricilla, in honor of the wife of Purdue president Frederick Hovde.

It all started with a little garden at a cottage on Lake Oscawana, New York, where his family summered when he was a boy. There, Janick tended tomatoes and discovered his love for horticulture. During World War II he worked on New York state farms as part of the Victory Farm Corps program.

His mother and stepfather did not have a college education, but at the age of sixteen Janick graduated from high school and entered Cornell, graduating at the age of nineteen. In January of 1951 he entered graduate studies at Purdue and in 1954 was offered a job as a Boilermaker fruit breeder.

"Purdue is wonderful," Janick said. "Purdue is the best decision I made. Purdue let me expand my career from breeding to editing and horticulture history and art. I've traveled the world and I've enjoyed every minute of it. And my son Peter is a doctor who grows orchids and my daughter a lawyer who grows grapes."[26]

Janick is approaching his eighty-eighth birthday during Purdue's sesquicentennial and he continues teaching with no plans to slow down. His latest book, *Unraveling the Voynich Codex*, was published in the fall of 2018.[27]

27

The Spring of Discontent

1968 to 1971

My generation . . . came out of the Depression and went into World War II. We came home, got married, had kids, and said, "by golly, our kids are going to have it better than we had it." . . . We were amazed that they were different from us. And we made them that way.

—JOHN HICKS, SPECIAL ASSISTANT TO THE PRESIDENT OF PURDUE

Following a student uprising at Columbia University in New York City that captured national headlines in the spring of 1968, Purdue University dean of students Helen Schleman was quite blunt.

The same thing could happen at Purdue, she said.

Schleman had already brought about social changes at Purdue, including elimination of curfews for female students. In an age of in loco parentis when universities operated as substitute parents to students, there were women's "hours." At Purdue, female students had been required to be in their residence hall by 10:30 p.m. on weeknights and midnight on weekends. This often resulted in mad dashes to Windsor Halls and sororities as the clock ticked near the deadline.

During the war and her work with the SPARs in Washington, D.C., Schleman had seen women of college age handling their own lives quite nicely without male administrators giving them "hours." When she returned to campus in the late 1940s,

240

Schleman wanted to give female students keys to the residence halls and allow them to come and go as they pleased. This set off quite a debate. Trustees feared parents would not allow their daughters to attend a university without "hours."[1]

It took about twenty years for Schleman to prevail, proving once again she was way ahead of her time. In 1966 the trustees eliminated "hours" for sophomore, junior, and senior women. In 1969 the curfew was removed for freshmen.

But many inequalities remained. Equal rights for women protests were sweeping the nation, along with a growing sentiment on university campuses against the Vietnam War and racial discrimination.

"Open housing is not open enough here," Schleman said in a 1968 interview. "It is too difficult for a negro family or single student to find proper accommodations." She also said there was not enough dialogue between African American and white students, and minority students were not being given leadership opportunities.[2]

Civil Rights

Civil rights leader Martin Luther King Jr. was at Purdue on August 21, 1958. He spoke to a United Church of Christ national conference on Christian education, and then held a one-hour news conference in the Memorial Center (Stewart Center). He spoke against segregation in the United States and said he and other civil rights leaders had received death threats. "But we believe that this is a cause that is greater than life itself and that God is with us," King said. "It is a struggle to make the world better. God wants men to be brothers. If physical death of one man will free a people from psychological death, then to die is Christian."[3]

African American enrollment at Purdue remained small. In 1965 there were 20,176 students on the Purdue West Lafayette campus and only 129 were estimated to be African American. But opportunities were emerging. During the late 1940s and 1950s Purdue intercollegiate athletics had been integrated—something that had not taken place since before World War I. In 1947 Willard Ransom, an African American attorney and Purdue graduate, had challenged segregation in the University's football program, and an African American player was allowed on the team.

The first African American pictured as a member of the team in the *Debris* was in 1951. His name was Herman Murray. There were no others until 1955, when Lamar Lundy became the first African American scholarship athlete on the football squad. In the 1955 *Debris* he was also the only African American on the basketball team. Lundy was the most valuable player on the football and basketball teams in his senior year and was drafted in both the NFL and NBA. He went on to a professional football career. As years passed, African Americans continued to participate in Purdue intercollegiate athletics in larger numbers. In 1962 Forest Farmer, a football player, was named co-captain of the Boilermakers.

There were African American academic successes as well. In 1950 Phillip Hammond became the first African American to earn a PhD at Purdue. His field was pharmacology. In 1955 Dolores Cooper Shockley became the first woman to receive a PhD from Purdue and the first woman in the United States to earn the degree in pharmacology.

The Fire Next Time

On April 4, 1968, King was assassinated in Memphis, Tennessee, leading to demonstrations nationwide, some peaceful and some violent. At Purdue, African American students decided to make their positions known peacefully, said Eric McCaskill, who was a student at that time.[4]

Lamar Lundy, first African American scholarship athlete at Purdue. He played football and basketball. (Photo courtesy of Purdue Athletics Archives)

On May 16, 1968, about 150 African American students gathered at Lambert Fieldhouse and walked in single file to the Administration Building at the center of campus. As they passed the Pharmacy Building that was under construction they paused, and each person picked up a red brick—a symbol of the University. Purdue student James Bly drove a borrowed car at the back of the marchers providing protection from anyone coming from behind. They had decided on a silent protest and the marchers stopped before the steep steps at the front entrance to Hovde Hall. A list of demands was given to the dean of men, O. D. Roberts, who came out to meet them. Each of the young men and women carried a single red brick, and silently they walked up the steps, setting their bricks down in a pile. When all the bricks had been deposited, they left a sign that read ". . . or the fire next time." *The Fire Next Time* was the title of a book by James Baldwin that focused on race in the United States. The line comes from an African American spiritual song.

The demonstration lasted about forty-five minutes during the noon hour, and about five hundred people stopped to watch. "It was a peaceful and orderly meeting," Dean Roberts said.[5]

Among the marchers were some of Purdue's sports stars, such as basketball player Roger Blalock, track star McCaskill (who had set national high school records in the hurdles), football legend Leroy Keyes, and more. McCaskill and Keyes both came from competing and segregated high schools in Newport News, Virginia. In taking part in the demonstration, student athletes were putting their scholarships at risk. No actions were taken against them, but their participation added to the credibility of the student concerns. All had faced incidents of personal discrimination on campus. They had been called racial names, people had spit at them, and more. Keyes said that when a group of female students saw him and several other African American athletes walking together to a campus restaurant one evening, they became frightened and called campus police. Police arrived, found the African American male students in the restaurant, and demanded that Keyes leave with them. He refused, saying he had done nothing wrong. Police were called in from other departments in the county and the street was filled with squad cars. It took the arrival and intervention of football coach Jack Mollenkopf and Hovde before police left.[6]

Also among those who took part in the 1968 demonstration was Mamon Powers Jr., who was named to the Purdue University Board of Trustees in 1996, retiring in 2011. In 2014 he was given an honorary doctorate in engineering.

The demands from the 1968 demonstration were that the University pressure its departments to recruit qualified black professors for the next school year; that the history department integrate the content of its U.S. history courses; that student organizations be integrated immediately; that courses be started or expanded to deal with black culture; that black artists be incorporated into music and art appreciation courses; that the University compile a public list of discriminatory housing; that the administration show more than token integration; that the University see to it that black professors did not experience discrimination in procuring housing; and that a class dealing with diversity be instituted as a requirement for all students.

In his book *A Century and Beyond: The History of Purdue University*, Robert Topping said Hovde "responded positively to the student demands, not because of implied threats but because they addressed some valid points and listed some areas which, as Hovde admitted, the University had neglected."[7]

A year later students believed their demands had not been met.

The year 1968 was significant for African Americans on the Purdue campus. At homecoming that year cheerleader and African American Pam King gave a Black

Panther bowed head, raised fist salute during the National Anthem. It was not noticed in the large stadium. On December 5 she did it again at a basketball game. It was noticed and caused a controversy. She called it "The black man's salute. A symbol of pride and dignity."[8]

The cheerleading squad, which was part of the Athletic Department, was ordered to create a code of conduct, which no doubt included unsubtle hints to ban the controversial salute. By the next home game the cheerleaders had not completed the code, and Athletic Director Guy "Red" Mackey refused to allow them on the floor during the national anthem. During halftime King resigned from the squad.

The year 1968 was also the year Helen Bass Williams became the University's first African American professor. Purdue education professor Mary Endres had worked with Williams on community action efforts in the southern United States. She wanted Williams at Purdue and persuaded Hovde assistant John Hicks to find money and a position for her. "Helen was worth her weight in gold," said Betty M. Nelson, then an assistant in the Office of the Dean of Women.[9] At Purdue Williams was an instructor in French and a counselor in the School of Humanities, Social Sciences, and Education. Before arriving at Purdue she had played a significant role in the American civil rights movement. She was active in the southern United States at a time when King was leading peaceful protests. In Mississippi, Williams was beaten, gassed, and jailed. Her stature was short and her voice was high-pitched and tiny. But what she had to say was powerful, and she would stand in the way of any injustice. She worked very actively on civil rights during her time at Purdue. "Helen Bass Williams had a lot of wisdom, she was very wise," McCaskill said. "She said 'we're going to do this the Dr. King way, peaceful protests.'"[10]

A focus on African Americans was added to courses in the Departments of Speech, History, and Sociology in 1968, and it was the year the Krannert School of Management launched the Business Opportunity Program to increase diversity and give all students access to world-class business management education. It was the first program of its kind at Purdue. In the first year it was headed by Professor Dan Schendel on an interim basis. In its second year Cornell A. Bell was hired to head it. Bell went on to mentor students

Helen Bass Williams, first African American member of the Purdue faculty.

Cornell Bell, director of the Krannert School of Management Business Opportunity Program.

for thirty-seven years until his retirement in 2006, becoming a father figure to countless students. The program continues at the University's sesquicentennial with a 90 percent five-year graduation rate. In its first fifty years nearly fifteen hundred students participated in the program.

Also in 1968, the first African American sorority was founded at Purdue, Delta Sigma Theta.

On Monday, April 14, 1969, about a hundred Purdue African American students sang "We Shall Overcome" as they marched from campus to Lafayette City Hall in support of track athlete McCaskill—again with Bly following them in a car for protection. McCaskill had grown a mustache that was against the Purdue Intercollegiate Athletic Department's clean-shaven policy. McCaskill said mustaches were part of African American male culture and the policy was racially biased. He said there were seven African Americans on the track team. Four of them shaved and three didn't. They had close-cropped, trimmed mustaches.[11]

On Saturday, April 12, as the track team boarded an airplane at the Purdue Airport bound for Iowa City and the University of Iowa, McCaskill was told he could not make the trip because of his mustache. Although he was injured and could not run, McCaskill wanted to support his teammates. Angered over the policy and other issues, he remarked that there "ought to be a bomb on the plane." The pilot panicked and refused to fly.[12]

McCaskill's remarks resulted in a bomb squad being called in from Fort Benjamin Harrison to search the plane. Nothing was found. McCaskill was then questioned by an FBI agent, who said no federal law had been violated. Finally, Indiana State Police officers arrived and arrested McCaskill on a charge of disorderly conduct. The story made national news; his mother first learned about it while watching television in New York City. McCaskill's bail was set at $300. He would have spent the rest of the weekend in jail but the dean of the School of Humanities, Social Sciences, and Education—who knew him to be an award-winning student—bailed him out.

During the Monday morning court hearing at Lafayette City Hall, which was packed with African American Purdue students, Tippecanoe County prosecutor

David Crouse said no law had been broken, and charges were dropped.[13] The students marched back to campus and entered the Administration Building to talk with Hovde. One of their top concerns by this point was the creation of a place on campus where they could gather. Many of these students had come from racially segregated communities and were not accustomed to being around so many people of a different race. They wanted a place of their own where they could feel comfortable.

Hovde's secretary informed the students that the president was out of town, but she reached him by telephone and Hovde spoke with McCaskill. There had been previous conversations with the administration about African American concerns. Now a new idea was brought forward—the creation of a Black Cultural Center. A committee was established that recommended creation of the Center, with the needed funds coming from the Purdue Research Foundation.

McCaskill graduated that spring of 1969 and in the fall came back to campus to work on a master's degree from Krannert. He was the first Business Opportunity Program graduate counselor, along with fellow students Dutch Ross and Joe Knox. In 1969, creation of a Black Cultural Center was approved. It opened in 1970 in a large house on University Street, across from the Armory. Its first objective was to serve the University's six hundred African American students, but the Center's administrator, Professor Singer Buchanan, called it a place for sharing the black experience with everyone on campus and in the community. (More on the Black Cultural Center in part 3.) McCaskill was on campus to see the Black Cultural Center open.

And he kept his mustache—while at Purdue and for many more years after.

McCaskill became an ordained minister and ultimately returned to his Newport News, Virginia, hometown. He is part of the Alpha and Omega Network of Ministries, which its website says works to "restore virtues in the culture/village; rebuild broken families and relationships; renew hope in the American ideal; and empower at-risk youth to realize their maximum potential!"

Keyes returned to Purdue in the mid-1990s and held a variety of coaching and athletic department positions. At the sesquicentennial he worked with the John Purdue Club, whose members support Purdue intercollegiate athletics. He was among the most popular and loved people on campus by both students and alumni.

In 1980 the Black Cultural Center celebrated its tenth anniversary. Tony Zamora, director of the center, wanted all African American alumni to return to campus for the events, but the University did not have a complete list of graduates by race through its history. So several students went through the *Debris* yearbooks looking at senior photos and found about twelve hundred African American graduates. Invitations went out to all of them and about two hundred returned.

Hovde, who was then president emeritus, was invited to take part and was presented with an award for his role in starting the center. He was moved by the honor. Privately, he told Zamora, "I didn't think anyone would remember."

The 1980 reunion became a catalyst for the creation of the Purdue Black Alumni Organization as part of the Purdue Alumni Association.

In 1968 Helen Schleman retired, and the University lost a tireless advocate for women, minorities, and progressive thinking, change, and progress. She urged the administration to hire one of the assistants in her office as the new dean of women. Schleman won her last campaign at Purdue and M. Beverley Stone replaced her. Stone was born in Norfolk, Virginia. In 1936 she graduated from Randolph-Macon Woman's College in Lynchburg with a degree in chemistry. She received a master's degree in student personnel administration from Teachers College of Columbia University where Edward C. Elliott and Dorothy Stratton had also studied. For seven years Stone had worked in the dean of women's office at the University of Arkansas and was dean of women from 1954 to 1955. She arrived at Purdue in 1956.

She had plenty of challenges awaiting her.

The Spring of '69

Some of the largest demonstrations on the Purdue campus came in the spring of 1969 when the University raised tuition and fees for the next academic year. Fees for in-state students went from $400 to $700 a year. The increase for out-of-state students went from $1,200 to $1,600. Since room and board costs also went up $100 per year, estimated total costs including expenses, books, and room and board were going from $1,950 to $2,350 (20 percent) for Indiana students and from $2,750 to $3,250 (18 percent) for out-of-state students.

Hovde believed the University had little choice but to raise fees and tuition when the Indiana General Assembly did not increase state support for higher education in its two-year budget. He said part of the increase went to need-based student support.

On April 17, 1969, about six hundred to eight hundred students marched on the Administration Building. An estimated three hundred of them entered the building and asked to see Hovde, who was out of state. The students left a series of demands, including an end to the fee and tuition increase, and promised to come back when Hovde returned. They also announced a boycott of food and other items sold in the Memorial Union.

On April 21, about two hundred students entered the Administration Building while University officials were meeting there behind closed doors. The students were not allowed in the meeting but remained in the building. They sat against walls and lined both sides of stairways, leaving the middle open. Some studied. Hicks, assistant to Hovde, said the students were not disruptive and as long as they were not, they could remain until the building was closed at 5 p.m. *Journal and Courier* reporter Larry Schumpert noticed a cartoon hanging on a wall in Hovde's reception area. It featured a university president in an office filled with sitting demonstrators being asked by a reporter: "As president of a great university, sir, what would you say was the most significant change you have observed in the last twenty years?"[14]

Evelyn Emerson, a World War II marine sergeant, was a file clerk in the Office of the Dean of Women in the Administration Building. During the sit-in Hovde's receptionist felt intimidated by the noisy, milling students and asked to be relieved. Emerson volunteered to take her place. "Like an old drill sergeant she was formidable and quickly organized the students to clean up their trash every hour," Betty Nelson said. "She and the students got along famously."[15]

When the workday was coming to a close, the students were reportedly becoming disruptive, and they were asked to leave. Emboldened by the arrival and participation of Keyes, they refused. With that, University police arrested forty-one students, including Keyes. As the arrests were made, the students sang "Hail Purdue." Within an hour the students were told that if they would leave the building, everyone arrested would be released without charges. This time they did leave and the arrested students were released. A group of students stayed behind to clean up the building.

That evening as many as five thousand demonstrators marched through campus.

On Tuesday, April 22, six thousand students packed the Elliott Hall of Music and demanded that Hovde, who arrived late, meet with Governor Edgar D. Whitcomb to ask for a special meeting of the General Assembly to deal with the issue. Although retired, Helen Schleman attended the meeting. She introduced herself, shook hands with students, and asked what they were protesting and what they really wanted. Hovde said he would contact the governor, but several days later announced he could not do that. Whitcomb favored a much tougher stance against protesting students in his state. During a closed-door meeting with faculty, Hovde asked for their support and received a standing ovation.

But the situation did not get better. It got worse. First, on Wednesday, April 30, several hundred students met on the mall and marched to the Memorial Union for what they called a "live-in." They went in with pillows, blankets, guitars, and books. The protest centered on the tuition increase, but it also had to do with anger concerning the Vietnam War, concerns about racism and sexism, and a rebelliousness by students against the perceived wrongs committed by everyone more than thirty years old.

A pair of representatives from the offices of the dean of men and dean of women were in the Union at all times during the "live-in," including Betty Nelson. She remembered: "One night the lounge just east of the Great Hall was filled with students who were socializing and playing guitars. They were not being ugly or abusive of the place."[16] Two naked students did streak (run naked), barely missing members of the board of trustees, who were leaving a black tie Union party. There was even naked horseback riding outside on State Street.

The students were still there on Friday, May 1, when about three hundred students opposed to the war in Vietnam met on the mall, listened to speeches, then marched on the Administration Building. They went to Hovde Hall and were met by Hicks. The students had three demands: an end to ROTC on campus (military training was required by the Morrill Land-Grant Act of 1862), an end to all classified military research on campus, and an official University statement against the war. Hicks said he would not negotiate with a mob and rejected the demands.

The students left the Administration Building and went to the Armory, where preparations were underway for an ROTC awards presentation. They took over the microphone and speaker's stand and were told by Associate Dean of Men William Brown that they were violating University policy. He asked them to leave, and as they did they tore down an ROTC sign.

A short time later a nonstudent demonstrator was arrested at the Armory, and in response the previous group marched back, walking five abreast and chanting. They broke an overhead door and marched into the Armory, this time while the ceremony was underway. They sat in the middle of the drill floor and a student grabbed the microphone.

At this point, on instructions from Hicks, police entered and removed the students. Some resisted and there were skirmishes. The police had clubs and used them. Some of the students left bleeding. Four students were taken to the health center and treated for cuts and bruises. One of the students arrested was a relative of Associate Dean of Women Barbara Cook. Cook was not pleased.[17] The Hovde administration decided that all the students who participated in the demonstration during the ceremony would be suspended if they were identified.

Meanwhile the students remained in the Union. On Monday, May 5, they expressed concerns that other students who were opposed to what they were doing had threatened to physically assault them and asked for police protection. It was decided that the Union should be closed and the students asked to leave. The order to leave was given to the students at 2:39 a.m. and 2:59 a.m. on May 6. Police waited almost an hour, until 3:55 a.m., to announce that the Union was closed, and then the doors were locked. Two hundred twenty-nine students who had remained inside were peacefully arrested.

Later that day when Hovde stood to make his remarks at a celebration of Purdue's centennial in the Hall of Music, one thousand students got up and walked out silently. It was a difficult moment for Hovde at a time of celebration. It didn't get any better for him when U.S. health, education, and welfare secretary Robert Finch, who had been invited to speak, said that nationwide, two-year community college programs would eventually replace land-grant universities as "democracy's colleges."

At 2 p.m. that afternoon about eight hundred students marched on the Administration Building and began knocking on the door to Hovde's office, without getting a response. They sang "Happy Birthday" to the University.

Students returned to the building and at 5:30 p.m. were told that the governor had called in the state police. That resulted in more students entering the building. After another warning to the students, state police marched in precision from the Armory parking lot and entered the Administration Building at about 6:40 p.m. They did not wear helmets and they did not carry clubs. Police said they were spit at, kicked, and punched, and rocks were thrown. A rock shattered a glass door. Police used Mace to "ward off belligerent students." About two hundred students were in the building, and once all the state troopers had entered, police announced, "You are free to leave the building if you care to."[18] The students left. Many marched to the Union, where they planned to spend the night. Nelson was on duty that evening, and she and a representative from the Office of the Dean of Men were called to a meeting in the office of Union director John Smalley. The University Cabinet was meeting in Hovde's Seventh Street home in Lafayette. Vice President Lytle Freehafer brought the message that the cabinet had decided to tell police to arrest students who had come back to the Union. Nelson was among those who spoke against the decision and convinced the cabinet to ask the students to leave by midnight. The students were asked and they left.[19]

The rest of the school year remained relatively calm compared to what had transpired. There were also demonstrations against tuition hikes at the state's other public universities that spring, and students at many of Indiana's public and private universities held protests against the Vietnam War.

The staff in the Office of the Dean of Women met with each female student arrested that spring to establish a point of support within the University administration and to discuss continuing their education. Nelson said numerous strong and lasting relationships began at that time.[20]

There were more demonstrations against the war in the next several years, but there was never anything like the spring of 1969 at Purdue.

"This whole period was one of the most dramatic of President Hovde's twenty-five years here," Hicks said. "I think it bothered him more than anything. I've often thought about my particular generation. We came out of the Depression and went into World War II. We won the battle over the forces of darkness. We came home and said, 'Well, we've beaten the forces of evil. Now we can get on with our lives.' So we got married, had kids, went about our work, and said, 'By golly, our kids are going to

Student protesters on Purdue campus, 1970.

have it better than we had. . . . We were amazed that they were different from us. And we made them that way."[21]

President of the board of trustees Maurice Knoy said of Hovde, "In the brutal situation of the student unrest, the thing that hurt him the most—really shook him to his foundation—was that any of his students would turn against him. He just couldn't believe that it was possible."[22]

Seven months after the spring of 1969, on December 19, Hovde stepped to the podium at a routine faculty convocation and announced his retirement. He would serve until June 30, 1971, to give the trustees time to find his replacement. The faculty were stunned. He had not yet reached Purdue's mandatory retirement age. But he had completed a personal goal of becoming Purdue's longest-serving president. His tenure would be twenty-five years. Hovde was Purdue's seventh president spanning one hundred years since the University's founding. And he was the first president to leave on his own terms. The others had

John Hicks, special assistant to the president, later acting president.

either died in office or were forced by the board to step aside.

Hovde emphasized that his retirement had nothing to do with the student unrest and what he considered a lack of faculty support the previous spring. Still, some people wondered.

As the board of trustees set about searching for a new president, a number of Purdue people made suggestions. Among them was Schleman, who nominated ten people—all women.

On June 24, 1971, a retirement dinner was held for Hovde in Indianapolis. President Richard M. Nixon attended. Nixon wondered how Hovde would most be remembered:

U.S. president Richard M. Nixon at a retirement dinner for
Purdue president Frederick L. Hovde, Indianapolis, 1971.

Those who remember football will remember him for that. Others who think
of his service in the war will remember him for that. But I think on a night like
this that the nation will remember him for the monument that he really deserves, and that monument is 80,000 men and women. Eighty thousand men and
women have graduated from Purdue in the twenty-five years that he has served
as president. They have gone forth through this land, throughout this world, as
a matter of fact, serving the cause of progress, the cause of understanding. They
have been men and women with brains, good educations, but more than that
they have been men and women with character and it is that, that Fred Hovde
stands for. . . . That is the tradition he leaves with Purdue and with America.[23]

Hovde and his wife remained in Lafayette. She died in 1980. Hovde died of
emphysema on March 1, 1983.

He was eulogized in a University publication, *Perspective:* "Suffice it to say, it
mattered not at all to him whether we were students or staff members, deans or department heads, or even United States presidents or drivers of campus mail trucks,
Fred Hovde's generosity of spirit brought out the best in each of us.

"And what else could we ask of any man."[24]

Roland G. Parrish
(Photo by Mark Simons/
Purdue University)

Roland Parrish was a senior at Indiana's Hammond High School in 1970 when he had a visitor.

"There's a man named Cornell Bell and he wants to talk with you," Parrish was told. That conversation changed his life.

The meeting with Bell, the son of a steel worker and a stay-at-home mother, led Parrish to Purdue University, where he received two degrees from the Krannert School of Management and ultimately became the owner of twenty-five McDonald's restaurants in northern Texas, with annual revenues of more than $66 million.

"Krannert and Cornell Bell are the cornerstones of my life," Parrish said.[25]

Bell was director of the Krannert Business Opportunity Program, the school's most important catalyst in recruiting and retaining a diverse student body.

Parrish was the first person Bell recruited to Purdue. But other Boilermakers also influenced him, including a fourth-grade student teacher. And as an all-state athlete in track, Parrish was impressed to see Purdue football star Leroy Keyes on the cover of *Sports Illustrated*.

Purdue track and field coach David Rankin offered Parrish a full scholarship to attend the University. And when Bell put Rankin together with Parrish's father during the state track and field regional finals, the deal was sealed. The following weekend Parrish won the State Championship in the 880-yard run.

Bell not only recruited students to Purdue. He worked with them to help them succeed.

"Dr. Bell at first was an academic counselor but he got to know us on a personal basis," Parrish said. "It wasn't long before he dropped by my parents' home to get to know them. He stayed for dinner. He came to family reunions and weddings. He became a very important person in our lives."[26]

After receiving a master's degree in 1976, Parrish worked thirteen years for ExxonMobil and applied to become a McDonald's franchisee in 1987. He wasn't immediately accepted. While he continued working fifty hours a week for ExxonMobil, he worked without pay twenty-three hours a week at a McDonald's restaurant—and had to commute ten hours weekly for two years. This taught him the importance of hard work and long hours.

In addition to his academic success, Parrish won six Big Ten medals in track, was twice named the Boilermaker MVP, and was voted team captain his senior year.

He gave $2 million to the University to renovate what was named the Roland G. Parrish Library of Management and Economics. He believes in giving back for all he received on the West Lafayette campus.

"Purdue prepared me for life," he said.[27]

Dolores Cooper Shockley

Dolores Cooper Shockley lived a life of firsts. But all of her accomplishments came with obstacles she had to overcome. She was an African American and she was a woman.

In 1955 Shockely became the first African American woman in the United States to receive a doctorate in pharmacology, which she earned at Purdue University.

Shockley was born in Clarksdale, Mississippi, in 1930. Her formal schooling was difficult in the town's segregated schools. "School supplies were very limited for blacks," Shockley said. "We got leftovers from the white schools. Most of what I did was learned at home."[28]

Despite the challenges, there were educated people in her family, some of them in healthcare fields, and they inspired her. Shockley studied pharmacology at Xavier University of Louisiana, in New Orleans—a Catholic and historically black school. She graduated first in her class. Faculty at Purdue wrote many of the textbooks she used. So she selected Purdue over the seven other graduate programs that accepted her.[29]

She found some discrimination at Purdue and even more in the community. "In Mississippi . . . there were signs in bold letters WHITE and COLORED," Shockley said. "In West Lafayette discrimination was covert and insidious. There were no signs, but a refusal to serve you or to rent you a room. This was extremely hurtful . . . I went to my room and cried several times."[30] She succeeded through her own determination and with help from some faculty.

She first lived in a Purdue residence hall and later struggled to get an apartment in West Lafayette. Finally a physics professor agreed to rent her an apartment.

After Purdue, Shockley used a Fulbright Fellowship for postgraduate research at the University of Copenhagen. She later accepted a position on the faculty of Meharry Medical College in Nashville, Tennessee, where she was offered a "ridiculous" salary. She demanded to be paid the same as men.[31] Ten years later she became chair of pharmacology at Meharry, the first African American woman to chair a pharmacology department in the country. She also was a visiting professor at Einstein College of Medicine in New York.

"The education I received at Purdue was well worth the subtle and not-so-subtle indignities that I suffered," Shockley said, adding that things have now changed "dramatically."[32]

"I am proud to be a graduate of Purdue. Purdue prepared me for the rigor of teaching and research at a medical school. I consider myself blessed to have degrees from Purdue."[33]

28

Purdue Astronauts

1946 to Present

One of the first things you see when you're out of the spacecraft
is the earth and all its beauty and splendor, the blues of the
ocean and the whites of the snow and the clouds.

—Eugene Cernan

Between 1950 and 1957 four young men who would play central roles in the race
to the moon graduated from Purdue University: Virgil "Gus" Grissom, Neil
Armstrong, Eugene Cernan, and Roger Chaffee.

They attended classroom lectures, studied in the library, and sat under tall trees
that had shaded students in the horse and buggy days seventy-five years before
them—long before the Wright brothers flew at Kitty Hawk, North Carolina. They
imagined their future and what it might hold. They dreamed about flying, but not
in space. In the 1940s and early 1950s the word "astronaut" was an obscure science
fiction term that wasn't even part of the vocabulary of most people.

But eleven years after he graduated from Purdue, Grissom became the second
American and third human launched into space. Fourteen years after he graduated, Armstrong became the first person to walk on the moon. Sixteen years after
Cernan graduated, he was the last Apollo astronaut to leave his boot print on the
lunar surface.

As Purdue approached its sesquicentennial, twenty-four of its graduates had been selected as astronauts. Four of Purdue's graduates were part of the Mercury, Gemini, and Apollo programs that conquered the moon. Two of the twelve men who walked on the moon were Purdue graduates. Eighteen Purdue alumni flew in the NASA space shuttle program between 1981 and 2011, and two more graduates became astronauts in the post-shuttle era. In 2018 two Purdue astronauts worked aboard the International Space Station, Drew Feustel, BS '91, MS '92, Science, and Scott Tingle, '88, MS Mechanical Engineering.

Only about 11 percent of all people selected as astronauts in the world have been women. Two Purdue women have flown in space, and a third was selected by NASA as an astronaut in training in 2017.

Purdue's first female astronaut was Janice Voss, who graduated from Purdue in 1975 with a degree in engineering science. During her undergraduate years she worked in a co-op program with NASA at the Johnson Space Center in Houston. She received a master's in electrical engineering and a PhD in aerospace engineering from MIT.

As a young girl she loved to read science fiction, and one of her favorite books was *A Wrinkle in Time*, the story of two children who travel in space helping their father, who is trying to save the universe. She thought space travel sounded pretty neat. She was selected for astronaut training in 1990, launched on five shuttle flights, and spent more than forty-nine days in space.

Voss was very involved with Purdue students and faculty during her career with NASA, working especially to encourage female students in engineering. She died in 2012 after a battle with cancer. In 2015 Purdue dedicated the Visiting Our Solar System interactive exhibit in Discovery Park. The acronym for Visiting Our Solar System is VOSS, and it celebrates her contributions to space exploration and the University. The exhibit includes to-scale models of the sun and eight planets in our solar system.

The second Purdue female astronaut was Mary Ellen Weber. She graduated from Purdue in 1984 with a degree in chemical engineering and took part in a co-op program with Ohio Edison, Delco Electronics, and 3-M. Part of what led her to Purdue was its reputation in space. Weber earned a PhD in physical chemistry from the University of California, Berkeley, in 1988 and a master's of business administration from Southern Methodist University in 2002.

She is a licensed pilot with an instrument rating and flew jets with NASA. "When I was in graduate school I saw the space program as a fantastic melding of the two passions that I had—science and aviation," Weber said.[1] She was selected by NASA in 1992 and flew on two shuttle missions, logging more than 450 hours in space. She is a skydiver and also enjoys scuba diving.

Loral O'Hara, who received her master's degree in aeronautics and astronautics from Purdue in 2009, was selected for a two-year Astronaut Candidate Class in 2017.

Purdue's history in flight goes back to the earliest days. Wilbur and Orville Wright are acknowledged as the first people to fly a controlled, heavier than air machine on the morning of December 17, 1903. After graduating from Purdue in mechanical engineering in 1908, Cliff Turpin joined them. He helped with their design and became a Wright Exposition Flyer, demonstrating flight to people around the nation. He had the twenty-fourth pilot's license issued in the United States.

Aviation developed slowly at Purdue, although there had been an airshow on campus during the Winthrop Stone administration in 1911. Around the world, interest in flight intensified during the Great War, and in 1919 Purdue graduate George Haskins, '16, flew from Dayton, Ohio, and landed his airplane on Stuart Field. He became the first alumnus to land an airplane on campus, and he brought with him a proposal from alumni in Dayton to launch a School of Aeronautics at Purdue. Courses in aeronautics were added to the Mechanical Engineering program during the 1920–1921 school year.[2] Haskins was a World War I pilot.

Haskins was recruited to Purdue in 1929 and placed in charge of the aeronautics courses in mechanical engineering that he had helped create. As head of aeronautic engineering within mechanical engineering, Haskins played a key role in building its national reputation. With the start of World War II he returned to military service and never returned to Purdue.

On the evening of June 20, 1969, Haskins, who was born before the Wright brothers' first flight, sat in his living room in Pasadena, California, and watched television as Armstrong, who went through his Purdue aeronautical engineering program, became the first human to touch the moon. The second man on the moon that day was Edwin "Buzz" Aldrin. Aldrin was the son of Haskins's lifelong best friend.

Two months later Haskins died, but he lived to see history that he helped to create.[3]

Two early pilots from Purdue who played a role in the advancement of aviation and probably would have become astronauts if they had survived were George Welch and Iven Kincheloe.

Welch came to Purdue from Delaware in 1937. During his junior year he was accepted into the military and left campus, finishing his degree in the army. He was stationed at Pearl Harbor on December 7, 1941, and was a hero during the attack by Japan, one of the few pilots able to get his plane into the air. He shot down four Japanese planes that day (some say six), more than any other U.S. pilot.

After the war he became a test pilot and is officially credited with being the second person to break the sound barrier. Some claim he was the first. He died in an airplane crash on October 12, 1954.

Kincheloe came to Purdue from Dowagiac, Michigan, in 1945. He graduated in 1949. He served in Korea and by the mid-1950s he was flying the Bell X-2, a rocket-powered research airplane. In September 1956 he became the first person to fly one hundred thousand feet above the earth. For that accomplishment he was not called an astronaut. He was called a "Spaceman." He was one of three pilots selected to fly the X-15 aircraft, but he died in a crash testing another airplane on July 26, 1958.

Grissom, Purdue's first astronaut, came to Purdue in the fall of 1946 on the GI Bill. He had graduated high school in the small southern Indiana town of Mitchell in 1944. He immediately went into the military and hoped to fly, but there was an abundance of pilots during the last year of World War II and he didn't get an opportunity. He was released from the army in 1946. Grissom returned to Mitchell, where he married his high school sweetheart, Betty. He worked for a short time at a company outside of town making school buses. But he was restless. He decided to go to Purdue on the GI Bill.

Grissom returned to the military after graduation, received flight training, served in the Korean War, and in 1959 was named one of the original Mercury Seven NASA astronauts. His first space launch was July 21, 1961. On March 23, 1965, he commanded the first manned Gemini space program flight.

Armstrong was born in Wapakoneta, Ohio. He became a licensed pilot at the age of sixteen. He was a good student and skipped a grade in elementary school. By his junior year of high school he was making plans to attend the Massachusetts Institute

Purdue alumnus and astronaut Virgil "Gus" Grissom signs autographs for boys outside a West Lafayette home during a May 1964 visit to campus.

Neil Armstrong waving a Purdue flag from the Ross-Ade Pavilion during a football game, 2007.

of Technology. He wanted to design airplanes. But as much as he loved airplanes and flying, he also loved football.

On October 20, 1945, he was in the Ohio State football stadium with more than seventy-three thousand other people for the game against Purdue. The Buckeyes were ranked fourth in the nation. They had won twelve straight games going back to the previous season. Only one team had scored six points on them all year. A self-described skinny freshman quarterback named Bob DeMoss led Purdue that day. The Ohio State fans were expecting a big win. But when play started, Purdue scored twenty-two points before Ohio State made a first down. The Boilermakers won the game 35–13. Armstrong was so impressed he decided to take a closer look at Purdue.

In 1947 Armstrong showed up on the West Lafayette campus, a seventeen-year-old surrounded by World War II veterans in their twenties. He was very nervous. Sixty years later he stood in front of a Purdue building that had been named for him and looked at a bronze image of himself as a teenage student. "Only Purdue would put a statue of a scared freshman in front of an engineering building," Armstrong said that day.

Armstrong flew in the NASA Gemini program before landing on the moon on July 20, 1969. The fiftieth anniversary of the moon landing coincides with Purdue's sesquicentennial year.

Cernan arrived at Purdue in the fall of 1952 from Bellwood, Illinois, a Chicago suburb. He received a naval ROTC scholarship and planned to study at Purdue. But the navy said the Purdue program was already full. They wanted to send him to the University of Illinois. His father wouldn't hear of it. He insisted that Cernan go to Purdue and sent him to West Lafayette without a scholarship. Cernan participated in ROTC and did receive a naval scholarship for his junior and senior years. "My life started on the fast track at Purdue," Cernan said. "We had to do a lot of studying."[4] Cernan flew one mission in the Gemini program, did everything but land on the moon on Apollo 10, and in December 1972 became the last person to walk on the moon.

Chaffee, from Grand Rapids, Michigan, transferred to Purdue in 1954 from the Illinois Institute of Technology. He also was on a navy ROTC scholarship. At Purdue he and Cernan knew each other. Cernan also knew Chaffee's girlfriend and future wife, Martha. Cernan was in charge of picking a queen for the ROTC Ball. Martha, a homecoming queen, interviewed for the title and she enjoyed teasing him about it. "Years later when we were reunited in the astronaut program I reminded him of that interview," Martha said. "I said, 'Thanks a lot. You didn't pick me.'"[5]

Among the young people who followed the space flights of those four was Jerry Ross from Crown Point, Indiana. He wanted to fly into space, so he went to Purdue. It had worked for Grissom, Armstrong, Cernan, and Chaffee. Maybe it would work for him. It did. He graduated in 1970, received a master's degree in mechanical engineering in 1972, found his wife, Karen, at Purdue, and shot into his future—which was space. Ross was launched in a space shuttle seven times—an accomplishment matched by only one other person in the world.

As the University approached its sesquicentennial, Purdue graduates had made more than sixty space flights, about one-third of all NASA human missions. In addition to its astronauts Purdue graduate Robert Foerster was one of ten finalists to be a "Teacher in Space," a mission that went to Christa McAuliffe and ended in tragedy in 1986. Many Purdue graduates have worked for NASA and private corporations in the space industry.

Grissom, Armstrong, Cernan, and Chaffee were together at Purdue's January 2, 1967, Rose Bowl game. Ross was also there, as a freshman Purdue student. Grissom and Chaffee died in an *Apollo 1* test accident twenty-five days later. Armstrong died in 2012 and Cernan in 2017. They were both eighty-two.

Cernan loved to talk about his experience on the moon. "You look all around and you're standing in sunlight and the blackest black you can conceive in your mind surrounds you. Not darkness—blackness," he said in a 2003 interview. "One of the first things you see when you're out of the spacecraft is the earth and all its beauty and splendor, the blues of the ocean and the whites of the snow and the clouds. It is three-dimensional out there in that blackness. . . . You could put your hand around

Left to right: Purdue graduate and Indiana senator Birch Bayh, Purdue president Frederick L. Hovde, first man on the moon and Purdue alumnus Neil Armstrong, and Indiana governor Edgar Whitcomb at a news conference on the Purdue campus in 1970.

the earth and almost focus on the endlessness of time, the endlessness of infinity. Can I show it to you? No. Can I draw you a picture of it? No. But I can tell you it exists because I have seen it with my own eyes."[6]

Purdue, Lafayette, and West Lafayette celebrated Armstrong's moon landing on June 20, 1969. "Man Sets Foot on Moon," the Lafayette *Journal and Courier* headline said on July 21. "Neil Armstrong, Purdue Graduate, All-Time Hero."[7] In an editorial the paper said, "To Purdue University came the great thrill of seeing one of its graduates make that first big step onto the lunar surface, fulfilling a dream that has beset man since he first gazed toward the sky. Neil Armstrong performed just as all who knew him were certain he would. . . . The Apollo 11 mission provided all humans who watched it with one of the greatest thrills of their lifetime."[8]

After the moon landing *Journal and Courier* editorial cartoonist Dave Sattler drew Purdue Pete with his arm around Armstrong. "That's my boy," Pete said.

Armstrong did not often talk publicly about his moon landing, although he provided full details of the flight to space historians and to NASA, which made his interviews available to the public. A number of astronauts, including Armstrong, have donated their papers and memorabilia to Purdue's Barron Hilton Flight and Space Exploration Archives. On the moon landing's tenth anniversary, during a news conference at the University of Cincinnati where he was teaching, Armstrong said he did enjoy, sometimes, looking up at the moon shining in the nighttime blackness of space. It brought back memories.

He said he could see a place where he had been.

Frank Bass

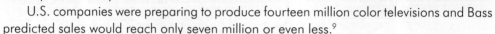

In the mid-1960s American color television sets hit the market, and everyone expected an endless boom in sales that would continue for many years—everyone except Frank Bass of the Purdue University Krannert School of Management.

Using a differential equation, Bass determined sales would peak in two years and then decline.

U.S. companies were preparing to produce fourteen million color televisions and Bass predicted sales would reach only seven million or even less.[9]

"I got lots of phone calls and mail from people about this, and some of them were angry," Bass recalled years later. "Some Wall Street people in particular were especially nasty. It involved a differential equation and a certain amount of mathematics, and they just didn't understand that sort of thing at all."[10]

Bass held his ground. It wasn't long before he was proven correct. In 1968 color TV sales peaked.

Bass is considered the "Father of Market Science."

He was among a group of young, aggressive faculty recruited to Purdue by Emanuel Weiler, who has head of economics and then dean of the new School of Management.

Bass joined the Purdue faculty in 1961. Two years earlier he was one of forty management faculty members from around the nation selected by the Ford Foundation to improve business education. They were sent to Harvard. That's where Weiler recruited a number of faculty, including Bass.

"One of the reasons that Frank Bass came here was because [Weiler] told him that he wanted him to do the best marketing done anywhere in the world. Weiler realized that he was a talented person and turned him loose," said Dennis J. Weidenaar, a former dean of Krannert.[11] That was Weiler's style with all his faculty members.

The mathematical model Bass used to make his forecasts became known as the Bass Model. It caught on around the nation. Computer programs were later written for it.[12]

Another milestone in his career came in 1974 when Bass published a paper stating "that consumers' choices cannot be predicted easily but that they vary depending on mood and many other factors." This went against the prevailing theory that marketing could predict "with certainty" what individuals would do.[13]

Bass, the Loeb Distinguished Professor of Management at Purdue, left for his home state and the University of Texas at Dallas in 1982. He died December 1, 2006.

Bass's son, Douglas, said his father "took marketing, which was really like a black art when he came along, and basically transformed it almost single-handedly."[14]

It was exactly what Weiler hoped he would do.

Scott and Nikki Niswonger
*(Photo by Mark Simons/
Purdue University)*

Scott Niswonger grew up in Van Wert, Ohio, took his first flight lesson at twelve, and had a pilot's license before he could drive a car. His love of flying led him to Purdue.

In 1981 Niswonger created Landair Transport, a high-service trucking operation for the air cargo industry. Recently he sold Landair to Covenant Transport Group. Proceeds will support the Niswonger Foundation. He is also chairman emeritus of Forward Air (NASDAQ FWRD), a public company he founded in 1990. Today, those two businesses have annual sales exceeding $1.4 billion. Scott has lived a life of incredible philanthropy. He also chairs the Niswonger Children's Hospital.

It all started when he was five years old in Van Wert, near the Indiana border. "I was riding a tricycle and watching in the sky as DC-6 airplanes landed at Baer Field in Fort Wayne," he said. "I decided that's what I wanted to do.[15]

"When I was growing up my mom knew where to find me," he said. "I was at the local airport begging to wash an airplane for a flight." He soloed at sixteen and had his private license at seventeen.[16]

Niswonger never knew his father. His parents divorced when he was very young and his father died soon after. The father figure in his life was his maternal grandfather, a magistrate and county employee in Van Wert. Niswonger's mother worked as a bank teller.

In 1965 Niswonger graduated from high school. It was an exciting time at Purdue with Aviation Technology, already emerging as a leader in flight programs and providing students with many flying experiences.

Niswonger said he was an average high school student. His grandfather took him to an interview at Purdue with Professor Charlie Holleman, who was then an instructor in professional pilot technology. Holleman saw Scott's potential and Niswonger was admitted to Purdue. He graduated in 1968 with an associate's degree, later earning a bachelor's degree from Tusculum College. Since then he has been awarded four honorary doctoral degrees—from Purdue, the University of Tennessee, Tusculum University, and East Tennessee State University, where he is chairman of the Board of Trustees.

After Purdue, Niswonger moved to Greenville, Tennessee, and flew as a corporate pilot for the Magnavox Company until he saw an opportunity to launch his own air cargo business. Within five years of graduation he founded General Aviation, an airline with several hundred employees flying cargo every night. With the deregulation of surface transportation in 1980, he entered the trucking business.

"Purdue means everything to me," Niswonger said. "The wonderful professors and instructors we had, like Charlie Holleman, made a real difference in my life."[17]

Niswonger has given generously to Purdue. Two buildings are named for him: the Niswonger Aviation Technology Building and the Holleman-Niswonger flight simulation facility.

Part Three

1972 to 2018

29

Arthur Hansen:
The Students' President

1970s

Walking with him on the way to a meeting across campus
was a remarkable experience. He was the Pied Piper of
Purdue. Students were attracted to him like a magnet.

—Betty M. Nelson, associate dean of students in the 1970s

In the fall of 1946, twenty-one-year-old Arthur Hansen began graduate studies in mathematics at Purdue University. He earned money by joining a labor union and helping to build temporary Quonset huts on campus to help with the huge influx of GI Bill students. Twenty-five years later Hansen returned to Purdue—this time as president of the University. Some of the temporary Quonset huts were still there.

Described as erudite, friendly, and supportive, in 1971 Hansen became the first alumnus to be elected president of Purdue. He quickly became known as the "students' president." While other college administrators could not understand the baby boom generation and the campus unrest they created, Hansen spoke at length about the social changes that were taking place and the forces behind it. He not only opened his office to students, he opened his home to them, and his calming presence ultimately quieted the West Lafayette campus. Students still protested the Vietnam War and fought against rules they thought were archaic. But students dealt with issues more positively through an administration that was very open, understood them, and listened, and during the 1970s the turmoil of the 1960s gradually drifted away.

"President Hovde didn't understand all the student unrest," said Carolyn Gery, who went to work in Purdue University fund-raising in 1972. "He had a difficult time accepting that people were challenging and showing disrespect for the institution that meant so much to him. Hansen had an easier time understanding the students. He was amenable to sitting down and talking with them."[1]

Hansen was a renaissance man. He looked young. He was handsome. He smoked a pipe. He received his bachelor's degree in electrical engineering and his master's and PhD in mathematics. But his hobby was philosophy.

President Arthur Hansen at commencement wearing Oxford University cap originally worn by President Frederick L. Hovde.

He was born February 28, 1925, in Sturgeon Bay, Wisconsin. The family moved to Green Bay after the start of the Great Depression in 1929. His father worked at a paper mill and later owned and operated a neighborhood grocery store where Hansen helped stock shelves and delivered food. He picked fruits, had a large newspaper route, and enjoyed hiking, fishing, and hunting—activities he pursued his entire life. His mother died when he was eight years old and he was raised by his father, whose education had stopped at the eighth grade. His father read to him every evening, and a teacher who challenged him sparked his interest in math.[2]

The class valedictorian, he graduated high school in 1943 during World War II. Knowing he would quickly be drafted, he came across a Marine Corps recruiting office and liked the uniforms he saw in photos in the window. He enlisted and was sent to a V-12 program at Purdue to study engineering. He knew nothing about Purdue.[3]

After receiving his bachelor's degree, Hansen started work on his master's degree at Purdue in the fall of 1946. With the war over, he was reassigned to the Marine Corps Reserves. At Purdue he was a teaching assistant in mathematics, and he was quite surprised when Nancy Tucker, a freshman he had already started dating, walked into his class the first day of school. In a short time he recommended she drop the course "because I wasn't getting it down too well," Tucker recalled years later. She stuck it out. Hansen gave her a C.[4]

Tucker left Purdue and finished her bachelor's and master's degrees at Butler University in Indianapolis. Hansen remained on campus. He met and married Margaret Kuehl, who was a dietician in the Women's Residence Halls. The couple had two daughters and three sons. He finished his master's degree in 1948 and before leaving campus went to the Placement Office, where he met a recruiter for the National Advisory Committee for Aeronautics (NACA)—the predecessor of NASA. Hansen spent ten years at NACA and midway through began work on his PhD at what is now Case Western Reserve University.

In 1958 he became an associate professor of mechanical engineering at the University of Michigan. From 1964 to 1966 he was department chair. By 1966 he was at Georgia Institute of Technology in Atlanta, first as dean of engineering, then in 1969 as president.

When his children look back at that period of his career, they most remember his heart for social justice. He picketed a barbershop that refused to cut the hair of African Americans.[5] He wrote newspaper letters to the editor denouncing McCarthyism. For six months he taught at Tuskegee Institute (now University), an African American school in Alabama. There he put his children into the family car, drove them to the civil rights march from Selma to Montgomery, and parked close enough that they could hear racial slurs coming from people who lined the street. "He told all of us kids, some day you will remember how important this is," his son Geoff said. Geoff also remembered his father crying after National Guardsmen at Kent State University shot four students in 1969. "He just couldn't understand how something like that could happen on a university campus," Geoff said.[6]

On June 7, 1970, when Hansen received an honorary doctorate in engineering from Purdue, he was already on the University's shortlist to replace Hovde, who would retire the following year. While on campus Hansen talked with Maurice Knoy, president of the board of trustees. Students, faculty, and the board had all compiled lists of people worthy of consideration. Hansen was on all the lists.

Knoy personally called about twenty-five of the top candidates. Hansen's phone in Atlanta rang in early 1971 and when he answered, it was Knoy, who wanted to know again if he was interested in the Purdue presidency. They talked. Hansen said no. It had been too soon after starting his George Tech position. Several weeks later Knoy's phone rang late in the evening. It was Hansen: "Is the Purdue presidency still open?"[7]

On April 27 the board announced that Hansen would be the next president. The offer was approved at their meeting on June 12, 1971. On April 28 the *Atlanta Constitution* editorialized: "Dr. Arthur G. Hansen is leaving Georgia Tech and we are sad to see him go. As President of that institution he deserves special praise for holding the university together during those times when other campuses seemed more intrigued with shrill confrontations than calm collective reasoning. . . . Although Dr.

Hansen has only been at Tech since 1966, he leaves a warmth and respect that will long be remembered by the Tech community, the city and the state."[8]

Hansen came to Purdue alone. He and Margaret divorced shortly before he accepted the position.[9] When he was named Purdue's president, he received many congratulatory notes. Among them was one from Carolyn Burres, who was Nancy Tucker's sister. "How is Nancy?" Hansen responded. "Tell her I have by no means forgotten her. She still has a soft spot in my heart." Nancy was teaching in the Philippines at the time. She and Hansen started a correspondence by mail that lasted eleven months before they finally met again. Hansen announced their engagement four weeks later. Nancy said she knew they would marry as soon as they began corresponding a year earlier. "It was just one of those things you know," she said. "It took him awhile, though."[10] Hansen and Tucker were married on July 26, 1972. She took a very active role in the University.[11]

The Hansens opened a new presidential home at Purdue—Westwood on McCormick Road at the west edge of campus. In June 1971, Purdue vice president and treasurer R. B. Stewart and his wife, Lillian, had given the home to the University. It underwent a major remodeling, and the Hansens lived in the Seventh Street presidential home formerly used by Edward C. Elliott and Hovde until work at Westwood was completed.

In his book on Purdue history, Robert Topping said, "Over the next ten years, Westwood was not only the presidential manse, but also a sort of second Union Building. The Hansens opened it to students." Organizations met in the presidential home, and students gathered there for wiener roasts or to join the Hansens watching an out-of-town Boilermaker basketball game while munching on popcorn.[12]

In addition to students, the Hansens invited faculty and staff groups to events in their home and even offered tours. An *Indianapolis Star* reporter touring the home asked Nancy Hansen who owned the furniture. "Anything you can sit on or eat from belongs to the University," she said. "Anything pretty, weird, or moveable belongs to us."[13]

Purdue president Arthur Hansen and Nancy Hansen at the front door of Westwood.

At his office Hansen had an open door policy and liked to lunch with students, sometimes helping with their homework—especially if it was engineering, math, or philosophy. As he walked through the campus he stopped and talked with students, introducing himself and asking them questions. And he listened to what they had to say.

Betty M. Nelson, who served as assistant dean of students and then associate dean of students during the Hansen years and was later named dean of students, said he was the perfect choice for president of Purdue at that time. "It was popular at the time for students to assert that 'if you are over thirty you can't understand me,'" she said in a 2010 interview. "President Hansen's personality was so open, positive and inclusive that he was disarming. Even students who were being very cranky found it difficult to sustain that demeanor with him. Walking with him or behind him on the way to a meeting across campus was a remarkable experience. He was the Pied Piper of Purdue. Students were attracted to him like a magnet. His easy gait, ready smile, and genuine interest in students made him an attractive companion for a walk between classes."[14]

President Arthur Hansen talks with students on campus.

D. William Moreau Jr. was a freshman when Hansen was named president, and two years later Moreau was elected editor of the *Exponent*. He became a teaching assistant and graduate student at Purdue in 1975–1976 and served on the Purdue University Board of Trustees from September 1991 to June 2003. "Art has rightly gone down in Purdue history as the 'students' president,'" Moreau said. "He was only forty-six when he arrived, full of physical and intellectual energy. His first convocation for freshmen featured an admonition to relax and enjoy themselves. President Hovde would tell the freshman convo attendees to say hello to the guy on the right and the girl on the left. Only one of three would graduate!"[15]

Hansen did not shy away from taking stands on controversial issues. In the spring of 1974, the *Debris* yearbook hit the streets with a hammer and sickle on the cover. It made national news, and an Indiana state senator accused Hansen of "not having the wisdom to recognize the infiltration of a great university by Communists." The state senator called for an investigation by the FBI.

"Art had lived through and spoken out against the Red Scare of the 1950s," Moreau said. "His response was, 'This takes me back to the days when Senator McCarthy reigned supreme.'" The *Debris* editor was an industrial engineering student from Indianapolis and a Republican, Moreau said, who simply wanted to increase sales. "Only Art and a handful of others knew that the student was a finalist for the most coveted award a graduating senior man could receive, the G. A. Ross Award," Moreau said. "Art could have given the award to someone else, and no one would have been the wiser." But the honor went to the *Debris* editor.[16]

In the summer of 1973, in advance of the regular *Exponent* back-to-class edition, Hansen sat down with Moreau for a taped—and then transcribed—interview that spanned three editions of the newspaper. "He insisted we do the interview at my house, and he arrived at my broken-down apartment with a six-pack of beer," Moreau said. "We smoked cigarettes and drank beer for two hours."[17]

Moreau did learn, though, that Hansen's tolerance had limits. When he wrote an article for the newspaper that criticized Nancy Hansen, Moreau's telephone rang at 5:30 a.m.—not an hour when most university students are awake. But it is an hour when a university president awakes and reads the newspaper. Moreau put the phone to his ear: "Bill, Art Hansen here. Write anything you want about me, but leave Nancy alone." He hung up. "We did leave Nancy alone after that," Moreau said. "Believe me, we did."[18]

Hansen had disagreements with the *Exponent*, but in 1974 when the newspaper was in financial difficulties, he agreed to increase the University's advertising without asking for any editorial control in return. "Art was a personal hero," Moreau said. "I miss him to this day."[19]

Hansen relished good-natured humor. Purdue's annual Ag Alumni Fish Fry had been ongoing since the 1920s. Mauri Williamson, a 1950 Purdue graduate, was executive secretary of the Ag Alumni Association from 1953 to 1990. Williamson loved to poke fun at people and institutions, and the Fish Fry became a time for everyone to enjoy running gags and good-natured roasts of individuals—including Williamson, who often dressed as a woman for the event that drew huge crowds that filled the Armory.

Williamson was also very active in the Indiana State Fair. In 1961 he founded the Pioneer Village at the fair, allowing people to see and experience what farm life was like in the 1800s. Year after year he was present at the village in his blue coveralls and red checked shirt, doing what he loved—talking to people about agriculture.

Hansen played along with the gags at the Ag Alumni Fish Fries. One year he was Dirty Harry in a Wild West Show. Another time he was the Great Hansini with a circus theme, and Mark Twain for a riverboat program. The event sometimes drew a governor, other state officials, and U.S. senators and congressmen. Indiana University president John Ryan always supported a close relationship between the

Student athlete Larry Burton, winner of the Intercollegiate Conference Medal of Honor, talks with Purdue president Arthur Hansen, May 1975.

universities, and he took part in the fish fries. When a critic said Hansen was trying to create a Camelot atmosphere at Purdue, Hansen dressed as King Arthur and rode into the fish fry on a huge Belgian draft horse, followed by "King" Ryan, who rode in on a donkey. Hansen had never before ridden on a horse, and Ryan had probably never ridden on a donkey.

But Hansen's main focus was on accomplishing directives he had received from the board of trustees. The trustees did not want the University to continue the days of student unrest, protest, and constant confrontations with the administration. They wanted to improve the morale at Purdue. The board also wanted to move forward and find a new source of revenue that would be vital to Purdue's future. In 1975 the Indiana General Assembly gave a boost to student concerns when it approved a requirement that one of the ten members of state university boards of trustees had to be a student with full voting rights.

John Hicks, who was special assistant to both Hovde and Hansen, said the two men were very different. In a 1972 interview he said:

President Hansen probably leans a bit more in the direction of the general faculty having a larger role than President Hovde. President Hovde during the last several years of his presidency kept the [faculty] Senate very much at his arm's length. He chose not to attend Senate meetings. He chose to have as little to do with the Senate as he could. I think he had real purpose in this. He did not want to see the Senate become a particularly powerful body. . . . Hovde did not enjoy many aspects of the external roles of the president and he didn't do that as well as other things.[20]

Under Hovde and previous presidents the faculty reported to the deans and the deans reported directly to the president. Hansen changed that. On July 14 Hansen named Harold "Cotton" Robinson to a new position—provost. A native of North Carolina, Robinson was a vice chancellor of the University System of Georgia before coming to Purdue. Only two people reported directly to Hansen: the provost, Robinson, and the University vice president and treasurer, who was Lytle Freehafer. Unfortunately, disagreements quickly erupted between Robinson and Freehafer.

By the end of 1973, Freehafer took early retirement. Robinson left in 1974 to become chancellor of Western Carolina University in Cullowhee, North Carolina. Frederick R. Ford, a native of Kentland, Indiana, and a Purdue graduate, replaced Freehafer. In addition to his undergraduate degree, Ford had a PhD in management from Purdue. Robinson was replaced as provost by Felix Haas, a native of Austria, a Massachusetts Institute of Technology graduate, and a mathematician who had been dean of the School of Science since 1962.

Early in Hansen's term medical education returned to Purdue for the first time since 1907. After former president Winthrop Stone dropped Purdue's program, Indiana University had the only school of medicine in the state. By the 1970s there was political pressure to establish another school of medicine to meet the need for more doctors, especially outside of Indianapolis. As an alternative, IU launched a system of regional, or satellite, two-year medical school programs on eight campuses around the state, including West Lafayette. Among the key figures in creating the new locations was a man who would become the next president of Purdue—Steven Beering. During the twenty-first century the regional medical schools became four-year programs.

Many other changes came during the Hansen years, including merging the dean of men and dean of women into a single dean of students office. Purdue Dean of Women Beverley Stone and others in her office fought the consolidation. When it had taken place at other schools, a man had always been chosen as the new dean of students and the former dean of women was placed in a subordinate role. Stone was surprised when vice president for student services Bill Fischang named her to the new position. Purdue was the last Big Ten university to merge the two offices, but it was the first to name a woman dean of students.

Stone wasn't sure she wanted the job. She had another offer. But friends encouraged her, including Purdue professor of physics Anna Akeley, who told her, "you could not let the women at Purdue down. The men would have said, 'See, we give women a chance and they won't take it.'"[21] As a girl in Europe, Akeley had met Kaiser Wilhelm of Germany and Tsar Nicholas of Russia. She was Jewish and escaped from Vienna during World War II, arriving in West Lafayette in 1942. She did it by traveling west through Russia, Korea, and Japan.

"President Hansen supported the Dean of Students in decisions that were important 'firsts' for our campus—decisions that confirmed the rights of students in protected categories," Nelson said. "He knew he would receive vigorous criticism for some, but he also knew the decisions were the right ones."[22]

It was Hansen who approved the formation of the first gay student organization on campus. When the proposal was brought to him, Hansen articulated all the pros and cons for recognition. The next day he said, "Aw shucks . . . of course Purdue has to recognize this group."[23]

In 1980 Stone retired as dean of students. She was replaced by her colleague Barbara Cook, continuing a female line from the first part-time dean of women, Carolyn Shoemaker, in 1913, through Stratton and Schleman and Stone.

Cook was born in Memphis, Tennessee. She graduated from the University of Arkansas in 1951. In 1954 she received a master's degree in student personnel from Syracuse University and then worked as assistant dean of women and director of women's residence halls at Arkansas, where she became friends with Beverley Stone. She joined the Office of the Dean of Women at Purdue in 1956.

Among the first challenges Cook faced was an on-campus housing shortage. When she started as dean, about 45 percent of the thirty-two thousand students lived off campus. "It used to be that off-campus students lived that way by choice," she said. "But now, it's not by choice. Many are lonely, not part of the campus environment and have no one to relate to."[24]

There were lean financial times for the University in the 1970s, and Hansen immediately moved into another area in which the board of trustees had told him to concentrate: fund-raising, called development. In 1972 he appointed Purdue alumnus Stanley Hall, of New York, to head a new development area. Hall had graduated in engineering.

"President Hovde was very hesitant to get deeply involved in private fund-raising," Hicks said in his 1972 interview. "At that time I think he probably didn't really need it."[25]

In 1972 Hall hired Carolyn Gery. "President Hansen came in with a directive to start real fund-raising," Gery said. "We were ten years behind other Big Ten universities." There was some resistance to it at first, but success brought acceptance and eventually enthusiasm.[26]

An annual student Phonathon was launched. A Purdue Annual Fund was created along with a Parents' Association. One of the biggest steps was the President's Council, created in 1972. A minimum gift to the University of $1,000 per year was required to join the organization. As of the University's sesquicentennial, the amount had never been increased. Gery was director of the President's Council from the beginning.

President's Council events were held on campus. Hansen and other administrators, trustees, and faculty attended, giving donors an opportunity to meet and get to know the people who ran the University.

Among the biggest events were football pregame programs launched in 1973. President's Council members gathered before the game, ate, and talked with Hansen. The Glee Club or another Purdue Musical Organization group sang. Hansen traveled, bringing the campus to alumni in their hometowns. "President Hansen, being the wonderful, bright, caring guy he was—people loved him," Gery said. "We traveled around the country, even the world, often piggybacking with the Alumni Association." It was a huge success.[27]

Attendance at football pregame events was helped by the success of the team under Coach Jim Young. He was coach from December of 1976 through 1981 and as the number of bowl games sponsored around the nation increased, Young took the Boilermakers to postseason games for the first time since the 1967 Rose Bowl.

A standout freshman from Carmel, Indiana, north of Indianapolis, arrived on the Purdue campus in the fall of 1977. His name was Mark Herrmann. Young didn't want to start a freshman at quarterback the first game of the season. But Herrmann took over in the first half and went on to start forty-five of the next forty-six games. An injury kept him out for one game. During his career Herrmann led Purdue to victories in three bowl games—the Peach Bowl, the Bluebonnet Bowl, and the Liberty Bowl. He threw for 9,946 yards, an NCAA record. He was elected into the College Football Hall of Fame in 2010.

In basketball, Gene Keady came to Purdue in 1980. In a career that lasted until 2005, he had more victories than any other Purdue coach—493. He had the second most victories among Big Ten basketball coaches, with Indiana's Bobby Knight taking the top spot. Keady beat Knight in Purdue–IU games, 21–20. Keady led the Boilermakers to six Big Ten Championships and eighteen NCAA Tournament appearances. In 1978 he was named Big Ten Coach of the Year.

At the end of the 1970s, Hansen launched a Plan for the Eighties Fund Drive and named John Day, former dean of the Krannert School of Management, as vice president for development. Hansen talked about the plan during the campaign kickoff in March 1980.

"Fundamentally, our efforts in the 1980s will concentrate on pure research, applied research, and people," he said. "We will need special buildings, equipment, educational facilities, and new faculty. This is where we will need your help and support. . . . Our estimate for what we wish to achieve from private support is $34.2 million. This is not a great goal, but it is the largest ever set for Purdue and will take hard work to reach."[28]

They reached it. And more.

GIANT LEAPS IN RESEARCH

Seymour Benzer

S eymour Benzer was like every other child in Bensonhurst, Brooklyn, in the 1920s. But everything changed for the son of Polish Jewish immigrants after his Bar Mitzvah when he was given a microscope. "And that opened up the whole world," he said.[29]

Benzer, who received his PhD, taught, and researched at Purdue University, became a founder of modern behavioral genetics, revolutionizing our understanding of genes and their effect on behavior. His work led to speculation concerning whether aging can be manipulated.[30]

As a boy with his new microscope, Benzer retreated to the family basement, where he dissected flies. He read books about atomic physics during synagogue. At the age of fifteen he graduated from high school and studied physics at Brooklyn College on a scholarship. In 1942 he enrolled at Purdue to work on a PhD. Benzer joined the Purdue faculty but drifted from physics to genetics.

During the 1950s and 1960s, thanks to Benzer, Purdue was the site of some of the most significant findings in the new field of microbial genetics. Benzer devised a plan to map mutations in a gene, nucleotide by nucleotide. "This brilliant, yet simple, system made Benzer a star in research and brought much renown to the campus," said Professor Louis Sherman of Purdue Biological Sciences.[31]

Sherman said Benzer's research helped lead to the formation of Purdue's Department of Biological Sciences.

Benzer left Purdue for Cal Tech in 1967. But Sherman said his legacy remained, and Purdue is considered a significant institution in the development of microbial genetics.

In awarding him its $500,000 2004 Neuroscience Prize, the Gruber Foundation stated: Benzer is credited with demonstrating that a gene can be split into hundreds of components, each able to mutate. . . . He is recognized as a pioneer in the field of molecular biology. In the mid-1960s he moved into neurobiology and began to concentrate on behavioral genetics, beginning a lasting and spectacularly successful affiliation with an unpresuming ally: the fruit fly. . . . By painstakingly altering genes in individual [fruit flies], creating mutants, and observing how these changes affected behavior, he and his students not only established the molecular component in behavior, they identified numerous genes with specific functions, including genes involved in memory and learning, in vision, in sexual conduct, in determining time periods of activity and in aging. They . . . inspired myriad researchers to use the fly to study everything from alcoholism to the effects of psychotropic drugs.[32]

Benzer died in 2007 at the age of eighty-six.

H. Edwin Umbarger

H. Edwin Umbarger grew up in Mansfield, Ohio, fascinated by geography, history, and archeology. Although his parents never went to college, he planned to pursue an education majoring in Latin and Greek and to go on for a PhD in archeology.

Umbarger ended up a pioneer during the early days of molecular biology thanks to a high school Latin teacher who forced him to take biology, chemistry, and physics. He later observed that the choice of majors was obvious to him. "It was only in biology and chemistry that I received A's," he said.[33]

Following the success of Seymour Benzer's research and the creation of the Purdue Department of Biological Sciences, there was an explosion of faculty hiring in the field on the West Lafayette campus. Among those recruited to Purdue was Umbarger in 1964.

He had graduated from Ohio University in 1943, served in the U.S. Navy during the final years of World War II, and gone to Harvard University for graduate school, receiving his PhD in 1950. For the next nine years he taught and researched in the Harvard Medical School Department of Bacteriology and Immunology.

Umbarger left Harvard in 1959, worked in England, and then was appointed senior staff investigator at the Cold Spring Harbor Biological Laboratory in Long Island, New York. In 1964 he accepted a full professorship at Purdue and remained there the rest of his career. In 1970 he was named the Wright Distinguished Professor of Biological Sciences.

It was an exciting time in his field. A worldwide race was underway to learn how the main constituents of living cells were made.[34]

Louis Sherman, a professor in the Purdue Department of Biological Sciences, said research in Umbarger's lab demonstrated the relationship between enzymes and their metabolites (the byproduct of metabolism) in metabolic pathways (a linked series of chemical reactions in a cell). "The work began in bacteria, but was soon replicated in many organisms, including humans," Sherman said. "This work has led to important insights into nutrition and into the way that medically important drugs are metabolized."[35]

Arthur Aronson, who is retired from Purdue Biological Sciences, knew Umbarger. "His major contribution was in elucidating how a certain group of amino acids were synthesized, and most importantly how that synthesis was regulated so that just the right amount was produced for the particular growth conditions of the bacteria," Aronson said. "This regulation is called 'feedback inhibition' and was Umbarger's crowning achievement."[36]

Umbarger died in 1999. He was seventy-eight.

30

Opportunity, Equality, and Respect

1970s

If people, and I mean all people, can get to know a bit more about each other it will go a long way toward mutual understanding and appreciation.

—Singer Buchanan, first head of the Purdue Black Cultural Center

It was an incredible evening.

From his seat in Elliott Hall of Music, Tony Zamora, director of the Black Cultural Center at Purdue University, could turn, look behind him, and see the main floor and two balconies filling with an excited audience. Students, faculty, staff, people from the community, and people from surrounding communities all hurried into the University's signature theater for a Black Cultural Center program.

When the lights went down, a man walked onto the huge stage and up to the podium to roaring applause. Just two years after he had arrived on campus, Zamora had brought together six thousand people of all races to hear a talk by an African American. That had never happened before.

African American singers and musicians had drawn big crowds in previous years. But African American speakers had drawn smaller, mostly African American audiences.

Tony Zamora, director of the
Black Cultural Center.

On the evening of November 20, 1975, Elliott Hall of Music was packed to hear the lecture. The man speaking was Muhammad Ali—"the greatest," the heavyweight boxing champion of the world. Among the topics Ali talked about that evening was friendship. Zamora's center, created to share the African American experience with all people—black and white, students and faculty, Boilermakers and members of the community—was all about creating friendship.

The Black Cultural Center had been established in 1969 and its first building dedicated on December 4, 1970, during the administration of President Frederick Hovde. Singer Buchanan, coordinator of Purdue's black student programs and first head of the center, said: "We want to provide a facility where anyone feels welcome to come in for discussions, readings, social events—or to just sit down and talk for awhile. If people, and I mean all people, can get to know a bit more about each other it will go a long way toward mutual understanding and appreciation."[1]

Buchanan left Purdue in 1972, and direction of the Black Cultural Center was placed in the hands of a PhD graduate student, Johnnie Houston, until the arrival of Zamora in July 1973. Houston put off his studies for one year to lead the center.

Born in Chicago, Zamora moved to Champaign, Illinois, in the 1950s. A widely respected musician, he had a jazz band in Champaign. He lived near campus and interacted with the School of Music and with international students from Africa at the Union. With the help of Illinois students, Zamora created a program to provide free music lessons and instruments for children in the community. The local American Federation of Musicians union provided funds for instruments. As his work became publicly known, he was asked to be director of an African American cultural program at the University of Illinois. He became the second director of the program in 1970. But Zamora believed Illinois, at that time, was not committed to the program and resigned after one year.

Soon after he was contacted by Purdue's Black Cultural Center and hired as director, becoming a respected leader at the University.

Floyd Hayes III, a Purdue professor, said Zamora was "a jazz musician, community and university leader, institution builder, teacher, mentor/motivator, risk taker, colleague and friend." Black cultural centers on university campuses, Hayes said, were new when Zamora arrived. "Without written scripts or scores, Tony employed an historical Africanist tradition of improvisation as a foundation for leading the BCC," Hayes said. "Improvising upon conventional Purdue culture, Tony embraced the university's central mission of educational excellence in developing the Black Cultural Center, emphasizing cultural creativity, risk taking and inclusiveness."[2] The Purdue Center became a model for others around the nation.

Zamora started the Black Voices of Inspiration, a choir that had up to one hundred members. He also created the Jahari Dance Troupe, the New Directional Players, and the Haraka Writers, in addition to hosting many lectures and programs. During his tenure Zamora grew the center's library into a major collection of books, periodicals, and videos with a focus on African American culture and history. He initiated an artist-in-residence program and led an effort to create a University scholarship named for Helen Bass Williams, Purdue's first African American professor.

During the 1970s at Purdue, African Americans were gaining a voice through the Black Cultural Center, and the University had a stronger focus on bringing minority students to the campus. The 1970 *Debris* yearbook reported that 144 African American freshmen had enrolled at Purdue in the fall of 1969—a larger number than in any previous class. That brought the total African American student enrollment to above 500. By 1971 there were 631.

President Arthur Hansen hands flowers to Homecoming
Queen Kassandra Agee in the fall of 1978. She was
Purdue's first African American homecoming queen.

President Arthur Hansen served as chair of the National Research Council's Committee on Minorities in Engineering. In addition to the Black Cultural Center, he supported the Society of Black Engineers at Purdue. In 1971 Purdue undergraduates Edward Barnette and Fred Cooper had an idea for this society to improve recruitment and retention of African American engineers. They received vital help from their academic counselor, Arthur Bond. Without his help the society might not have been possible. In 1974 Anthony Harris, president of the Purdue group, contacted every accredited engineering program in the nation, suggesting the formation of a national organization. It was launched in 1975. The six founding members of the organization from Purdue, now known as the Chicago Six, were Edward Coleman, Anthony Harris, Brian Harris, Stanley Kirtley, John Logan Jr., and George Smith.

"Both Dr. Hansen and his wife, Nancy, were extremely supportive," Zamora said. "And their support came at a critical time when some people did not see the need for a Black Cultural Center on campus."[3]

One day in 1994 Zamora came home from work and told his wife, Betty, he had two things to tell her. The first was that he was going to retire. The second was accomplishment of the goal they had been working toward throughout their years at Purdue.[4] The Black Cultural Center would no longer be housed in a University Street converted home. Purdue planned to build a new building for the center that would advance everything Zamora had been working to achieve. (See part 3, chapter 34.)

The 1970s saw the beginning of more programs focused on underrepresented students. The Purdue University Minority Engineering Program was created in 1974 and focused on recruitment and retention of historically underrepresented students. Marion Williamson Blalock retired as director of the program in August of 2008 after serving Purdue for more than thirty-four years. She transformed it into a national model and mentored thousands of minority students at Purdue and across the country. Blalock assisted more twenty-three hundred students with their engineering degrees.

In 1972 Richard Weaver was hired as director of minority programs in Pharmacy, but the program was discontinued in 1981 when a grant ended. In 1992 under Pharmacy dean Charles "Chip" Rutledge, who supported programs for underrepresented minorities as well as cultural awareness, the Office of Minority Programs was created to increase diversity and enhance success. A major reason for the success of these programs was the director, Jackie Jimerson, who retired from Purdue in 2012 after working thirty-two years supporting many multicultural efforts on campus. The number of minorities who graduated from Pharmacy doubled under her tenure with the Office of Minority Programs.

In the twenty-first century a number of Purdue colleges had programs for underrepresented minorities and women. In addition to the Black Cultural Center, Purdue had an Asian American and Asian Resource and Cultural Center, a Latino Cultural Center, a Native American Educational and Cultural Center, and a Lesbian, Gay, Bisexual, Transgender, and Queer Center.

Title IX Opens Doors for Women

In 1972 when the Higher Education Act of 1965 was up for reauthorization before Congress, a U.S. senator and Purdue alumnus introduced an amendment to it. The thirty-seven-word amendment stated, "No person in the United States shall, on the basis of sex, be excluded from participation in, be denied the benefits of, or be subjected to discrimination under any education program or activity receiving federal financial assistance." It sounded simple. In fact, it was a game changer. Literally. The amendment, known as Title IX, was offered by Birch Bayh, who graduated from Purdue in 1951 with a degree in agriculture. He is called the "Father of Title IX."

The 1964 Civil Rights Act prohibited discrimination based on race, color, religion, sex, or national origin. But even with the act, the number of women teaching at universities lagged far behind men, and there was discrimination against female students in college admissions, organizations, programs, and athletics.

Title IX extended protections for women in all areas of higher education.

Bayh had help learning about discrimination against women from his wife, Marvella. She had been denied admission to the University of Virginia based on her gender.[5] "As we went through life together, from time to time she would remind me what it was like to be a woman in a man's world," Bayh said in 2014.[6]

On the Senate floor Bayh called the Title IX Amendment "an important first step in the effort to provide for the women of America something that is rightfully theirs—an equal chance to attend the schools of their choice, to develop the skills they want, and to apply those skills with the knowledge that they will have a fair chance to secure the jobs of their choice with equal pay for equal work."[7]

Passage of Title IX was not the end. Regulations had to be formulated, and that work continues into the twenty-first century. Kate Cruikshank of Indiana University, where Bayh's papers are archived, wrote that after passage of the amendment: "Bayh rode herd on the Department of Health, Education, and Welfare to get regulations formulated that carried out the legislative intent of eliminating discrimination in higher education on the basis of sex."[8]

Title XI launched athletic intercollegiate competition for women, and much of the initial resistance came from men. A representative of Purdue athletics told Bayh that Title IX would ruin the football program. It did not.

Athletic opportunities are what the public noticed most. But Title IX was focused on giving women equal opportunities in academics and participation in leadership organizations. Betty M. Nelson, an assistant and associate dean of students in the 1970s who became dean of students in 1987, said that without Title IX she would not have lived long enough to see the changes that took place as a result of the act.[9]

Nelson said the dean of students office worked to help implement Title IX at Purdue, and it had a large impact on campus organizations and academics.

Title IX opened previously all-male Purdue organizations such as Iron Key to women. It also opened the previously all-female Mortar Board to males. Some thought Mortar Board might be discontinued nationally after Title IX, but Dean of Students Barbara Cook fought successfully to keep it going for male and female students. The Purdue Mortar Board chapter is named for Cook.

In 1981, twelve hundred women were enrolled in Purdue engineering. Ten years earlier there had only been about two hundred. A major factor in this was the Purdue Women in Engineering Program, created in 1969 during the Hovde years and supported by Hansen. It was the first of its kind in the nation and served as a model for other universities, recruiting women and helping them succeed.

Indiana senator Birch Bayh, a 1951 Purdue graduate in agriculture, exercises in the Purdue Fieldhouse with female student athletes. Bayh was the "Father of Title IX," which gave women equal opportunities in public education, including athletics. This photo is part of the Birch Bayh Senatorial Papers, Modern Political Papers Collections, Indiana University Libraries.

Bayh served three terms in the U.S. Senate. Growing up, he always wanted to make a difference. He succeeded. "I guess those thirty-seven words made a difference beyond my wildest dreams," he said.[10]

A. Leon Higginbotham Returns to Purdue

As the decade of the 1970s drew near to a close, there had been many changes in American society and at Purdue.

When he left Purdue after being refused the right to live in a University residence hall in 1944, A. Leon Higginbotham vowed never to return, saying the experience would be "too painful, too dreadful." But in November 1978, as a distinguished federal judge, he was invited to Purdue to lecture to African American students. He accepted and spoke about his meeting with then president Edward Elliott that had led him to withdraw from Purdue.[11] (See part 2, chapter 25.)

Six months later he returned to the Purdue campus again. Hansen and the board of trustees had offered him what he had earlier been denied. On May 12, 1979, in Elliott Hall of Music, Hansen placed a hood on Higginbotham's shoulders and presented him with an honorary PhD.

The honorary degree did not symbolize what Higginbotham had been denied at Purdue thirty-five years earlier. He was a brilliant man who could have received a degree at Purdue or any university where he wanted to study. What he received from Purdue in 1979 that he had been denied in 1944 was something much more important. It was a basic human right.

Respect.

Beth Brooke-Marciniak
*(Photo courtesy of Beth
Brooke-Marciniak)*

At the age of thirteen, Beth Brooke-Marciniak injured her hip and a doctor told her she might never walk again.

"I looked at the doctor and said, 'Not only will I walk again, I'm going to be the best athlete you've ever seen,'" Brooke-Marciniak said.[12]

After two surgeries and more than nine months of exercise, she became the Taylor High School (Indiana) MVP in basketball, golf, tennis, and softball. She averaged an amazing double-double (points and rebounds) per game her sophomore, junior, and senior seasons and scored forty-four points and had thirty-four rebounds in one game her senior year. With the last name Millard, she was on the Indiana High School All-Star basketball team. When Indiana University said she was too short, she was among the first women to receive a basketball scholarship to Purdue following passage of Title IX, playing for Coach Ruth Jones. She is in the Indiana Basketball Fall of Fame.

And that's nothing compared to what came next. In a career with EY (Ernst & Young, LLP) that has extended thirty-seven, years Brooke-Marciniak has risen to global vice chair of public policy and is the global sponsor of diversity and inclusion for the company with more than 260,000 employees in more than 150 countries. Its 2018 global revenue approaches $35 billion. As a top executive at EY, *Forbes* named her ten times to its World's 100 Most Powerful Women list. In 2017 she received the NCAA Theodore Roosevelt Award, its highest individual honor, presented for national recognition and outstanding accomplishments. Brooke-Marciniak is married to Michelle Brooke-Marciniak, a former All American collegiate and professional basketball player who led the Tennessee Lady Vols to their fourth national championship and was named MVP in the 1996 Final Four.

Brooke-Marciniak came to Purdue unsure of what she would study. A counselor at Krannert School of Management advised her to double major in industrial management and computer science and to take every accounting course she could find. She did. He was correct.

"Purdue taught me how to think analytically and how to problem-solve," she said. "Many of the accounting majors I've known from other schools were taught rules. They were never taught to think. Purdue was also a cultural fit for me. I was serious about academics and athletics. I was on a mission. I wasn't there for a good time and Purdue was filled with students who were like me. Purdue fit my values and my entire mindset.[13]

"I love Purdue. The further and further I get the more I appreciate the experience I had there."[14]

Carol Pottenger
(Photo courtesy of Carol Pottenger)

Looking back on her thirty-six-year military career, Admiral Carol Pottenger said, "I was always proud I was able to rise to the occasion and do what the navy needed and expected me to do."[15] It wasn't a boastful statement. It was an understatement. She had a career filled with firsts.

A 1977 Purdue University graduate, Pottenger was among the first women the U.S. Navy assigned to sea in 1978. She was the first female admiral to command a major combat strike group and the first female admiral to lead a combatant force–type command charged with the manning, training, and equipping of more than forty thousand expeditionary sailors in preparation for deployments to Iraq and Afghanistan, as well as global security assistance operations. Her final navy post was with NATO as deputy chief of staff for capability development at Supreme Allied Commander Transformation, Norfolk, Virginia. She was the first female officer to hold that position.

Pottenger said the navy has blinders when it comes to gender and she was always encouraged to use her full potential. "I had outstanding mentors and they were all men," she said. "Those men—those leaders—opened doors of opportunity for me. Now in turn I am helping others."[16]

Born in Oak Park, Illinois, Pottenger grew up in Pittsburgh, Pennsylvania. Her family moved to St. Petersburg, Florida, in her senior year of high school. It was there that she determined she wanted a navy career, and from that point deciding to attend Purdue was easy.

At that time, Purdue was one of only four navy ROTC programs in the nation that accepted women. "Purdue was leading the way, as it has in so many things," Pottenger said.[17]

At first she thought she might be a navy nurse. But the Naval ROTC program opened her eyes to more possibilities as she learned about ship navigation, engineering, and leadership. She majored in history and has retained a deep interest in the subject to this day.

In 1976 Pottenger was selected as the Purdue Naval ROTC Midshipman of the Year. Tradition called for presenting the recipient with a navy sword. But navy regulations at that time did not permit women to wear a sword in uniform. Purdue Naval ROTC considered giving her a ship's chronometer. She told them she wanted the sword—which women were later approved to carry. She wore it during several ceremonies for changes of command and at her retirement in 2013.

Purdue has awarded Pottenger an honorary PhD and she is in the ROTC Hall of Fame. "Purdue gave me a fabulous grounding in academic, social, and ethical behavior," Pottenger said. "Purdue has always been in my heart."[18]

GIANT LEAPS IN LIFE

Anthony Harris
*(Photo courtesy of
Anthony Harris)*

Anthony Harris received his Purdue University bachelor's degree in mechanical engineering in 1975. He earned an MBA from Harvard and was named a Purdue Outstanding Mechanical Engineer and a Distinguished Engineering Alumnus.

He is president and CEO of Campbell/Harris Security Equipment Company in Alameda, California, a manufacturer of contraband, explosives and, "dirty bomb" detection equipment with sales in seventy countries.

He's come a long way in life from his childhood on South Side of Chicago.

"I was born in the 'hood,'" Harris said. "It was a tough area. We were surrounded by gangs."[19]

His father worked as a longshoreman, and while his parents did not even finish high school, they told him from the time he was able to understand that was expected to go to college.

He was placed in a Chicago magnet high school for college-bound students. But he had to hide his academic interests in the neighborhood.

"I had two sets of books," he said. "I kept one set at home and the other at school so no one would see me as a nerdy kid walking around with books."[20]

In 1971 Purdue offered Harris financial support to enroll in engineering. In his sophomore year he won a Shell Oil fellowship.

Coming from an African American community, Purdue was a "culture shock," Harris said. But he found support in the Black Cultural Center, and the School of Engineering provided courses to help students succeed.

Still, "it was a tough time," he said. "There were very few blacks admitted and fewer matriculating. There were twenty-five of us in my freshman engineering class. Only three of us graduated on time."[21]

During the 1960s, 80 percent of African Americans in Purdue freshman engineering dropped out.[22] In 1975, at the suggestion of two students, Edward Barnette and Fred Cooper, and the help of Academic Advisor Arthur J. Bond, the university launched the Black Society of Engineers to help with recruitment and retention of African Americans. Harris became president of the group and renamed it the Society of Black Engineers.

He then wrote a letter to the nation's 288 accredited engineering programs and proposed a national organization. The first meeting was held in April of 1975 and the National Society of Black Engineers was born.

Today it has more than seventeen thousand members.

"I didn't understand the true impact of Purdue until after I graduated and realized the breadth and scope of my experience was better than my peers," Harris said. "Purdue changed me a bit. I changed Purdue a bit. And we're both better off."[23]

31

Purdue Celebrates a Nobel Prize

1979 to 1982

[West Lafayette is] an ideal climate for students.
There's so little to [distract] them from studies.

—PROFESSOR HERBERT BROWN

When he was a student at the University of Chicago, Herbert Brown's girlfriend wrote a message in his yearbook: "To the future Nobel winner."

Some forty years later the Purdue University professor was summoned to Stockholm to accept the most prestigious award in science. He was the co-recipient in chemistry. His girlfriend had long before become his wife, so Sarah Brown wasn't the least bit surprised when the early-morning call came from the Nobel Committee on October 15, 1979. "It took awhile. But it happened," she said matter-of-factly.[1]

Brown had first become interested in his Nobel-winning research focusing on the chemical element boron when Sarah gave him a $2 book on the subject as a graduation gift. "I still have it," Brown said about the book when he appeared at a news conference with his wife beside him. "You should [still have it]," she said. "I suspect it cost me five days' lunch."[2]

Brown, the Wetherill Research Professor Emeritus in Chemistry, became the first current member of the Purdue faculty to receive a Nobel Prize.[3] His professorship was named for the same Richard Wetherill, a Lafayette physician and philanthropist, who had enrolled in Purdue's preparatory school in its first years of existence.

Clearly pleased and honored by the prize, Brown added that it was not his life's goal. "The greatest joy is enjoying yourself in your work, having fun, helping to solve the world's problems through research," he said. "If awards and honors come, that's great, they're extra gravy."[4]

Purdue president Arthur Hansen shared in the excitement. "I am absolutely delighted," he said of the award. "The result of his basic research has had tremendous impact on the entire field of applied chemistry. [It is] tangible evidence of the importance of basic research being done in a research-oriented campus like Purdue."[5]

Brown was born to Ukrainian Jewish immigrants in London, England, and was a University of Chicago alumnus. He was an instructor at the University of Chicago before moving to Wayne State University in Detroit. In 1939 he received a government request to explore new volatile compounds of uranium. It would become part of the research into the development of a nuclear bomb. "We were not advised what these were to be used for, but we were told it was important for the national defense," Brown said in a 1970 interview. His group discovered a new compound, uranium borohydride—a backup for uranium hexafluoride used in the separation of uranium isotopes. It was never needed in development of the nuclear bomb but it did find its way into pharmaceutical research.[6]

"I think it is an interesting point," Brown said, "that research that was carried out to find a material for use in [a nuclear bomb] would have found its major application in the pharmaceutical industry for the manufacture of drugs to alleviate human suffering."[7]

He was recruited to the Purdue faculty from Wayne State University in Detroit.

When I came to West Lafayette and saw what a lovely little community this was, without the problems of the large city I decided that for my family's sake it would be better to move [here]. Many people have indicated that they consider Lafayette to be a relatively quiet—perhaps dead—place without a great many distractions. But I found that to be very pleasant. I think it's an ideal climate for students—there's so little to [distract] them from studies. The time of graduate work should be a time when a man tests himself in his ability to produce new original things. He has his best opportunity for demonstrating this in a place like Lafayette where he can concentrate without being continually distracted by either physical things, such as boating on the ocean, or swimming, or hiking, or political agitation. I personally believe that Lafayette offers students an ideal environment to develop themselves.[8]

Brown was one of many successful researchers on campus during the Hansen years. In the 1990s Hansen was asked what had been his proudest accomplishment at Purdue. He answered immediately. "Buying Michael Rossmann that supercomputer," he said. It was the first supercomputer at Purdue, and Rossmann put it to great use.[9]

Rossmann heads a team of biologists who are unraveling the secrets of viruses. He and his team were the first to solve the structure of a common cold virus, and subsequently made other breakthroughs. Purdue professor Richard Kuhn and Rossmann have studied flaviviruses for nearly twenty years. They were the first to map the structure of any flavivirus when they determined the dengue virus structure in 2002. In 2003 they were first to determine the structure of West Nile virus and also the first to do so with the Zika virus in 2016. Once the structure of a virus is known, researchers can identify sites on the virus where antibodies can attach and disable it. The research could lead to the development of antiviral drugs.

Everything was going particularly well for Hansen as the decade of the 1970s came to a close. His fund-raising campaign, the Plan for the Eighties, would reach $40.7 million. The campaign ended in the spring. Hansen had accomplished the goals the trustees had set out for him when he arrived on campus ten years earlier.

The success of the campaign was overshadowed by a surprise announcement in November 1981. Immediately after a board of trustees meeting, the University held a news conference. Hansen announced that he was retiring from Purdue.

He told incoming trustees president Donald Powers the day before the announcement. "It was a surprise to me," Powers said.[10]

The next day at the news conference, Hansen said that when he took the job, he told trustees he would stay seven to ten years, although many had forgotten that statement. The previous three presidents had all served more than twenty years. "We have reached what I see [as] the right time," Hansen said.[11]

In late March 1982 Hansen accepted a position as chancellor of the Texas A&M University System. He left Texas A&M after four years.

Hansen returned to Indiana. He served for a time on the Commission for Higher Education and led the Hudson Institute, an Indianapolis-based policy research organization. In 2002 he surprised his wife, Nancy, by donating $1.8 million to build the Nancy Tucker Hansen Theater in Yue-Kong Pao Hall. Nancy died in 2003.

Hansen died July 5, 2010, in Fort Myers, Florida. He was buried with Nancy in McCormick Cemetery just north of his former home, Westwood.

Immediately upon Hansen's resignation, the trustees set out to find a new president. But by June 30 they had not yet named anyone. John Hicks, special assistant to Frederick Hovde and Hansen, was named acting president. Many people believed

Hicks would have been a great choice for president, and former *Journal and Courier* publisher Jack Scott pushed for it. But Hicks, at sixty years old, didn't want it. "I'm too old and too undignified to be president," he said.

Hicks had strong support at Purdue. "John had probably the greatest knowledge of anybody at the University about the institution," said Joseph L. Bennett. In 1982 Bennett was Purdue's director of public information. He would retire in 2008 as vice president for university relations emeritus.

In addition to being a wonderful administrator, Hicks had a great sense of humor. Among the duties of a president or acting president was speaking to new freshmen students packed into Elliott Hall of Music. At the fall of 1982 convocation Hicks was introduced to the students as their president. He walked on stage and immediately went into a very dramatic presentation of the poem "Casey at the Bat," his favorite. When he was finished, he walked off without saying another word. He received a standing ovation.[12]

"John was a brilliant man, but also very self-effacing," Bennett said. "He told me as he was leaving Elliott Hall that after the poem he heard two students talking and one of them said, 'Who was that guy?' The other one said, 'I don't know. But he was pretty good.'"[13]

Powers and a search committee, including faculty and trustees, went to work finding a successor to Hansen, but they struggled to find the right person.

Powers finally did find a man everyone liked. There was only one problem.

He was an IU guy.

GIANT LEAPS IN RESEARCH

Michael Rossmann
(Photo by Mark Simons/
Purdue University)

Purdue president Arthur Hansen considered his proudest accomplishment buying a supercomputer for Professor Michael Rossmann. It was an even bigger moment for Rossmann.

"That was a very good thing," said Rossmann, Purdue's Hanley Distinguished Professor of Biological Sciences. "It enabled us to determine the structure of the common cold virus in 1984 using ideas I had been developing since I was in Cambridge and had been working on ever since."[14]

A preeminent Purdue researcher, Rossmann is known and respected by structural biologists worldwide for his discoveries showing how viruses work.

He was born in Frankfurt, Germany, in 1930, when Adolf Hitler was rising to power. Rossmann's parents divorced when he was young and he lived with his mother and maternal grandmother, who were Jewish.

"I had to go to public school and I had a lot of problems because of my Jewish family," he said. "I was terrified. The teachers were unkind. The kids were constantly chasing me home, beating me up."[15]

In 1939 he went to a boarding school in Holland, and when his mother managed to emigrate to England he joined her. He attended a boarding school in Essex, fifty miles north of London, and experienced the nightly German bombings that terrorized the nation. "Every morning you woke up and saw which houses had been destroyed," he said. "We could see the fires in London."[16]

Rossmann received his undergraduate and master's degrees in physics and mathematics at the University of London and earned a PhD at the University of Glasgow. He spent two years at the University of Minnesota doing postdoctoral work with William Lipscomb, who would later receive a Nobel Prize. In 1958 Rossmann returned to England and the University of Cambridge, working with Max Perutz, who won the 1962 Nobel Prize in chemistry.

Perutz was actually skeptical of Rossmann's work in the new field of structural biology, but he gave him free rein.[17] In 1964 Rossmann was recruited to Purdue.

"I had to find a job," Rossmann said. "I had many job offers from the United States but no offers from Britain. The offer from Purdue was good. I liked the head of the biology department. He made things attractive to me.

"Purdue has been very good to me," he said. "I was given the facilities, computers, everything I needed."[18]

His research at Purdue has been heighted by eureka moments when everything comes together. But Rossmann doesn't linger over his successes.

"As soon as the 'eureka moment' has passed," he said, he immediately asks, "What's next?"[19]

Carolyn Woo
(Photo courtesy of
Carolyn Woo)

Carolyn Woo came to Purdue University from her native Hong Kong in 1972 with only enough money for two semesters of study. She took twenty-one hours of coursework each semester. It started her on a career as a worldwide force for good.

Her next six years of study in the Krannert School of Management were supported by a Purdue scholarship and she worked as a teaching/research assistant and a residence hall counselor to earn her PhD in 1979. She then worked two years in the private sector before joining the Purdue faculty.

"I got so much out of my Purdue education, Woo said. "Purdue was important to me not just because of academic and professional opportunities, but because it gave me mentors, people who wanted me to succeed."[20]

During her Purdue career she became a full professor, was placed in charge of the MBA program, and was named associate executive vice president for academic affairs.

Woo left Purdue in 1997 to serve as dean of the University of Notre Dame's Mendoza College of Business, taking it to top national rankings. From 2012 to 2016 she was president of Catholic Relief Services, assisting more than one hundred million people annually worldwide. She was named by Foreign Policy one of the five hundred most powerful people on the planet and one of only thirty-three in the category "a force for good."[21]

Woo grew up in Hong Kong and was educated by the Maryknoll Sisters of Ossining. "It was an extraordinary education, not only in terms of academics but in terms of values, the expectations of service to others, and growth in my faith," she said.[22]

In Hong Kong, Woo met an economics professor from Purdue who advised her to become a Boilermaker. She had no idea where West Lafayette was and knew nothing about Purdue. She didn't even know what she wanted to study, but she took his advice. At Purdue she took a course in economics from Professor Dennis Weidenaar. "He was so invigorating." She said. "I decided that's what I wanted to study.[23]

"Purdue gave me so many opportunities," Woo said. "Vice President for Academic Affairs Bob Ringel and Executive Vice President and Treasurer Fred Ford mentored me. President Steven Beering nominated me for corporate boards. That was the generosity I received at Purdue."[24]

In 2017 Woo returned to Purdue in an advisory role as a Distinguished President's Fellow for Global Development, helping to connect Purdue innovation with the needs of people worldwide.

"Purdue is distinguished by its ethos of contribution," Woo said. "At Purdue we roll up sleeves and get the job done. No drama. We don't walk away before it's done. It doesn't matter who gets the credit. These are the values I learned at Purdue."[25]

32

Steven C. Beering:
The Years Before Purdue

1932 to 1983

When I think about it, I didn't have a real childhood.

—Steven C. Beering

It was July 1948. Just one month short of his sixteenth birthday, Steven C. Beering stood on the main deck of the *Britannic* with his hands on the railing as the ship cut through choppy water in New York harbor and passed the Statue of Liberty.

The deck of the 1929 ship, which had been used as a troop transport during World War II, was filled with people like him—immigrants straining to get a look at the green statue lifting its lamp "beside the golden door."

Steve was standing beside his father, who placed a hand over his son's, and the two watched as a ray of sunlight broke through the clouds shining on Lady Liberty. The Beering family had survived bombings and war. They had lost friends and relatives. They had lost their home, all their possessions, and for several years they had lost one another. But now the terror, devastation, and sorrows were behind them. They had survived and they were arriving in a new world.

When Steve saw the statue he understood what the struggle to reach the new world had been all about. "It was early morning," he remembered, many years later.

"There was fog on the water and mist and then there was a shaft of sunlight that hit the statue just as we were coming up to it. It was magic, just absolute magic, like in a movie. And my father said, 'There she is, the Statue of Liberty. Never forget this moment. This is a signal for you to make something out of yourself.'"[1]

It would be thirty-two years before the father and son held hands on a railing again—this time on the railing of a hospital bed. As the father lay dying, he looked up at his son standing beside him and said, "It worked out okay, didn't it."[2]

It wasn't a question. It was a statement of fact.

It had worked out more than "okay" for Steven Claus Beering—a lot more than okay. A teenage immigrant from Germany whose young life had been torn apart by World War II, he went on to become a medical doctor who treated a president of the United States, a physician to astronauts in the early days of the race to the moon, a medical school dean, a president of Purdue University, and finally a chairman of the U.S. National Science Board. It was more than enough to fulfill a lifetime of accomplishments for most people. But when it was all said and done, Beering was nowhere near ready to quit. He wanted to do more.

On July 1, 1983, Beering became the ninth president of Purdue, a position he would hold for seventeen years leading up to the twenty-first century. He arrived at Purdue with a partner, his wife, Jane, who took an active, upfront role as "First Hostess," a Boilermaker "friend maker," and a trusted adviser to her husband. They were a team. It was not unusual for one to finish the other's sentences. Jane and Steve Beering loved their work so much they did not take a single vacation away from work during their seventeen-year tenure.

During those seventeen years annual donations to the University increased by more than four times; the endowment went from $121 million to more than $1.3 billion; twenty new buildings and facilities were dedicated; diversity and international programs increased. Those are all great accomplishments. But when asked what he did for a living, Beering said, "I build people."

Students, faculty, staff, alumni, and friends of Purdue were family to the Beerings. He was known for walking the corridors of Hovde Hall and stepping inside offices to talk with staff, checking on the lives of administrative assistants who kept the University running. Even some faculty who had early disagreements with Beering would ultimately use words like "kind," "decent," and "friend" to describe him.

What is known of Beering's early life is sketchy. He told different parts of the story in different interviews, and there were some details he never explained. He said there were things he just couldn't remember or had blocked from his mind. "You know what happens to people if there's a painful episode in their life," he said in a

2009 interview. "I'm part of that. I tend to repress it. I just exclude it from my memory and I don't deal with it."[3]

Beering was born Klaus S. Bieringer[4] in Berlin, Germany, on August 20, 1932. His father, Stefan,[5] was a marketing manager for a commercial furniture company, Thonet, headquartered in Vienna, Austria. As a successful businessman and a Catholic, Stefan was respected and the family enjoyed a prosperous life. Klaus's mother was named Alice. After Klaus was born, the Bieringers moved to Hamburg, where Alice's parents, Theodore and Emma (Oppenheimer) Friedrichs, lived. The Friedrichs operated clothing stores in Germany and in other countries.[6] Alice was their only child. She was artistic, wrote and illustrated children's stories, and spoke several languages including German, French, and English.[7] Klaus's brother, Hans, was born in 1936.

Bieringer family in Berlin, 1937: Alice, Klaus, Hans, and Stefan. (Beering family photo)

The Friedrichs were Jewish. In Judaism, a person born to a Jewish woman is Jewish, making Alice and her children, Klaus and Hans, Jewish. But Alice and her children did not practice the Jewish faith. She had converted to Catholicism.[8] At a party rally in Nuremberg in September 1935, German Nazi leaders announced new laws concerning Jewish people. They defined a Jew as anyone with three or more Jewish grandparents, no matter what faith they practiced. So by Nazi law, Alice was a Jew. Her children were not, but it was a fine line and they were classified as *Mischlings* (mongrels or half-breeds) of the first degree. Nuremburg laws banned marriages between Jews and Aryans after 1936.

World War II in Europe began in September 1939 when Germany invaded Poland. By that time the Friedrichs had fled Germany. They traveled to England and eventually to the United States, where they settled in Pittsburgh. Before they left they tried to persuade their daughter and her family to join them, but the Bieringers stayed in Germany believing they would be safe.

Beginning in September 1941 all Jews in Germany six years of age and older were ordered to wear a badge in public. The badge included a yellow Star of David on a black field with the word *Jude* (German for "Jew") inscribed in the star. The law would have applied to Alice.

According to the Beering family, early in the war mixed faith marriages from before 1936 were ignored by the Nazis, and the resources of the Bieringers allowed them to evade scrutiny, at least for a time.

The year 1943 was a horrible turning point for the family. In the summer of 1943, Stefan traveled to Berlin, reportedly on business, and left his family behind in Hamburg. Beginning the last week of July, Hamburg was bombed day and night in Operation Gomorrah—an Allied air attack named for one of the biblical cities God destroyed with "brimstone and fire." It created one of the largest firestorms of World War II. Forty to fifty thousand civilians were killed and thirty-seven thousand were injured. A huge part of the city was incinerated. Alice and her sons spent the week huddled in a shelter with other German citizens.

When the bombing finally ended, they crawled out through the debris. All they saw was death, destruction, rubble, and fire. The building where they had lived was destroyed. The city was ablaze. Alice believed her husband had been killed in Berlin. He was actually alive, and he thought his wife and children had been killed in Hamburg.

Beering always said authorities put his mother and the two boys in a truck and took them away. The authorities were most likely police or soldiers. Because of their mother's Jewish heritage, they were taken to and confined in a work farm in Bavaria, probably because of Alice's Jewish background. There they were put to work from early in the morning until late at night. It was very hard work, but they were very fortunate they were not killed. They were ultimately freed by advancing U.S. soldiers.

When the war ended the Friedrichs in Pittsburgh worked through the International Red Cross to find their family and helped to reunite Alice and the boys with their husband and father, who had lost his hair and forty pounds—but survived.[9]

"On my mother's side of the family, there weren't many of them to begin with," Beering said in a 2009 interview. "But there were a number of them who were lost. Those were not the best years, I'll tell you. When I think about it, I didn't have a real childhood—not the kind of childhood that my kids had and my grandchildren have. It's okay. I've overcome that. But I had lost five years of school during the most formative time of a person's life."[10]

In 1946 the family was able to move to London and stayed with Alice's relatives while they waited for an opportunity to go to the United States. They immigrated through Ellis Island and settled in Pittsburgh where Alice's family was established, and they changed their names to sound less German. Bieringer became Beering. Klaus

Stefan became Steven Claus. Beering's father went to work with Alice's parents, who had a very popular ladies' ready-to-wear shop in the Jewish neighborhood, Squirrel Hill. He did bookkeeping. Young Steven also worked at the shop doing marketing. "My grandmother gave me a sum of money to grow the business," he said. "If I lost it I would be responsible for it." He grew the business and she allowed him to keep the money to pay for his education.

His horrific wartime experience pointed him toward a career in which he could help people. He wanted to go to medical school and become a physician. "I saw so much disease and so many problems during those war years," Beering said. "My mother was a nurse and she influenced my thinking."[11]

He enrolled in a Pittsburgh high school where he was placed with students his age, because they weren't sure in what grade level he should be placed.

His academic work went very well. He received straight A's and graduated in three years. After high school he enrolled at the University of Pittsburgh to study biology and chemistry. He was given a job as a lecturer in German and French, and that provided him with tuition remission. He also worked during the summers driving a cab, working in his grandmother's dress shop, and mowing lawns. He practiced speaking to get rid of his German accent.

At the beginning of his sophomore year Beering met a young woman named Jane Pickering, who was a freshman at Pittsburgh. He thought she was beautiful. She hardly noticed him. They talked a bit. He offered her a peppermint. She accepted. They didn't see one another again until she invited him to a dance sometime later, and after that evening they began dating regularly.

Beering finished his undergraduate degree, graduated summa cum laude, and entered the Pittsburgh School of Medicine with a fellowship. Jane graduated from the University of Pittsburgh the year after Steve. They were both inducted into Phi Beta Kappa academic honorary. She majored in political science and Far Eastern studies and after graduation went to Washington, D.C.,

Klaus Bieringer going to first grade in Germany, 1937, with a traditional *Schultüte* or school cone filled with school supplies, toys, and candy. (Beering family photo)

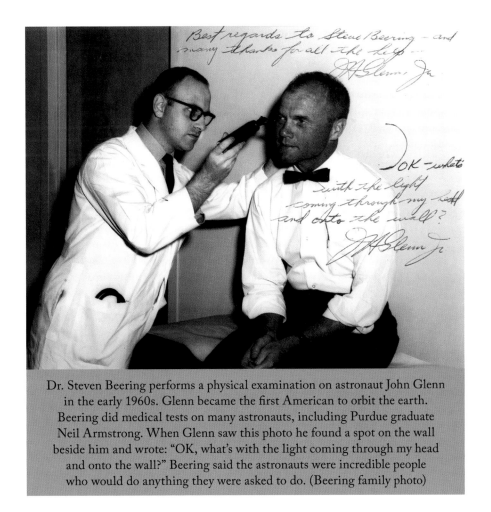

Dr. Steven Beering performs a physical examination on astronaut John Glenn in the early 1960s. Glenn became the first American to orbit the earth. Beering did medical tests on many astronauts, including Purdue graduate Neil Armstrong. When Glenn saw this photo he found a spot on the wall beside him and wrote: "OK, what's with the light coming through my head and onto the wall?" Beering said the astronauts were incredible people who would do anything they were asked to do. (Beering family photo)

where she worked for the CIA. They married in 1956. In 1957 Beering joined the U.S. Air Force Medical Corps, and he remained in the service for twelve years, reaching the rank of lieutenant colonel.

Beering graduated from medical school in 1958 and did an internship at Walter Reed Army Medical Center. During his internship he was part of a medical team that treated President Dwight Eisenhower, who had suffered one of his heart attacks.

His next assignment was to Wilford Hall Medical Center at Lackland Air Force Base in San Antonio, Texas, where he became chief of internal medicine and performed medical examinations on astronauts. "The general who hired us told us they were going to put a man on the moon," Beering said. "We thought this man had a screw loose. As we got involved in it and met these people who had been selected as astronauts they were extraordinary human beings. Anything seemed possible. It became an expectation that had to be done."[12]

The Beerings' three sons were born in Texas—Peter, David, and John. When his service in Texas was coming to an end, the air force wanted to send Beering to Washington, D.C., to work in the Surgeon General's office. "I came home and told Jane, I'm not going to work in administration," Beering said. "That's the last thing I'm going to do. I'm going to continue with my medical practice and research and teaching. I've done my duty. I've fulfilled my obligation. I'm going to resign from the Air Force." The Indiana University School of Medicine in Indianapolis was recruiting him and he accepted the offer.

He arrived in Indianapolis on July 1, 1969—just weeks before his former patient, Neil Armstrong, commanded the *Apollo 11* moon landing. Beering became assistant dean and then dean of the Indiana University School of Medicine, a position he held for ten years. Some people thought he would become president of Indiana University.

But Beering's future would take a different course.

Don Thompson
*(Photo courtesy of
Don Thompson)*

Don Thompson went from Chicago's Cabrini-Green Housing Project to president and CEO of McDonald's Corporation. An education at Purdue University helped give him the confidence to pass through the doors of opportunity.

Thompson is founder and CEO of Cleveland Avenue, a Chicago-based venture capital firm focused on new food, beverage, and restaurant concepts.

Thompson, who was raised by his grandmother, moved with her to Indianapolis in the early 1970s for better education opportunities and a safer environment. He entered Purdue in the fall of 1980 and immediately met a young woman named Liz who would later become his wife. They had grown up in Chicago on the same street, Cleveland Avenue, only four blocks apart, but never met.

Both Thompson and his future wife took part in Purdue Minority Engineering. "While we were at Purdue the influence of the Minority Engineering Program was so great in our lives," Liz said. "It helped us make a giant leap from being high school kids who were really nervous, having never been in as diverse an environment as Purdue. It helped us in ways we didn't realize at the time, but we now know helped us transition to the adults we have become."[13]

Thompson majored in electrical and computer engineering and went to work with Northrop Grumman after graduation in 1984. He loved his work and grew within the organization. But one day a call came from a recruiter he thought was with McDonnell Douglas, an aerospace company like Northrup. The caller was actually with McDonald's, and in 1990 Thompson began a series of leadership positions with the company. He also worked at a Chicago franchise to learn from the ground up. From 2012 to 2015 he served as president and CEO of the company.

Thompson said his career has been one of opportunity and risk-taking. "What Purdue taught us was that every opportunity was also a problem to be solved," he said. "That's something we talk to students about. Thanks to Purdue there have been doors of opportunity opened for me and my family."

Liz is a Purdue engineer who decided she would rather engineer the careers of people. She is president of the couple's Cleveland Avenue Foundation for Education.

"Thanks to Purdue we have a platform to help countless people who are on the same journey we were," Liz said. "We're now able to say we can help you start this journey."[14]

33

The Steven Beering Years

1983 to 2000

He didn't like the idea that some of our programs
were second class and had no hope of being first class.
He wanted everyone to aspire to first class.

—Frederick Ford

The arrival of Steven and Jane Beering on campus February 4, 1983, capped a presidential search by Purdue University Board of Trustees president Donald Powers, vice president Bob Jesse, and a University committee that had been ongoing since shortly after November 1981, when Arthur Hansen announced he was retiring. They were struggling to find the right person.

"Finally," Powers said, "John Hicks, who was acting president, said, 'You know, there's a guy at IU you ought to look at. He's dean of the medical school in Indianapolis.'" President emeritus Frederick L. Hovde offered the same advice, but a lot more emphatically. He talked to Powers at a basketball game in 1982, jabbed his finger at Powers's chest, and said, "If you don't talk to Steve Beering, you're not doing your job."[1]

In late 1982 Powers called Beering, who thought a doctor/president would not be compatible with the engineering/agriculture curriculum at Purdue. And he wasn't sure the people at Purdue would accept a president from Indiana University.

The Beerings' second son, David, was a sophomore at Purdue at the time. He heard that his father had been contacted and had turned down the opportunity to interview for the position, and he asked his father to reconsider. So a meeting was arranged between Powers and Beering at the IU School of Medicine in Indianapolis.

It was a busy day when Powers arrived for the meeting. Beering suggested they get sandwiches out of a machine in the basement and then eat and talk in his office. Powers was impressed with Beering's practicality. The two found themselves in agreement on higher education issues and the future of Purdue. Powers decided immediately that he had found the new president. Beering was less certain. "This was the most talked over and prayed over decision he ever had to make," his wife, Jane, said. "He would be leaving medicine and going in a totally different direction."[2] There would be no turning back.

Purdue wasn't the only university that had contacted Beering. The University of Iowa was searching for a new president, and they had contacted him. He was not interested. At the same time John Ryan, president of Indiana University, had served more than ten years, and people were mentioning to Beering that he might ultimately be a candidate for that position.

By January 1983 Beering had decided he would accept Purdue's offer. It came after a dinner with the board of trustees at Westwood. Powers and Jesse asked Beering to join them in the lower level of the home. Powers said the entire board was in agreement. They wanted him to be the next president. "Steve was just flabbergasted," Jesse said.

Purdue president Steven Beering and Jane Beering.

"Couple of tears rolled down his cheeks. He said, 'I can't think of a higher honor. And I will pledge to you this, that I will devote all my energies, all my time and all my love to this university for as long as I'm here. You will not have anybody that tries harder at this institution.'"[3]

The board of trustees met February 4, 1983, to elect Beering. He and Jane arrived early and a rare winter rainbow arched over the campus. They considered it a good omen. After the board meeting there was a news conference and a reception at the Union. "I accept [this position] humbly because of the enormity of the task, but gladly because it is an opportunity to make a difference," Beering told reporters. What would he do with his red-and-white IU wardrobe, he was asked? "A garage sale has been suggested," he said.[4]

Director of Public Information Joseph L. Bennett watched Beering during the board meeting, news conference, and reception. "Both Steve and Jane were very pleasant," he said. "Steve as always [was] very gracious. But he gave you a sense of authority. You felt like he was somebody who was in command. You always had a feeling that this guy was smarter than anyone else."[5]

In February Hovde fell gravely ill with emphysema. Beering went to Home Hospital to visit the president emeritus. Hovde's two daughters were also in the hospital room. "It was the most meaningful encounter I have had with another man," Beering said. "He bared his soul. He said Purdue had been his life's work and that he wouldn't trust just anyone with it. . . . Tears were running down my face and his two girls were sobbing. It was quite a scene. I leaned over and hugged him."[6] Hovde said he could die peacefully knowing Purdue was in good hands.[7] He died on March 1, 1983, less than a month after Beering was elected president.

Beering continued as dean of the Medical School until Purdue's fiscal year began on July 1. During the interim, he drove up to West Lafayette from Indianapolis to talk with all the campus leaders—administrators, deans, department heads, faculty, students, as well as community leaders.

"My already positive impressions about Purdue were reinforced," Beering said. "Every one of these people was dedicated and informed and loyal to the institution." He was especially impressed with Robert Ringel, dean of humanities, social sciences, and education. Beering would soon promote him to executive vice president for academic affairs. "I wanted him in charge of all academics for the whole system," Beering said. "He was one of the superstars of the group, but there weren't any soft spots in this faculty and administration. It was an array of professional people who worked together beautifully."[8]

Initially some people were concerned about the selection of Beering. Some faculty noted that he was a medical doctor, as opposed to a PhD. One of the biggest questions centered on bringing the "IU guy" to the Purdue campus.

"There's no question that the idea of bringing someone in from IU was really revolutionary," Bennett said. "There was a feeling that—is this person going to identify with Purdue? I suppose there were some people who even wondered where his loyalties would be. But it didn't take very long for that to change. I don't think you could ever find somebody more loyal to Purdue. Steve became steeped in Gold and Black. He truly loved the University."[9]

Among Beering's first activities was traveling the state with IU president Ryan, who was a friend as well as his former boss. They would arrive in Indiana cities and towns together, meet with local officials and legislators, and promote the state's two major public research universities.

Astronaut Reunion

In 1985 Beering had the idea of bringing all the Purdue astronauts back to West Lafayette for a reunion. They had been on campus as students and many made return visits. But they had never all been on campus at the same time.

"This was Steve's idea," Bennett said. "He knew many of the people at NASA and he was very well respected there. This was something he was very excited about. When he unrolled it for me, I thought, this is going to be tough. But we started putting together contacts for these men and women and every single one of the astronauts returned for that reunion."[10]

At that time there were fourteen living astronauts who had Purdue degrees. Virgil "Gus" Grissom and Roger Chaffee had died in January 1967 in an *Apollo 1* training accident. Among those who returned were Neil Armstrong and Eugene Cernan, the

Purdue president Steven Beering with Boilermaker astronauts during the reunion in 1985. Also pictured is Robert Foerster (first row, right), an alumnus and one of ten finalists for "Teacher in Space." The position went to Christa McAuliffe, who died in the *Challenger* shuttle explosion in January 1986.

Purdue alumnus and first man on the moon Neil Armstrong hits the World's Biggest Drum during halftime of a football game in 1999.

first and last men on the moon. Armstrong agreed to do public appearances, but not media interviews.[11] It set the tone for more astronaut reunions that were to follow.

When IU's Ryan retired in 1987, Beering said he was approached about returning to lead his former school. But he wouldn't consider it.[12]

Faculty Senate Unrest

The spring of 1987 was an exciting time on campus when President Ronald Reagan visited, flying into the Purdue Airport on Air Force One. But the fall brought a low point for Beering. Nationally, the health insurance industry was instituting changes. The benefits of Purdue employees were going to be changed, and discussions were taking place among University administrators and trustees. But information about the changes was not reaching the faculty. The Purdue *Exponent* ran a story about what was taking place. "All of a sudden we had a very hot situation," Bennett said. "As president, Steve could not avoid taking responsibility. But it certainly wasn't Steve's doing."

After faculty and staff protests about decisions that had been made, the board of trustees approved a compromise insurance plan. But in the meantime, the issue came before the faculty senate.

At a senate meeting an assistant professor told Beering, "You seem to be confused by the reaction of the faculty to this. You seem to think we work for you. I may be a little old fashioned about academia, but I always thought the faculty carried out the main business of the university. We do in fact, run the university and you work for us." Beering answered calmly: "We do feel we're servants of the university. . . . But I believe the trustees of the university selected me as president because I have a degree of expertise in administrative matters that I can use so you can be free for research and teaching."[13]

Margaret "Peggy" Rowe was a member of the senate. She would later become dean of humanities, social sciences, and education and a vice provost. "Steve Beering got off to a very bad start," she said. "I think he wasn't listening and he got very bad advice. The University Senate was a hotbed at that time. We fought everything. There was a big change in healthcare and instead of bringing people into a discussion it was announced [what the University would do]."[14]

There was a call to take a vote of no confidence on Beering by the entire faculty. The Senate vote to send the no confidence issue to the faculty took place on October 19, 1987. Eighty senators were present and forty-nine voted to poll the entire faculty. Since the motion required a two-thirds vote, fifty-four senators were needed, and the motion failed. Some members of the senate wanted to continue pressing the issue. Beering, who attended the meetings, said, "My intent is that this University continue moving forward with the momentum we've established and that we put this unhappiness behind us."[15] He vowed better communication with the faculty and staff.

Rowe was part of a committee that met with Beering. One meeting in particular she never forgot. She arrived early. Beering was known for being very punctual, and when she arrived he was already sitting at the table by himself, waiting for the others. The two sat alone together until others arrived. "It was extraordinarily difficult," Rowe said. "I know it was for me and I sensed it was for him. But he did not make it more difficult. I was impressed by that." They talked and got to know one another. What struck her most about that day was his "civility." They became friends.[16]

Years later Beering said health insurance was an issue very important to the faculty. "You have to take that in stride," he said. "I didn't take it personally. I understood."[17]

Even when he was being publicly insulted, he never lost his sense of decorum, Bennett said. "He was always polite in his responses. But behind the scenes it angered him. He resented not so much that it was personal to him, but that people would behave in a way that was less humane to other people. It was contrary to his view of the world."[18]

Liberal Arts

Beering was credited with building up the liberal arts. He started in 1989 by separating the school it was in, Humanities, Social Sciences, and Education, into two schools: Liberal Arts and Education. The School of Liberal Arts that Dorothy Stratton and Helen Schleman had fought for beginning in the 1930s was finally realized. In 1993 the University built a $28.5 million facility for Liberal Arts and Education. After Beering retired, the board of trustees named it for him.

Executive vice president and treasurer Frederick Ford said presidents had seen the liberal arts at Purdue differently before Beering arrived:

> Under President Hovde they were a service for our engineers, pharmacists, people in agriculture, science, and so on. Art Hansen was a little more oriented in the direction of liberal arts. He had a broader [concept]. He was not just an engineer and mathematician. But Steve Beering put more priority in the liberal arts. . . . He didn't like the idea that some of our programs were second class and had no hope of being first class. He wanted everyone to aspire to first class. You can't have a university where everything is top rate. There isn't enough money in the world to do that. Steve invested a lot more money in the liberal arts and education and John Hicks supported that as well. He thought it was a good idea. On the other hand, some of the engineers didn't think that was a good idea.[19]

Gene Keady Stays at Purdue

Beering's office on the second floor of Hovde Hall had windows that looked out on the Purdue Mall, and nearby sidewalks were often crowded with students hurrying between classes. Beering used that view with great success on at least one occasion.

Campus crises take all forms, and a serious storm arose at Purdue in mid-February 1989, when ESPN-TV basketball color commentator Dick Vitale suggested the University's legendary and loved coach, Gene Keady, might be a perfect fit at Arizona State University. Arizona State turned out to be the one other job Keady had always wanted.

Arizona State did come calling for Keady, and the coach was thinking maybe it was the right moment for change. In his previous eight seasons at Purdue, he had received four offers from other schools. None had interested him, until Arizona State.

Purdue men's basketball coach Gene Keady.
(Photo by Tom Campbell/Purdue University)

Keady flew to Tempe, Arizona, on Sunday, March 12, when the basketball season was finished. He got little sleep the night before and on the night he returned home. His pay at Purdue was estimated at about $225,000. Arizona State offered him $325,000 to $375,000.[20]

On Monday morning, the day after his trip west, Keady met with Purdue athletic director George King, who was beginning to believe he would lose his basketball coach. But Keady had one more meeting before he would announce his decision. Beering invited him to Hovde Hall to meet in the president's office with its view of the campus and the students.

Beering loved intercollegiate athletics and he supported them. He showed Keady the students passing by beneath his windows. "I said forget the money; money isn't everything," Beering remembered years later. "What about the people here and what you have built here. You just can't walk away from that. Look what you have done for the students here. Look what you have done for the young men [on your teams] and for school spirit and for the University. Go home and rethink this. I really want you to stay."[21]

That same day, during an emotional media conference, Keady announced his decision to remain at Purdue. "The fact that we have great tradition is hard to find," Keady told reporters. "I was just not willing to give that up. [Arizona State is] the one school that I really thought I'd end up coaching at some day, if they ever wanted me. But because of circumstances here and the happiness of our family, we decided to stay."[22]

Years later Keady said he never really wanted to leave, and he remembered meeting and talking with Beering. "I really loved Purdue," he said. "I loved the people at Purdue. It was a great group of people, a lot of guys around me helped us win. It wasn't about me, it was us." He said he remembered other coaches who left successful positions for new opportunities that didn't work out. "The grass is not always greener on the other side," Keady said. "I just asked for a little raise, he [Beering] agreed, and I was happy."[23]

University Traditions

A supporter of all things Purdue, Beering resurrected the University Hymn and Seal. In 1968 Al Gowan, an assistant professor of the then School of Creative Arts, had designed the ninth version of the seal, which featured a griffin. When the Beerings arrived on campus, it was not being used, and he brought it back. He had small lapel pins created featuring the griffin, and they were given to his cabinet and to friends of Purdue. They were cherished gifts.

Beering also wondered why Purdue did not have a hymn. He was told the University did have one, written in 1941 and first performed in 1943 by the University Choir, but it was no longer being performed. In 1993 Beering had the "Purdue Hymn" officially named the Alma Mater of the University, and it is performed often at Boilermaker events. Alfred B. Kirchhoff, who was a teacher, principal, choirmaster, organist, and youth leader at St. James Lutheran School and Church in Lafayette from 1932 to 1948, composed the hymn.[24]

A New Kind of Athletic Director

Before Keady, one of the legendary basketball coaches at Purdue had been King. He came to the University in 1966, replacing Ray Eddy, after playing college basketball at Morris Harvey College (now the University of Charleston) in West Virginia, playing in the NBA, and coaching at West Virginia University. When King retired from coaching at Purdue in 1971 he became athletic director. It was common at that time for former star athletes such as King to become coaches and then to move into the position of athletic director. But when King retired in 1992, Beering and the board of trustees looked for something different. Intercollegiate athletics had changed as more games were televised. The hiring of coaches and the negotiation of television contracts had become much more complex.

Trustee J. Timothy McGinley, who became chairman of the board in 1993, was very much involved in the process of hiring a new athletic director. "In the past almost all the athletic directors were former coaches, usually football or men's basketball," McGinley said. He was a Purdue basketball star himself during his student days. "When the athletic director opening occurred, we studied the job requirements and concluded the skill and experience of coaches did not meet the need. Some required skills were balancing a multimillion-dollar budget, private fund-raising, maintaining and growing a sizeable physical facilities plant, managing eighteen coaches and other personnel, as well as negotiating television and other vendor contracts. Our conclusion was to pursue someone with a business, financial, or administrative background. We were one of the first universities to not hire or promote a coach. We were pioneers."[25]

In 1993 the board hired Morgan Burke as director of intercollegiate athletics. Burke had been captain of the swim team at Purdue when he graduated, Phi Beta Kappa, with a degree in management in 1973. He received a Purdue master's degree in industrial management in 1975 and in 1980 graduated from the John Marshall Law School in Chicago. He worked for Inland Steel in northern Indiana and advanced through thirteen positions in eighteen years. He was a vice president at Inland when Purdue hired him. He had the business, financial, and administrative background the board was seeking. His legal degree was a plus.

Burke found a tough situation at his alma mater. "We really had some terrible facilities issues," Burke said. "We had the worst [football] physical plant in the Big Ten. And I'm not talking about trying to keep up with Texas. I'm trying to stay up with Harrison High School. That's how bad it was. It begins to wear on you as a coach that you keep losing recruits because you are so badly overmanned in terms of facilities."[26]

There was more. "We were dramatically behind in compensation," Burke said. "We were probably 80 percent of the median. And we were barely breaking even."[27]

He was able to attract donors who made a difference with their financial support.

"The stadium [Ross-Ade improvements] came into being really in large part because of Fred Ford and his wife, Mary," Burke said. "Mary let me know in her polite but no uncertain terms that the restroom situation in Ross-Ade for women was deplorable. And it was. Fred said, 'We'll start the process of slowly but surely building restrooms around the concourse.' We had other issues. The concrete was crumbling. You had a little press box. It was a mess. I went to Dr. Beering at that time and I said we have to get a master plan."[28] They created one, built new athletic facilities, and remodeled existing ones.

In October 1997 Beering, and everyone else at Purdue, was stunned when Burke announced he was returning to Inland Steel as vice president for human resources. Beering did not want Burke to leave. He met with Burke, told him how much he appreciated all he had done, and encouraged him to reconsider. On Friday, December 12, Burke announced he was staying.

"I wanted him back," Beering said. "I didn't want him to leave in the first place. I had great confidence in him. I liked his style. He'd been good for Purdue and we had been good for him. He's highly regarded by everyone."[29]

Burke listed six reasons for his change of mind. The first one was his close personal relationship with Beering. During Burke's tenure, in addition to Ross-Ade renovations, tennis, soccer, softball, and baseball received new facilities. The golf complex and swimming and diving facilities became among the best in the country and hosted national championships. Mackey Arena had a nine-figure renovation.

In 1997 Burke hired Joe Tiller as head football coach. Tiller had been head coach at Wyoming when he was hired, but he had been an assistant coach at Purdue from 1983 to 1986. He inherited a program that had only two winning seasons in the previous eighteen years. In his first year at Purdue he was 9–3 and tied for second in the Big Ten. His team went to the Alamo Bowl. He would take the Boilermakers to bowl games in ten out of twelve years as head coach—including the 2001 Rose Bowl.

After the 1996–1997 women's basketball season, Coach Nell Fortner announced she was leaving to be head coach of the USA Women's Basketball team. Rather than go outside and bring in someone new, Burke elevated Assistant Coach Carolyn Peck. She became the third women's head basketball coach in three years.

In her first season, 1997–1998, Peck's team was 23–10 and made it to the Elite Eight in the NCAA tournament. Meanwhile, the WNBA expanded from ten teams to twelve, and Peck was offered and accepted the job of head coach of a new Orlando team. But her players asked her to reconsider, and she did. She delayed going to the WNBA for one more year.

In that 1998–1999 season Peck had three standouts—Katie Douglas, Stephanie White, and Ukari Figgs. The Boilermakers went 28–1, their only loss being in a one-point game against Stanford. They won the NCAA National Championship in 1999, winning all six of their tournament games by at least ten points. The Associated Press named Peck the women's basketball coach of the year. When the players cut down the net after the national championship, they took it to the stands and gave it to one of their biggest fans—Jane Beering.

Purdue Students

As the Vietnam War declined in the 1970s, so did the raucous student protests. But social consciousness continued to simmer as students urged resistance to apartheid, sweatshops, and coal-fired boilers. From hunger strikes to marches on the malls to an invasion of balloons in Hovde Hall, students drew attention to their concerns.

Knowing that connected students are successful ones, the University encouraged clubs and organizations, which grew to about a thousand at Purdue's sesquicentennial. The clubs changed with the time, as yesterday's model railroaders made room for video gamers who created and hosted national competitions.

Whether for credit or group projects, students created solar racecars, candles from soybeans, futuristic energy-efficient homes, and Rube Goldberg contraptions. Come spring, energies turned to the Grand Prix race and Spring Fest, drawing national coverage for its cockroach races and cricket spitting. Cricket spitting was a Spring Fest campus event created by entomology professor Tom Turpin. Contestants placed a dead cricket in their mouths and competed to see who could spit it out the farthest. The contest became so popular that other universities adopted it. Spring Fest drew more than thirty thousand people to campus each year for a variety of demonstrations and activities.

As campus organizations expanded, some students even honed their financial skills—and made money—in an investment club and entrepreneurship competitions.

Early-morning classes were, as always, unpopular, but the students didn't mind staying up all night to hit the bars in costume for Breakfast Club early on game days. Breakfast Club became a tradition as students dressed as if for Halloween went to West Lafayette bars that opened early on the morning of home football game Saturdays.

Purdue students weren't immune to society's struggle with drugs, but it was alcohol that drew the most concern. Purdue Student Government successfully lobbied the Indiana General Assembly to minimize penalties for Good Samaritans who called for help for those whose lives were endangered.

Always a magnet for international students, Purdue's rankings and reputation increased its draw. In 1998–1999, near the end of the Beering era, Purdue enrolled more international undergraduate students than any other public university in the nation.

Purdue was a favorite spot for recruiters, who were drawn to campus by graduates' reputation for a Midwest work ethic, teamwork, and preparedness for careers.

Nude Olympics

One campus event Beering was glad to end after years of coaxing students to stop on their own was the Nude Olympics. No one is quite sure when it started, but "streaking"—running naked, mostly by males—had a long tradition at Purdue and other universities. The Nude Olympics at Purdue was an event that featured naked men and eventually a few women running around Cary Quad on the coldest day of winter.

It dated to the late 1950s or early 1960s, originally as a late-night, clothed, quick run on a very cold night—which could be well below zero. It eventually evolved into an endurance run with no clothes, attracting thousands of spectators. In its later stages it took on a mob mentality, including broken windows, drunk and disorderly behavior, and mistreatment of runners. Students were treated for exposure at hospital emergency rooms. In the early years the University tolerated the event, never strongly taking a position against it. But in the late 1980s and early 1990s Beering directed the housing staff to phase out the event altogether. It took a couple of years of efforts by several campus agencies, but eventually it came to a close in the mid-1990s.

For the most part, students considered it a lark. "If you have friends in it you sure get to know them better," a female sophomore observer told a Lafayette *Journal and Courier* reporter in 1985. But she never intended to watch it again. Once was more than enough.[30] In 1981 there were reports of as many as two hundred students jumping into and out of the race. The longest runners continued for more than one hour and forty minutes.[31]

By 1985 Playboy Television was showing an interest in filming the event, and video of it had run on ESPN, CBS Morning News, CNN, and the Independent News Network. After having the event reviewed by a committee, in 1985 Beering said it was a health and safety hazard, and he ordered it stopped. There was property damage, fights, and at least one assault on a police officer in addition to the health hazards. In 1986 about eighty Nude Olympians faced University disciplinary action. Later in 1986 there were attempts to organize a winter carnival to replace the Cary Hall run. But the Nude Olympics continued. In 1991 twenty-two students were placed on probation for taking part in the event. In 1995 five students were disciplined after the race.

Tom Paczolt said it was his "unwritten duty to make certain the event didn't occur again and remarkably it never occurred during my tenure as Cary manager. The last time the run occurred was early spring semester in 1996. There were rumors about it happening for two or three years after that, but it never happened."[32]

A Makeover for the Purdue Campus

Beering's accomplishments at Purdue were many. But when people are asked about his most memorable contribution, they agree. Indiana University had long been known as one of the most beautiful campuses in the nation and it was. Purdue was not.

Beering changed the "look" of Purdue.

34

A Makeover for Purdue

1983 to 2000

We need to plant trees and create parks where
we can sit and relax and study.

—Steven Beering

The first time Steven Beering took a seat behind his desk in Hovde Hall, he could turn and look out the office windows at an asphalt road, a parking lot, and an old smokestack that was shedding bricks from more than two hundred feet above the ground.

During the seventeen years that Beering was president, the parking lot became a park; the asphalt roads became walking paths; and the smokestack was replaced by a beautiful tower that housed the Heavilon Hall bells chiming the hour, half hour, and the beloved end of classes.

It was the greatest period of campus beautification in the history of Purdue. It was stunning. As newly planted trees grew, grass replaced asphalt, and fountains gushed water, people who worked on campus every day suddenly saw that their functional, practical University had transformed from the ugly ducking into the swan.

And it was all done in the face of criticism from people who were actually very fond of Purdue's traditional roads, parking lots, and brick-dropping smokestack.

In the 1870s Purdue began on a cornfield—a treeless expanse of flat ground. As years passed, most of the buildings were designed more for function than beauty.

Space wasn't set aside in buildings for airy atriums. Space was used for classrooms and laboratories, and that space was used all day, Monday through Friday, and at least half a day on Saturday. There were efforts to make the campus more appealing. Founding fathers, such as trustee Martin Peirce, provided Purdue with a tree nursery and ten thousand plantings. The Purdue founders planted shrubbery. Elm trees once lined State Street and Memorial Mall. But Dutch elm disease claimed those trees over the years, and the other plantings gave way to the need for new buildings and parking lots. In the latter half of the twentieth century, campus growth was being guided by a master plan from the 1920s that could never have conceived of the increases in automobiles and enrollment.[1]

The smokestack behind Loeb Fountain. Some people protested taking the smokestack down.

John Collier graduated from Purdue in 1984 with a degree in landscape architecture and went to work in the University's Office of Facilities Planning. He said the original Purdue campus included green space, but that was pushed aside by the need for buildings and to accommodate automobiles. "After World War II in 1950s and 1960s the automobile became the dominant thing and it had to [be] accommodated," he said. "Green space was lost." The University became a parking lot. "It was austere. People described it as neo-penitentiary."[2]

The campus simply wasn't considered a thing of beauty. People were proud, instead, of its Purdue practicality. Ford, whose office oversaw campus planning and building, was aware that people were complaining. "Campus facilities are a major challenge," he said. "At one point in time the University Senate got very concerned about the appearance of Purdue and all the red brick buildings. They thought it was boring."[3]

There had been a lot of other pressing problems in the twentieth century that had priority. Campus beautification took a back seat.

But Ford listened to the concerns. He talked with his staff about what could be done to make the campus more attractive. They couldn't change the buildings. But they came up with the idea of changing the landscape, adding green space. The plan found huge supporters in Beering and his wife, Jane.

"Some of the long-term faculty told me about remembering the trees," Beering said. "I heard about the elm trees dying, and I said we need to do something about that. We need to plant trees and create parks where we can sit and relax and study."[4]

By 1985 the University had a new master plan done by Sasaki Associates in Watertown, Massachusetts. The general plan, without details, cost only $30,000. It called for replacing roads and parking lots with grassy malls and parks. Trodden grass was replaced by cobblestone walkways. Automobiles were parked in garages or in parking spaces on the periphery of campus. People began talking about a pedestrian campus.

Many people were involved in what took place. The board of trustees supported the plans and approved funding. Ford was a key player, along with people in his office, including Ken Burns, who was vice president for physical facilities and later vice president and treasurer. Thomas Schmenk, director of facilities planning and construction, helped lead the way.

In the end, it was done on Beering's watch. "What it boils down to is when you have someone like Dr. Beering in charge who has a feeling for the aesthetics of the campus, then things are going to happen," Schmenk said in a 2000 interview. "It was really exciting for us to realize we had such a man in charge. Twenty years ago people talked about 'the brick factory' and 'the prison.' I got sick of that. There were obviously so many things that could be done, but without someone at the top to say it was a major commitment, it wasn't going to happen. We had made some strides to making it an attractive campus. But when he came, it really changed."

The first project was the Purdue Mall, surrounded on two sides by engineering buildings with Hovde Hall at the west end. When Beering arrived a horseshoe-shaped street ran from Northwestern Avenue, curved in front of Hovde Hall, and returned to Northwestern—with parking all the way around. It was the main entrance to the campus and provided a vista of Hovde Hall and its Loeb Fountain for people driving along Northwestern Avenue.[5]

Meanwhile, Purdue needed a Materials and Electrical Engineering Building, and it was decided that the best place to put it was along Northwestern Avenue between, and in fact connected to, the Physics and Electrical Engineering Buildings. That blocked the traditional gateway to campus and the view of Hovde Hall and the Loeb Fountain from Northwestern Avenue. In fact, the Loeb Fountain would have to go, too. It was too small for the setting. So it was put into storage with the intention of placing it elsewhere.

There were some unhappy Boilermakers who thought their view of the fountain and Hovde Hall from Northwestern Avenue was being destroyed—until they saw the park that replaced the parking lot. The Materials and Electrical Engineering Building was constructed with a large atrium and windows so people could see through to Hovde Hall. And then came a fountain that became a focal point for the University.

Students returning to campus in the fall of 1989 were struck by how their University had been transformed by the wide-open Purdue Mall and what was called a "water sculpture" built through a gift of $350,000 from the Class of 1939, which was celebrating its fiftieth anniversary.

The fountain was turned on the evening of July 17, 1989. Thousands of people from the community showed up to see it. There were "oohs" and "aahs" from the crowd as if they were watching Fourth of July fireworks as the water reached its maximum height of thirty-eight feet, with colored lights forming a rainbow of colors.[6]

Designed by Robert Youngman, it represented technology, engineering, and ideas. Some people said it resembled a space shuttle. People loved it, and the fountain quickly became a popular place where students and community visitors could run through the cooling water on hot summer days and evenings.

A Bell Tower near Hovde Hall was dedicated at Homecoming, October 14, 1995. One hundred sixty feet high, it replaced the old smokestack that some people had declared a campus icon that could not be replaced. The redbrick tower with a white base and top housed the Heavilon Hall bells, which had been in storage since 1956. In addition to chiming the time, "Hail Purdue" played from the top of the tower along with other recorded songs, including carols before and after the Purdue Musical Organizations' Christmas Show in nearby Elliott Hall. It also played Purdue songs during commencement. The Class of 1948 provided the leadership gift for the $1.45 million tower that quickly became the symbol of the campus. Soon, no one remembered the dangerous, deteriorating smokestack.

The Beerings' son David, who was a student at Purdue and a member of the band, had the great idea of using the old smokestack as a fund-raiser. When it was taken down, the bricks were preserved and numbered, with #1 going to the president. The others were sold, raising $40,000 that the Department of Bands used to purchase new sousaphones.

People liked their "new" campus. And more was happening. If the Loeb Fountain wasn't the "right fit" for the Purdue Mall, it was perfect for the new Founders Park in front of the new Liberal Arts and Education Building. The redesigned fountain eliminated the basin and flattened the look—also making it more accessible for "fountain runs." The park was dedicated April 23, 1994, during Purdue's 125th anniversary celebration. More than two acres, the park featured walks, trees, tables, lighting, and a Loeb Fountain redesigned to look and function better than ever. Founders Park had been a green space, a park, when the campus was started with buildings surrounding it. So, in effect, it returned the space to what was originally intended.

Loeb Fountain, Founders Park.

In 1997 the University got rid of a ground-level parking lot behind the Memorial Union and turned it into Academy Park, where professors could lead outdoor classes on beautiful spring and autumn days. The park honored teachers at Purdue. A Book of Great Teachers, located in the west foyer of the Purdue Memorial Union, is an extension of Academy Park. In the park, etched into a granite obelisk, are the names of the six professors who taught at Purdue when the University began classes in 1874.

As parks and malls were created on campus, Beering also commissioned outdoor art. A number of buildings went up, including the Hansen Life Sciences Research Building, Knoy Hall of Technology, the Class of 1950 Lecture Hall, the Mollenkopf Athletic Center, Hillenbrand Hall, and the Nelson Hall of Food Sciences. New buildings during Beering's tenure totaled $162 million.

Academy Park behind the Memorial Union.

One of the most spectacular buildings was the Black Cultural Center, dedicated in 1999 with more than four hundred people spilling out of tents set up for the occasion. It expanded the previous center's square footage from five thousand to eighteen thousand. The cost was $3 million, with most of that coming from fund-raising efforts.

African American students and alumni had been pressing for a new building for a number of years. Black Cultural Center director Tony Zamora had requested a new center in his annual report every year. But there was also concern among African Americans that their building on University Street would be closed and the Black Cultural Center would be placed in an office in the basement of the Union or merged into a multicultural center on campus.

With pressure from African American students and alumni, a decision was made to build a new Black Cultural Center, a promise made to Zamora before he retired in February 1995. Students and alumni continued to push for the new center. In March 1995 about two hundred students went to Hovde Hall and stated that the building on University Street was too small, the water was undrinkable, the basement flooded, and the building contained asbestos.

Beering received a standing ovation when he reached the podium to speak at the dedication of the new center. "We recognize not just where we came from, but where we have yet to go," he said. "We are united by our diversity."

Designed by Blackburn Architects of Indianapolis, the center captured African American culture and the spirit of traditional African architecture.[7] "The center's appearance evokes a feeling of community, reminiscent of the bonds that form within a traditional African village," said Renee Thomas, director of the center. Thomas was assistant director in 1995 when Director Tony Zamora retired. She served first as interim director and in 1996 was named director. "More and more, the United States is becoming a global village," she said. "So this state-of-the-art facility is geared towards educating all students about the contributions of African Americans, and helping all students develop the skills needed for success in a global environment."[8]

Among the projects at Purdue shortly after the Beerings arrived was the renovation and restoration of Westwood. When it was owned by Purdue vice president and treasurer R. B. Stewart and his wife, Lillian, it had been a smaller home. It was enlarged for President Arthur and Nancy Hansen. But still more space was needed for the entertaining that would be part of fund-raising. The University considered constructing a second building next to Westwood for entertaining. "But Steve didn't want that," Ford said. "He thought the president's house is where people should be entertained."[9] During the Beerings' tenure, more than twenty thousand people ate at Westwood.

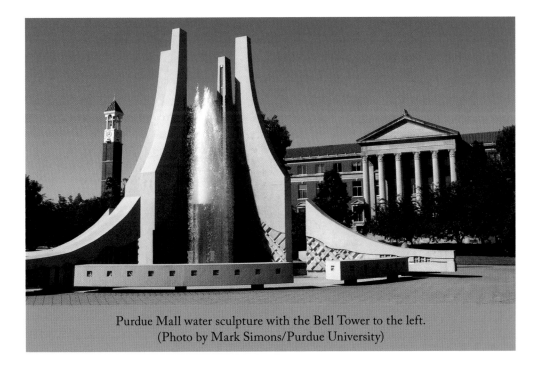

Purdue Mall water sculpture with the Bell Tower to the left.
(Photo by Mark Simons/Purdue University)

The University remodeled and enlarged the home for large groups. Ford said Purdue offered to lease a home for the Beerings during the construction period. But the Beerings turned it down and moved to a suite in the Memorial Union. Steve and Jane Beering ended up having to live there for about a year—longer than expected. "They were good troupers," Ford said. "They never complained."[10]

Throughout the physical beautification of campus, Beering never forgot what was most important to the University—people. "It is my belief that it is not the buildings, it is not the gadgetry, it is not the equipment, it is the people that have to be central in a university," he said. "It isn't the bricks and mortar you remember, it is that person who somehow spoke to you and touched you, and this experience even intensifies in the graduate years."[11]

Fund-raising was becoming increasingly important to the University. It was among the focus items Beering listed when he first arrived on campus, and Steve and Jane Beering worked on it as a team.

"One of the things that Steve should be given full credit for was raising the level of private fund-raising," Ford said. "At Purdue historically we hadn't become quite as sophisticated until they came. Jane should get a lot of credit in fund-raising. The donor relation work she did was amazing. She always had a camera."[12]

If a picture is worth one thousand words, Jane Beering wrote one hundred million of them.

35

A Photo Finish

1983 to 2000

Purdue is not a job to us, it's a life.

—STEVEN BEERING

I f you had been invited to Westwood for a Purdue University event in the mid- to late 1980s and throughout the 1990s, you would have met Steve and Jane Beering.

He would have spoken with you in a tone that was at once friendly and presidential, warm but formal. You would have come away thinking this was a very smart and dignified man.

She would have laughed with you and taken your photo. A week later, you would have received an envelope from Westwood. Inside would be a warm note from Jane and your photo, a memento of your evening with the Beerings at the presidential home. Jane estimated she took a hundred thousand of those photos, before digital cameras, in the days when she had to carry and load film and have it developed and printed. Those photos are in Boilermaker family albums all over the nation and the world. In a March 2000 interview Jane Beering said she didn't raise funds. She raised friends.[1]

Wherever Jane Beering went, her camera went with her. In addition to all the work taking and processing the photographs, she also catalogued them. "I can't imagine how many hours she spent doing this," Purdue vice president and treasurer

Purdue president Steven Beering celebrates the Vision 21 Campaign with alumni, astronauts Neil Armstrong (left), Eugene Cernan (to Beering's right), and ABC sportscaster Chris Schenkel, a Purdue alumnus.

Frederick Ford said. "A lot of people received notes and pictures from her and that helped to build relationships between donors and the University."[2]

Fund-raising had been among Beering's goals when he arrived on campus in 1983. President Frederick Hovde had one capital campaign near the end of his term. President Arthur Hansen raised the bar. When Beering arrived, Purdue was receiving about $20 million a year in private support. During his last year the University raised about $90 million. He increased membership in the President's Council from fourteen hundred member households to seventy-six hundred.

In 1985 he named Charles Wise vice president for development. A 1965 Purdue graduate, Wise had been vice president of business services and assistant treasurer. The appointment set the stage for the biggest fund-raising campaign in the University's history. It was called Vision 21, and its goal was to raise $255 million and launch Purdue into the twenty-first century. Neil Armstrong and Eugene Cernan, the first and last men to walk on the moon, were honorary co-chairs. The campaign ended in 1994, having raised $332 million—the largest amount raised by any public university in Indiana at that time.

Betty M. Nelson

In 1987 Betty M. Nelson was named dean of students, replacing Barbara Cook, who retired. She was Purdue's fifth female dean of women and dean of students following Carolyn Shoemaker, who served part-time beginning in 1913. Part of the tradition of a new dean was writing in a Bible that had been passed from Shoemaker, to Dorothy

Five women deans: (left to right) Helen Schleman, dean of women 1947 to 1968; Beverley Stone, dean of women 1968 to 1974 and dean of students 1974 to 1980; Barbara Cook, dean of students 1980 to 1987; Betty Nelson, dean of students 1987 to 1995; Dorothy Stratton, dean of women 1933 to 1947. (Photo by David Umberger/Purdue University)

Stratton, to Helen Schleman, to Beverley Stone, to Barbara Cook, and to Nelson. No other university had a heritage to equal that of Purdue's female deans.

A native of West Virginia, Nelson received a bachelor's degree from Radford University in 1957 where she studied psychology and sociology. Radford was a Virginia women's college where female students had "hours" during which they were required to be in the residence halls. If they went out after dinner, they had to sign out, and to ride in a car with a date, female students needed a letter of permission from their parents.

She earned a master's degree in student personnel administration from Ohio University and arrived at Purdue in 1965. She served Purdue for more than thirty years, beginning in the psychology department before joining Helen Schleman in the Office of the Dean of Women.

One of her tasks was helping to implement the federal Rehabilitation Act of 1973 that prohibited discrimination on the basis of disabilities. This meant making the Purdue campus accessible. People who used wheelchairs could not get to upper floors in classroom buildings that featured steep steps and no elevators. Getting over curbs to cross streets was nearly impossible.

Nelson said the University was slow in beginning the work, and student organizations played a major role in getting things going. Curb cuts costs $200, and twenty-five student organizations were quick to come forward, donating the funds in return for having their name placed at the locations. Students planned Handicapped Awareness Days that included asking people to negotiate the campus in wheelchairs.

Among Purdue administrators who were very active in accessibility issues was Ken Burns. Burns began his Purdue career as a trainee on the West Lafayette campus in 1967. Later that year, he became purchasing manager at Indiana University–Purdue University Fort Wayne. In 1973, he returned to West Lafayette as director of purchases and service enterprises, a position he held until taking over the University's physical facilities in 1981. In 1998 he was named executive vice president and treasurer, and he served in that position until he retired in 2004.

Nelson was sometimes described as an iron fist in a velvet glove.[3] She served with distinction and retired in 1995. She was replaced by Tony Hawkins, an associate dean of students who had worked at Purdue for twenty-one years.

Chairman of the Board

In 1989 Donald Powers, the board of trustees president who recruited Beering to Purdue, stepped aside. Bob Jesse replaced him. Four years later Jesse was replaced by one of the two longest serving board leaders in Purdue history—J. Timothy McGinley. The title was changed to chairman.

A Purdue basketball star as a student, McGinley won honors for both his athleticism and academics. He graduated with a degree in chemical engineering in 1965. McGinley went on to earn an MBA from the Harvard University Graduate School of Business, after which he was a White House Fellow under President Lyndon Baines Johnson. He was a special assistant to the secretary of labor, ran unsuccessfully for the U.S. Congress in 1970, and was president and founder of House Investments, a real estate investment banking and management firm that he led for more than thirty years. Governor Evan Bayh appointed him to the Purdue board in 1989.

"When Evan Bayh called me and asked me to go on the board, the first question I asked him was do you have an agenda or is there something you want to accomplish. And his answer was the appropriate one. He said no. He said I just want good people and I want them to use their best judgment about what to do. That was a wonderful principle he adhered to. And frankly, all the subsequent governors I've served under, Democrat or Republican, have pretty much been the same mold."[4] McGinley became the only board chair to serve with three Purdue presidents, and his sixteen-year tenure leading the trustees is only equaled by that of David Ross.

McGinley came to the board admiring Beering, and that never wavered. "Purdue has been fortunate; we've had great presidential leadership," he said in a 2008 interview. "Steve was very bright, very strong willed and those traits served him well. And you can't ignore the role that Jane Beering played. Jane was totally committed to the University and to her role in helping Steve."

In 1997 Beering was heading toward mandatory retirement age for a Purdue president. The rules at that time required top administrators to step down at the end of June after they turned sixty-five. Beering turned sixty-five in August 1997, meaning he would have been required to retire at the end of June 1998. He was in no way ready to retire. He wanted to continue and the board also wanted him to continue, but only for two more years. They granted him that extension.

McGinley said:

> There was a two-part conclusion [on Beering]. Steve certainly had the continued
> intellectual and physical abilities to do the job. Things were going well. And, if
> we had a defined period—two years—to decide on a process for selecting a new
> president, that would be very beneficial. We hadn't searched for a president for
> almost seventeen years. It was a combination of feeling comfortable with Steve,
> comfortable with the direction the University was going, and willing to have a
> lot of time to think things through and get a process in place that would lead us
> to the next president. The board was committed to a selection process that was
> very inclusive with communication and input from all stakeholders including stu-
> dents, faculty, staff, alumni, donors, the local community, and leaders of the state.
> That takes time, but is critically important for the success of a new presidency.[5]

Under McGinley's leadership the board started evaluating the University, its needs and potentials, and the kind of leadership it needed next.

The celebrations for Steve and Jane Beering began at Homecoming in October 1999, when eighteen of Purdue's nineteen living astronauts gathered on campus for a second reunion to honor the Beerings.

As of the spring of 2000, more than half of Purdue's three hundred thousand living graduates had a diploma with Steve Beering's signature.

At a board of trustees meeting on March 1, 1997, when his tenure was extended two years, Beering summed up his feelings. "Our time at Purdue has been the most important period in our lives," he said. "I love the University and its students. I deeply respect my faculty and staff and colleagues, and I look forward to meeting the challenges that remain during the rest of this century."[6]

Purdue president Steven Beering (center, first row) is surrounded by Boilermaker astronauts in front of the Memorial Union during a second campus reunion in 1999. Next to Beering are Betty Grissom (left), widow of Virgil "Gus" Grissom, and Martha Chaffee (right), a former Boilermaker homecoming queen and widow of Roger Chaffee.

Later he would add, "Purdue is not a job to us, it's a life." His wife and partner at Purdue, Jane, called their seventeen years "the greatest adventure of our lives."

When the Beerings retired, the Purdue University Retirees Association created a garden for Jane at the crest of Slayter Hill—one of her favorites places on campus. When she died in 2015 at the age of eighty-one, she was buried in the garden with a place left beside her for her husband. John Purdue, David Ross, Jane Beering, and eventually Steve Beering are the only people buried on the campus.

As the Beering years were winding down in West Lafayette, the president of Iowa State University received a phone call from McGinley, who introduced himself and said they were looking for a new president. "Thank you, but frankly I'm not interested," the Iowa State president said.

McGinley was persistent. And Purdue was about to start a seven-year sprint to preeminence.

36

Martin Jischke:
The Next Level

1941 to 2000

It is quite humbling when I think about myself, a
Midwesterner, a first-generation college graduate, assuming
the presidency of one of the nation's great universities.

—Martin Jischke

Martin C. Jischke and his wife, Patty, sat alone deep into the evening in the
Knoll at Iowa State University.

The Knoll was the 13,342-square-foot mansion where the Iowa State president
and first lady lived and entertained.

The Jischkes were waiting for guests—very late guests. They sat in the dining room where a meal was ready to be served. It had been prepared by Anthony
Cawdron—a man whose culinary and sommelier talents were so superb that people
would later donate $10,000 to charity to enjoy his dinners. The Jischkes finally decided
the only thing to do was to eat the meal themselves—along with Cawdron and Carol
Wightman, who also helped at the Knoll.

Jischke had expected his guests, the Purdue University Board of Trustees, to
offer him their presidency that evening. The chairman of the board had all but told

him so. But the trustees sat in an airport hangar in Indianapolis while a torrential downpour consumed the city and grounded flights. The Purdue presidency would have to wait.

Purdue's chairman of the board, J. Timothy McGinley, was not a waiting man. He had been working hard toward this moment for two years. Upon extending the contract of President Steven Beering to the year 2000, he immediately set the board to evaluating very throughly all that Purdue was and what it could become. They determined they were in an excellent position, but it was time to take the kind of step alumnus Neil Armstrong had accomplished on the moon. They picked the terminology themselves. They wanted to go to the "next level." It would be a giant leap.

That would require, among other things, a fund drive the size of which very few universities in the country had ever accomplished. It would take a strategic plan to prioritize focuses and investments. It would take a president who could administer a university, raise money, create a vision, sell others on his ideas, and accomplish them. It would take a president who could excite the campus community, its alumni, and its friends, and build partnerships with the state's business and government leaders. The Purdue trustees didn't know it when they started making these plans, but the man they were looking for all along was hard at work at Iowa State.

It was McGinley's persistence that brought Jischke to Purdue. The trustees hired a national search company to help them find their new leader. They had a campus committee and sought input from students, faculty, staff, and others. When the search began in earnest, they had a list of 120 names. They went through all the résumés, debated them, and whittled the number down to fifty. When they were down to about twenty individuals, everything got very serious. The search firm started calling candidates to find out who might actually be interested. Everything was done in complete secrecy, because candidates did not want people to know they were considering another opportunity. The search firm contacted Jischke by letter and did not receive a favorable response. Jischke was used to getting letters like that from "headhunters." Thanks, but no thanks, he responded.

"We got down to maybe ten candidates and Martin's résumé looked so perfect," McGinley said. "We thought, we ought to make sure he doesn't want to be a candidate—see if we could sell it to him. So, I decided to call him. He took the call."[1]

It was November 1999. McGinley and Jischke had a lively conversation. They discovered they had both been White House Fellows, one of the nation's most prestigious programs for leadership and service. The telephone conversation lasted about fifteen minutes before McGinley got to the point. Jischke was ready for it.

"Thanks for your call, but no thanks," Jischke said. "I'm happy. Things are going well here." McGinley told him they were down to ten candidates and he said he might call back. "I'll always take your phone call," Jischke said.[2] McGinley took that to mean that Jischke hadn't yet slammed shut and locked the door on the idea of moving to West Lafayette.

As the list of candidates dropped to five, McGinley called Jischke two more times. "I really don't want to be a pest," he said. "But why don't I fly to Iowa? Just meet me anywhere and let me talk to you about Purdue."[3] Finally, Jischke agreed, and the meeting was set for a Holiday Inn near the Des Moines Airport.

What McGinley did not know was that between the second and third calls Jischke had become concerned about the new governor of Iowa, Tom Vilsack, and his commitment to higher education. At a Big Eight basketball tournament in Kansas City, Jischke had received a telephone call from the chairman of the Iowa Board of Regents. The regents oversaw Iowa's public universities. Jischke had successfully worked with legislators to win funds for a new building at Iowa State. Vilsack wasn't happy about it, and the chairman of the regents sided with the governor. Jischke reminded the chairman that the board of regents had approved the building. After the "rather tense" phone call ended, Jischke turned to Patty and said, "Maybe this thing at Purdue might be attractive after all."[4] Within minutes the phone rang again. It was McGinley, and Jischke agree to meet some of the Purdue trustees.

McGinley, along with trustees Barbara Edmondson and Wayne Townsend, flew to Des Moines, where they had a three-hour breakfast with Jischke. From that point the dialogue moved quickly.

After more meetings, McGinley made it clear that they were going to make an offer, and the Purdue trustees made arrangements to visit Ames. That was the evening the Jischkes, Cawdron, and Wightman dined alone. Within a week the trustees visited Iowa State, had lunch with the Jischkes, toured the campus, and offered him the Purdue presidency. "I told them I was seriously interested," Jischke said. "We talked terms and conditions. That went pretty easy."[5]

The announcement came on March 23, 2000. The next morning the Lafayette *Journal and Courier* proclaimed at the top of page 1, "Jischke Crowned as University's Tenth President." He was pictured in a suit and tie holding up a gold Purdue sweatshirt with a big smile on his face and a Purdue cap on his head. As soon as he put on the cap, Patty Jischke took it off, bent the brim into the proper shape, and put it back. The new president and his wife worked as a team.

"Personally, it is quite humbling when I think about myself, a Midwesterner, a first-generation college graduate, assuming the presidency of one of the nation's great universities," Jischke said. "Perhaps this can only [happen] in America." Then he drew a laugh and applause from the audience when he added, "I think the trustees made a good choice."[6]

Martin Jischke with Patty Jischke as he is announced
the tenth president of Purdue in 2000.

McGinley was jubilant. "We love basketball here," the former Boilermaker basketball star said. "And [in Jischke] Purdue University has just signed our first-round draft choice." McGinley called it a "proud and historic moment."[7]

In a sense, Jischke's whole life had been building toward this moment.

In the 1890s his paternal grandfather immigrated to the United States as a boy with his parents from Silesia, which was then part of Germany and later part of Poland. The family settled in Sister Bay, Wisconsin, north of Green Bay.

Jischke's grandparents married and had a grocery business in Sister Bay and the couple had five children. The fourth born was Jischke's father. "He had ambitions to be a medical doctor," Jischke said. "But an unfortunate thing happened to him—called the Depression. So when he finished high school the family was very poor and the only thing he could do was leave home. One of the unfulfilled dreams he had for his entire life was that he never had the chance to go to college. It just couldn't happen. He went to Chicago looking for work."[8] In Chicago he met his future wife, who was a secretary for a broker. They married in 1940. Jischke, the oldest of six children, was born August 7, 1941.

In Chicago, Jischke's father made good use of his experience working in the family store in Wisconsin. He worked for large grocery chains such as Jewel Tea, the A&P, National, and IGA. For a while he had his own store. Jischke started going to work with his dad when he was about eight years old.

"He initially would take me along on Saturdays to help, sweeping and cleaning, not very glamorous stuff," Jischke said. "I graduated to slicing lunch meat and actually worked behind the counter. Eventually by the time I was in high school and college my father had moved on to becoming more of a sales manager for wholesalers."[9]

Jischke skipped a grade in elementary school. "I was always a good student," he said. "And my parents encouraged that. If I brought home straight A's I was rewarded. I remember one Christmas I got a chemistry set."[10]

He graduated from Proviso High School in Maywood, a suburb of Chicago. "When it came time for a college decision my parents were encouraging, but frankly didn't know much about it," he said. "They filled out all the scholarship application forms and they were excited about the idea of me going to college. But we never visited a college. I enrolled at Illinois Institute of Technology in Chicago based on information that came through the mail. It never dawned on me that you would go visit. My parents were enormously proud, but especially my father. It was the chance he didn't have."[11]

His father never took vacations and worked overtime whenever he could to make extra money. His mother worked when the children were older. Jischke won scholarship money and received Ford Foundation loans that would be forgiven if he became a teacher. His parents helped out with extra money when they had it.

Jischke also worked to pay his way through school. As an undergraduate, one summer he opened the meat market of a new supermarket in Oak Park. "My experiences in grocery stores turned out to be pretty decisive," Jischke said. "I learned a lot of powerful lessons about hard work, what it meant to get up in the morning and work all day and into the night, almost so tired you couldn't eat, and then get up the next day and do it all again. I was a young man and I was hiring and firing, I was setting prices, running this little organization. I remember having lots of conversations with some of the men I hired, about what their lives were like. A number of them hadn't finished high school, or they had married at a young age, had families at a young age. They had limited possibilities because of lack of education. It was a vivid lesson for me. Conversations ended with them saying, 'For God's sake get a good education.'"[12]

Jischke was a student leader at Illinois Institute of Technology, where he was president of his class. He joined Delta Tau Delta fraternity and was vice president of his pledge class, rush chairman, and social chairman. He was a radio sports announcer. He earned a reputation in the fraternity house for falling asleep at two in the morning at the desk where he'd been studying.

"I was the first person in my family to go to college," Jischke said. "So it was altogether a new experience. And it was daunting. I lived on campus and majored in physics. I chose physics because it was general. I wasn't quite sure what I wanted to study, but I was very good in mathematics and science."[13]

The race to the moon started while he was an undergraduate, and he was fascinated by the excitement, drama, and technology of the space program. He took courses in engineering, and the idea of being associated with NASA appealed to him. He received his bachelor's degree in physics in 1963. During his senior year he applied to ten graduate schools and was accepted and offered financial support to every one of them, including Stanford and Massachusetts Institute of Technology. He picked MIT.

He earned his master's degree in two semesters. He considered going to Stanford for his PhD, but stayed at MIT, received fellowships, and finished in 1968. Jischke enjoyed being a teacher during his graduate years at MIT and decided on a career as a university professor, but not before interviewing with Boeing, McDonnell Douglas, Bell Laboratories, and other companies. He also interviewed with a number of universities including Penn State, the University of Oklahoma, and Purdue. Penn State and Purdue decided not to fill their positions. But Oklahoma offered him an assistant professorship of aerospace and mechanical engineering. He accepted.

"I thought I'd enjoy being in a university environment," Jischke said. "There's more freedom on the research side, and I thought it would be a more interesting environment having colleagues who weren't all aerospace engineers—people who were interested in history, literature, and so on. And secondly, pragmatically, if teaching didn't work out it would be easier to go from a relatively lower paying academic job to a higher paying industry job than the other way around."[14]

With the exception of his books, which he shipped, he put everything he owned in a small Corvair Monza and drove to Oklahoma. Along the way he stopped in Chicago for a three-week visit with his family. He gave a copy of his PhD thesis to his parents. He had dedicated it to them. "They had absolutely no idea what I was giving them," Jischke said. "They had never held one, had never seen one. They couldn't understand it. It was very mathematical. But they were proud of the fact that I did it." There were tears in his father's eyes.[15]

He started in Oklahoma at the beginning of the 1968 school year. All went well—with his work, at least. In the spring of 1970 his department head asked how his social life was going. Jischke told him it wasn't going well. "I was working hard," he said. "I was doing well as a professor, good teacher evaluations, got my research program going. But I didn't have much of a social life."[16]

A faculty member overheard the conversation and came across a friend, Patty Fowler, as he walked through campus. She was a friend of his daughter and was enrolled as a graduate student in library science. Her father was a distinguished physicist at the University. Jischke's faculty friend asked Fowler if she'd like to meet a nice professor of engineering. She said no. He persisted and arranged a picnic for Jischke to meet Patty. "I showed up and Patty didn't—she didn't agree to come because she had a prior commitment," Jischke said.[17]

His friend continued to persist and Patty came to a second picnic at the Wichita Mountains Wildlife Refuge in southwestern Oklahoma. "Patty wasn't going to waste any time," Jischke said. "She was taking a course in invertebrate zoology and brought along nets to capture bugs and killing jars to put the specimens in because she had to make a collection as part of her course. So we spent some time traipsing around this Wichita wildlife refuge, catching bugs."[18]

Their next date was to a student engineering club annual banquet. Jischke invited members of the faculty to his apartment for a party afterward. "I made whiskey sours and Brandy Alexanders and I had potato chips and dip and cheese. We had a great party. She was quite impressed that this bachelor would reciprocate all the kindnesses that faculty colleagues had shown him. And could pull it off!"[19]

He sealed the deal quickly, they married on December 26, 1970, and had two children, Mary and Charles. Jischke's career advanced rapidly. He became a full professor and he had an opportunity for a sabbatical. He had offers to study at Imperial College in London or Cambridge University in England or at Harvard. He also applied for a White House fellowship, was accepted, and took it from September 1975 to September 1976. It was an amazing experience. President Gerald Ford introduced him to the national press corps during a White House ceremony in the Rose Garden. He spent a year in Washington as special assistant to secretary of transportation William Coleman.

"Coleman was an interesting guy, a very talented guy," Jischke said. "He was one of the young attorneys who worked with Thurgood Marshall on *Brown v. Board of Education*. As part of the White House fellowship we were encouraged to think about what we were going to do with our life. It was that year that I decided I wanted to be a university president."[20]

He created a personal strategic plan to accomplish his goal. He studied the large issues facing higher education. He learned about fund-raising. He worked on public speaking and took a class that helped him develop that skill.

By the fall of 1976, when he returned to Oklahoma, he was thirty-five years old. His goal was to be a university president by the time he was fifty—fifteen years. He was first a department head and in 1981 was named dean of engineering. In 1984 William Banowsky had resigned as president of the University of Oklahoma, and Jischke accepted an offer from the board of regents to become interim president. He was forty-three. He had achieved his goal in eight years.

He became a candidate for the permanent position, but the board was split concerning many issues, and their disagreements continued. He didn't get the job. "It was tough, but it was a marvelous learning experience," Jischke said. "I came to the conclusion that while being president of a university was an enormously heady thing, not all these jobs were good jobs. I became more discerning as to what circumstances you needed to be successful. The split in the board made the University of Oklahoma job impossible."[21]

Other offers came. He was considered for president of Kansas State University and chancellor of the University of Missouri–Rolla, later renamed Missouri University of Science and Technology. The offer from Missouri–Rolla came first. He took it and served from 1986 to 1991. At Missouri–Rolla he created a strategic plan. He improved fund-raising and external communications.

In 1991 he was hired as president of Iowa State. He started work in Ames on a Saturday morning by calling the affirmative action officer to his office. He signed equal employment opportunity and affirmative action policy statements. There had been no such policies in place, and he wanted everyone to know he had signed them. He had meetings and held a news conference. He let everyone know he had arrived and he was at work.

"I've done that everywhere I've been," Jischke said. "Whatever day I start, that day is full and very public and it is to say to people, 'We're here, we're at work, and this is serious business.'"[22]

People at Purdue would soon learn that for themselves. Before he arrived at Purdue he met with people who would report directly to him, and the Saturday before his first day at the office, he called together administrators for a mini-retreat at Westwood. He told them there would be changes; the intensity would notch up. He thought the University needed more energy and accountability. He believed people naturally wanted to contribute, and to do that they needed to know what the University was trying to accomplish. Jischke said some people loved the meeting, but it made others nervous.[23]

The next day he held a meeting with his cabinet in the Purdue Memorial Union Anniversary Drawing Room. Tables were set up in a "U" with a table for Jischke at the open end. One by one he spoke to each cabinet member with pointed, knowledgeable questions about their operation. Cabinet members came away understanding he had done his homework and knew about the University he was going to lead.[24]

During the next seven years, many people at Purdue would say they never worked so hard. They also said they had never accomplished so much and had so much fun.

GIANT LEAPS IN RESEARCH

Vernon L. Smith

In 1955, teaching his first class at Purdue University, Professor Vernon L. Smith involved his economics students in an experiment to understand why and how markets reach competitive equilibrium.

That experiment was the beginning of research that led Smith to receive the 2002 Nobel Prize in Economics.

Smith, the father of experimental economics, shared the prize with Daniel Kahneman of Princeton University. In addition to Purdue, Smith also taught at George Mason University and Chapman University.

"I taught Principles of Economics, and found it a challenge to convey basic microeconomic theory to students," Smith would later write. "Why/how could any market approximate a competitive equilibrium? I resolved that on the first day of class, I would try running a market experiment that would give the students an opportunity to experience an actual market, and me the opportunity to observe one in which I knew, but they did not know, what were the alleged driving conditions of supply and demand in that market."[25]

When the prize was announced, George Horwich, one of Smith's Purdue colleagues, said "This was the first time anyone had demonstrated experimentally how markets form. What Vernon . . . accomplished [was] an economic revolution. He took the theory of market behavior out of the ivory tower and showed us how to simulate what is going on in the real world." Smith continued the experiments at Purdue and left campus in 1967.[26]

Born in 1927 in Kansas, Smith grew up in a home without water or electricity. In 1945 he enrolled at Caltech, where he took a freshman course taught by Linus Pauling, who would receive the 1954 Nobel Prize in Chemistry. Smith studied electrical engineering but turned toward economics his senior year. He received a master's degree in economics from the University of Kansas and began working on his PhD in economics at Harvard in 1952.

In 1955 Emanuel Weiler, the new head of Purdue's Department of Economics, recruited Smith. Weiler looked for young people with strong potential, such as Smith.[27] Those same faculty would later form the nucleus of the Krannert School of Management.

In 1978 Smith wrote, "Many of us (in the economics department) had only one thing in common: a very subverting sense of considerable dissatisfaction with our own graduate education . . . and the state of economic knowledge. That allowed each of us to do our own thing.[28]

"Out of this menagerie many successful cultural experiments emerged (including), for me, experimental economics."[29]

And forty-seven years later a Nobel Prize.

37

Who Moved My Cheese?

2000 to 2001

The board talked about wanting to get to the next level,
and that language is the language of change.

—MARTIN JISCHKE

Among the many positives that attracted Martin Jischke to Purdue University
was its board of trustees. The trustees worked well together, they were united
in what they wanted to accomplish, and they were willing to take the steps necessary
to reach what they identified as "the next level."

"They kept saying they wanted to take Purdue to the next level and they said so
consistently," Jischke said. "They were together, united in this idea of taking Purdue
up a notch. They thought it required a strategic plan. They thought it required more
fund-raising. They thought it required a more public presence for the president. And
these were all things that I believed in and did well. The job description fit me. I
was very excited about it. The idea that a university of this quality and stature was
prepared to do better, I found very appealing."[1]

After Jischke's appointment was announced, but before he began work at Purdue,
a reporter asked him a classic question: What book had he recently read? He mentioned *Who Moved My Cheese?*, by Spencer Johnson. The 1998 book discussed change
in work and personal life and how two mice (people) reacted to it. The theme was
that change will happen, so prepare for it and go with it. It was a *New York Times*

business best seller for almost five years. When word got out that Jischke cited the book, Purdue people rushed to read it to gain insight into the new leader who was coming, promising change. "The board prepared the community for someone like me," Jischke said. "They talked about wanting to get to the next level, and that language is the language of change."[2]

When he arrived, for the most part Jischke found what he expected. But there were surprises. As he went from school to school, he learned that every one of them wanted a new building. "Second, I was surprised at how little interdisciplinary activity there was," Jischke said.

On the positive side, Jischke said he found the quality of the faculty to be high. He said people were committed to Purdue, and they were receptive to the changes he was introducing. "I think most people thought it was time for us to notch up," he said. "And the support of the alumni and the fund-raising was frankly breathtaking."[3]

Among Jischke's first moves was the reorganization of University Development, which was responsible for fund-raising. Carolyn Gery, a 1969 Purdue graduate and one of the first people hired to help the University launch fund-raising in 1972 during the administration of Arthur Hansen, was named interim vice president while a national search was conducted for a person to lead the University in a major capital campaign.

Jischke began presidential forums. These included deans, vice presidents, directors, and department heads—and even more people attended as time passed. The meetings, held monthly in the Union South Ballroom, starting with socializing over coffee at 7:30 a.m. The meetings began precisely at 8 a.m. Jischke knew the social time before the meeting would be valuable to people who were trying to get hold of one another. It also allowed people to meet and discuss work issues.

Jischke had only been in the job three months when, on November 16, he made his first big announcement—the Indiana Resident Top Scholars program. One hundred fifty full-tuition scholarships would be offered annually to in-state students based on academic merit. The goal was to keep the best and brightest students in Indiana.

"Everything's Coming Up Roses"

What happened next was a game changer. At the end of the 2000 football season, Purdue, led by Coach Joe Tiller, was headed to the Rose Bowl for the first time since January 1967. Quarterback Drew Brees was a Heisman Trophy finalist. He would go on to a certain NFL Hall of Fame career, first with the San Diego Chargers and then with the New Orleans Saints.[4]

Purdue went eight and four in 2000, was tri-champion for the Big Ten title, and received the Rose Bowl invitation. The October 28, 2000, game in Ross-Ade Stadium against Ohio State was among the great Boilermaker victories and one of the greatest comebacks by one of its quarterbacks. Purdue was heading to a rare victory against the Buckeyes and a key win in its effort to reach Pasadena. With the Boilermakers ahead 20–24, Brees dropped back to pass, slipped, threw the ball, and it was easily intercepted and returned to Purdue's two-yard line. Ohio State scored and went ahead 27–24 with two minutes and sixteen seconds left in the game.

A packed Ross-Ade Stadium was despondent. But not Brees. He fought back after the Buckeye kickoff. On the second play of the series he saw his receiver, Seth Morales, wide open and heading toward the end zone. Brees threw the ball, and to people watching, it seemed to take days before reaching the outreached hands of Morales for a sixty-four-yard touchdown and a Boilermaker victory, 31–27.

The following week Purdue lost to Michigan State in Lansing, but easily defeated Indiana University on a cold Saturday in West Lafayette, 41–13. As the final seconds ticked away, the Purdue "All-American" Marching Band played "California, Here We Come," and "Everything's Coming Up Roses."

"We're going to the Rose Bowl! We're going to the Rose Bowl!" Tiller shouted on the sideline as the game ended.[5]

A huge number of Purdue fans traveled to California for the Rose Bowl on January 1, 2001. Purdue lost the big game 24–35 to the University of Washington. But it won big time in getting its fund-raising off the ground. "Before we went I told

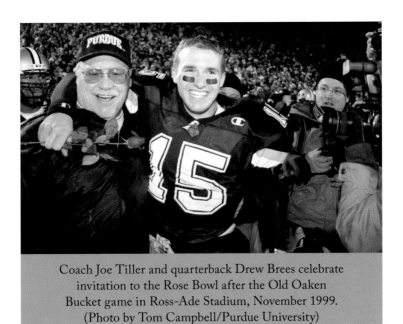

Coach Joe Tiller and quarterback Drew Brees celebrate
invitation to the Rose Bowl after the Old Oaken
Bucket game in Ross-Ade Stadium, November 1999.
(Photo by Tom Campbell/Purdue University)

Carolyn Gery this is not just a football game," Jischke said. "We're going to raise some money. So we went out there and I remember having breakfast with trustee Mike Birck and saying, 'We've got to build this nanotechnology building. I'd like to ask you for $30 million.' His response to me was, 'That sounds like something we might want to do.' I knew we were on our way."[6]

Tiller's final season at Purdue was in 2008. His eighty-seven victories gave him the most wins among all Boilermaker football coaches. He and his wife, Arnette, introduced the popular cheer "Boiler Up! and Hammer Down!" Tiller died September 30, 2017, at the age of seventy-four.

Women's Basketball and Student Vandalism

On the evening of April 1, 2001, another athletic event captured the focus on the Purdue campus. Students, faculty, and people throughout the area were locked to their television screens watching the Purdue women's basketball team battle for a national title for the second time in three years.

The game was a cliffhanger, and it was against Notre Dame, always an archrival of the Boilermakers. Played in St. Louis, the game was tied 66–66 when Ruth Riley, the Notre Dame All-American center, made two free throws. With 5.8 seconds left, Purdue got the ball and hustled down court. From the free throw line the team's star player, Katie Douglas, put up an off-balance shot. Boilermaker hearts dropped as the ball bounded off the rim and fell to the court.

Back home on the Purdue campus, students started coming out of their living units. The crowd was estimated at a thousand, most of them watching while a much smaller group set fires and vandalized property. When it was over, the vandalism was estimated at more than $76,000. As many as two hundred police were called in, and they had to use tear gas. There had been similar, but smaller incidents at Purdue after the women's National Championship game in 1999 and after a men's NCAA tournament loss in 2000. By 2001 student violence was becoming somewhat common nationwide after NCAA tournament basketball games. Such behavior had scarred the reputation of several other universities already. Jischke would not stand for it, and he was widely praised for his response.

The smoke had hardly cleared before police announced that five students had been arrested along with one former student. Two days after the event a straight A engineering student was packing his bags in Wiley Hall after being expelled. Other students faced suspension or probation in addition to court charges. Jischke made his feelings well known and called the conduct of the students "unacceptable. This is not what Purdue is all about," he said.[7] Police offered $5,000 rewards for information leading to arrests, and to help with identification, they posted photos of the vandalism as it was being committed on a website where they could be viewed by friends, acquaintances—and parents. The Internet was relatively new, and this was an idea that would catch on elsewhere.

"There will be consequences for this kind of behavior," Jischke said. He used his monthly radio show to lay partial blame on the students who watched as the vandalism took place. "Those watching provide a kind of anonymity and protection to the students who participated," he said. "They helped create the environment that makes this a difficult problem. They hindered police and the fact that they were there, cheering these people on, encouraged individuals to behave stupidly and commit acts of vandalism they might otherwise not have done."[8] Dean of Students Tony Hawkins said people who were spectators or agitators could face probation if identified through photographs.

In the years to come there were no more incidents of postgame vandalism at Purdue.

A Fence for the Fountain

A university president must wrestle with budgets and critical issues in higher education, and no matter what decisions are made, some people will be happy and some will not. But occasionally events take place on college campuses showing that what people are really concerned about are their traditions. And their fountains.

Two of the most popular features from the campus beautification project were the fountain on the Mall in front of Hovde Hall and the redesigned Loeb Fountain in Founders Park. People loved to sit and listen to the water. They loved to watch lights changing color on the Purdue Mall fountain. They loved it on hot days when a breeze blew a cooling mist of water.

But what some people really loved to do was run through the fountains and get soaking wet. There were all kinds of reasons to do this. It cooled a person on a hot summer day. People from the community brought little children to campus in bathing suits and let them get wet. Students did it on a dare. After commencement, celebrating graduates ran through the Purdue Mall fountain in full cap and gown. People ran through the water—just because it was there. It was fun. They believed the fountains weren't just to look at. They were to play in. They were interactive. Even dogs played in the cooling water.

During the summer of 2000 a child ran through the sculpture fountain in front of Hovde Hall. The spray of water that shot high into the air began at ground level.

The force of the water, which would stop and then come back on, knocked the child off balance and threw her into a concrete wall, breaking her arm. It was a wake-up call, University officials said.

In response, a decision was reached to place metal fencing around both the Purdue Mall and Memorial Mall fountains to keep people from running through the water. Three days after the announcement the Lafayette *Journal and Courier* carried a reasoned column by University vice president and treasurer Ken Burns, who explained what had to be done. Next to the column were letters from unhappy citizens. "Fountain fences: Dousing danger or is Purdue all wet," the headline stated. Burns said his own children enjoyed playing in the fountain. Administrators understood that people loved this activity that had became a tradition. Anything that continues for four years on a university campus becomes a tradition. But the fountains were not designed for recreational use, Burns said, and the decision to put up fencing came very reluctantly. "Injuries began to occur from the beginning and as more and more people engage in these activities, the danger becomes greater. If nothing is done, the question is not whether serious injury—or even death—will occur, but rather when." Many people were not convinced. One wrote a letter to the editor that read, "Another of my life's simple pleasures, gone."[9]

Jischke did not make the decision on the fencing and he didn't announce it. But when there is a strong president, people believe the top person is ultimately responsible for everything.

Five days after the announcement, a petition drive had been started, and Robert Youngman, the designer of the fountain, now retired from the University of Illinois, was brought in for advice. He sympathized with the concerns of the Purdue administration. But he said the fencing would spoil the design. He walked around, studied his work, and within thirty minutes came up with an idea—which he might well have had before he came. Instead of the water shooting out of the ground, why not have it spray into the air out of a tall, mirror-finished, stainless steel cylinder?[10] Administrators liked the idea and announced they would study it. It was later approved, to everyone's satisfaction.

May 13, 2001, was a Sunday and the date of the final spring commencement program. A former dean at Purdue who was then executive vice president for academic affairs was set to retire at the end of June. His name was Robert Ringel, and he was widely respected and loved as campus icons are.[11] He had been through 125 commencements at Purdue wearing his doctoral cap, gown, and hood. He looked quite the distinguished figure as he walked into Elliott Hall of Music for his final commencement. A couple of hours later, photographers captured him soaking wet, water dripping off his black robe and gold sash. The fountain wasn't altered yet. It had been turned on just for commencement. And Ringel couldn't resist. "I figured I'd never have another chance to wear a cap and gown and walk through the fountain," he said. "It was a good way to go out."[12]

Two More People Arrive from Iowa State

In 2001 two people were hired who would have a huge impact on the Purdue community. They came from Iowa State, where they had worked closely with Jischke, and they were among the most popular of his Purdue hires.

The first was Rabindra "Rab" Mukerjea. On March 1, 2001, Mukerjea was named Purdue's director of strategic planning and assessment. He was key to getting the Purdue strategic plan rolling and successfully completed.

The second was senior vice president for advancement Murray Blackwelder, a friendly, outgoing man who put fun in fund-raising and passed his enthusiasm on to others. The year before at Iowa State, which was smaller than Purdue, he had completed a $458 million campaign.[13]

Purdue's recent accreditation report said the University could probably accomplish a $1 billion campaign. It seemed staggering to many people at Purdue, but not to Blackwelder. He couldn't wait to get started on it. A $1 billion campaign, he said, was among the things that attracted him to Purdue.

"I'm a fund-raiser. I love campaigns,"[14] he said. From the beginning Blackwelder assured Jischke that Purdue could raise more than $1 billion. Jischke wisely believed him.

Left: Rabindra "Rab" Mukerjea, director of strategic planning and assessment. Right: Murray Blackwelder, senior vice president for advancement.

GIANT LEAPS
IN RESEARCH

Arun K. Ghosh
*(Photo by Steve Scherer/
Purdue University)*

In 2006 physicians gained their first FDA-approved drug to treat drug-resistant HIV thanks to a molecule created by Purdue University professor Arun K. Ghosh.

While there were many treatments for AIDS on the market, none were effective for the treatment of patients with drug-resistant HIV. "This [was] the first treatment effective against the growing number of drug-resistant strains of HIV, the virus that causes AIDS," Ghosh said. "The problem of drug resistance is widespread."[15]

The National Institutes of Health funded the research. NIH has now funded his work on multidisciplinary research projects in the areas of synthetic organic, bioorganic, and medicinal chemistry for twenty-four consecutive years.

Since 2006 his laboratory has expanded into many areas, including Alzheimer disease, cancer, and more. "We started the groundwork for treatment in Alzheimer's," Ghosh said. "We've done seminal work in this area. And from that we've expanded into possible new treatments for diabetes."[16]

Born in Calcutta, India, Ghosh grew up in a family of educators. "In many ways I have been fortunate," he said. "I had good teachers and they helped me open my potential and explore chemistry."[17]

He did his undergraduate studies at Calcutta University, graduating in 1979, and then received a master's degree in organic chemistry at the Indian Institute of Technology in Kanpur. He was partly inspired in organic chemistry by the work of Purdue University professor Herbert Brown, who won the Nobel Prize in Chemistry in 1979.

In 1981 Ghosh came to the United States and earned his PhD from the University of Pittsburgh, followed by postgraduate work at Harvard. He then went into industry as a research fellow with Merck Research Laboratories in West Point, Pennsylvania. He only stayed in industry for six years before joining the faculty at the University of Illinois at Chicago.

"I wanted to see briefly what kind of research was going on in pharmaceutical laboratories," he said. "They were doing cutting-edge research at Merck. They were coming up with incredible drugs for human diseases. But my goal was always to go into academia."[18]

In 2005 he was recruited to Purdue with a dual appointment as a professor in the Department of Medicinal Chemistry and Molecular Pharmacology and the Department of Chemistry. In 2009 he was named the Ian P. Rothwell Distinguished Professor at Purdue.

"Purdue is a very exciting place for chemistry," he said. "It's a terrific place to be. I think chemistry is the most coveted place to be at Purdue. The students are extraordinary and highly motivated. This is one of the best places to do innovative research."[19]

38

Preeminence

2001 to 2002

I believe this announcement is a turning point in the history of
Purdue. It's the start of an extraordinary chapter in our history.

—Martin Jischke

S ally Viparina arrived on the Purdue University campus in late summer of 1972.
She was twenty-two years old, entering graduate school to study biology, and
excited about her future. She was the first person in her family to go to college. Her
father had not graduated high school and now she was preparing to study for a PhD.
She believed she would have wonderful opportunities on the campus.

Shortly after she arrived at Purdue the graduate adviser brought together most of the
women students in science and told them he was all but certain they would not succeed.
But it wasn't important. The only reason they had been accepted at all was to fulfill a
quota of female students, he said. He told them that in two years they'd all be gone.[1]

In two years Viparina was gone—by her own choice after that meeting. She had
earned her master's degree, but instead of staying at Purdue to complete her PhD in
cellular, molecular, and developmental biology, she went to the University of Arizona
in Tucson and had a wonderful four-year experience.

In 2001 Jischke named Sally Mason provost at Purdue. She was the first woman
to hold the position. In discussing women's issues, Mason sometimes told the story
of Viparina. It was very personal to her. She was Viparina before taking her married

name. Mason would use her experiences and skills to help Purdue meet its goals, especially in diversity among students and faculty.

Mason loved working with Jischke. "He was very analytical in his approach to everything," she said. "He was like a freight train. You see him coming, you'd better hop on the train or get out of the way. I enjoyed being on the freight train with him. He was so full of energy and good ideas. He was tough. He was demanding. He expected high quality and he expected you to do what was asked of you and more. For me, that was enjoyable."[2]

Office of Engagement, Discovery Park

On August 1, 2001, Purdue created the Office of Engagement to address economic development in Indiana, and Don Gentry, former dean of technology, was named vice provost for engagement. Jischke had plans to use Purdue to help move the economy of Indiana into the twenty-first century, and he traveled all around the state selling his vision. On and off campus he delivered as many as four hundred prepared remarks each year.

He held "Discover Purdue" days in communities in every quadrant of Indiana. He visited with government and business leaders and toured industries. He held luncheons, often with Rotarians. Jischke was a powerful speaker with a deep, booming voice. When he stepped to a podium and began to speak, people sat up, took notice, and listened. He was confident and people believed in him.

He also visited Purdue schools and colleges every year and held forums with faculty, listening to their comments and answering their questions. In forums, meetings, wherever he went, he was always prepared. Faculty and administrators said he was open to debate if anyone disagreed with him. But they said that if you got into a debate with him, you had better be very well prepared. Because he would be. His strong leadership style made some people nervous, while others loved it. His Hovde Hall staff loved him. At meetings he would ask them questions about how people were reacting to various events and issues. And he listened to them.

Discovery Park

On September 7, 2001, at a news conference in the Purdue Memorial Union, Jischke announced plans for a $100 million interdisciplinary research, learning, and outreach area to be called Discovery Park. He also announced the first building in Discovery Park—a $58 million Birck Nanotechnology Center—thanks to that gift from trustee

Michael and Katherine Birck that Jischke secured at the Rose Bowl. In addition, Purdue alumni Donald and Carol Scifres gave $10 million for the center. The building opened in 2005.

Discovery Park received state and federal support. Indiana governor Frank O'Bannon and U.S. senator Evan Bayh were on hand for the announcement, and Jischke did not understate its importance. "I believe this announcement is a turning point in the history of Purdue," he said. "It's the start of an extraordinary chapter in our history."[3]

Charles O. "Chip" Rutledge, dean of the School of Pharmacy, Nursing, and Health Services, took on added duties as the first director of Discovery Park in November 2001. He later was named Purdue vice president for research.

By the time Jischke retired, Discovery Park had grown into a $350 million enterprise with eleven interdisciplinary research centers. As the University approached its sesquicentennial, Discovery Park had twelve centers on forty acres south of State Street on the Purdue campus. It had attracted more than $110 million in sponsored research and involved a thousand University faculty, students, and staff. The trustees had been enthusiastic supporters of the creation and growth of Discovery Park, and when Chairman J. Timothy McGinley retired in 2009, a plaza and fountain in the new area were named for McGinley and his wife, Jane, in recognition of their contributions to the University.

J. Timothy and Jane McGinley watch as water begins flowing at a fountain in the newly named McGinley Plaza in Discovery Park. President France Córdova is at the podium. President Emeritus Steven and Jane Beering are seated in the front row left and President Emeritus Martin and Patty Jischke are seated to the right.

Discovery Park was Jischke's vision, and to bring it to fruition he sought and received help from the Lilly Endowment, one of the world's largest private philanthropic foundations. On October 26, 2001, the Lilly Endowment announced a nearly $26 million grant to Purdue's Discovery Park. Purdue would receive more support from Lilly Endowment.

In an interview years later Jischke explained the reasons behind developing the research area. "First, there was this huge need for more space, particularly research space," he said. "Second, and much more fundamental, it was quite clear early on that the size of the research enterprise at Purdue was small given the stature of the University. We needed to grow our research program. Third, it was also clear to me that interdisciplinary research [at Purdue] was quite small and lacked investment. I also wanted to get this vision aligned with economic development. The trustees and I talked about this before I took the job. It became clear to me that arguably the biggest issue facing Indiana was the changing economy and the need for the state to move into a newer, more robust economy. I thought research could play a big role in this."[4]

Strategic Plan

Life at Purdue in the fall of 2001 meant moving forward with the strategic plan.

The board of trustees approved the plan on November 2, 2001. It called for hiring three hundred additional faculty and using fewer graduate teaching assistants, who would switch to research. It included funding for new facilities and programs, increased diversity, more internships, more study abroad, and community service programs. Research of all kinds, especially interdisciplinary research, needed to expand. The plan addressed economic development, increasing faculty salaries, and moving Purdue forward. Schools and departments also prepared their own strategic plans.

Tuition Increase

With state appropriations for higher education stalled, more revenue was needed to accomplish the goals. Purdue's tuition and fees were eighth among the schools in the Big Ten. The strategic plan called for a tuition and fee hike of $1,000 on new students beginning in the fall of 2002. Students already enrolled at Purdue would not be charged the additional $1,000. Because of state funding shortfalls, there was also a 10 percent increase in fees for all students beginning in the fall of 2002.

The fees for Indiana resident returning students jumped to $4,580 a year. Out-of-state students paid fees and tuition of $15,260. New students paid the additional

$1,000. The increase still left Purdue a long way from being the best funded in the Big Ten and among its selected peers. Jischke met with students, parents, legislators, state administrations, faculty, and staff—everyone—to explain the need for the increase. In the end, it received wide support.

All the money from the $1,000 strategic plan fee increase went to programs that benefited students and improved the quality of their education.[5] "When the strategic plan was approved we pledged to use the new student fee to fund initiatives that directly benefited the students. We will stay the course on that promise," Jischke said.[6]

More Announcements

Announcements kept coming. In February 2002 Purdue opened an Office of Engagement in the city of Indianapolis. Also in February 2002 Jischke announced a partnership with Indianapolis Public Schools to form Science Bound, a program that would mentor eighth- through twelfth-grade students in science, engineering, technology, math and science education, and agriculture. Students who completed the five-year program would receive a four-year full tuition scholarship to Purdue. Alumnus Bob Bowen of Indianapolis gave the lead gift for the program as well as sustaining gifts making it all possible. On September 9, 2002, the Purdue Research Foundation and the City of West Lafayette announced a $2.2 million development of fifty acres for the Research Park.

Campaign for Purdue

By September 27, 2002, Murray Blackwelder, Rab Mukerjea, and Sally Mason had all been in place for more than a year. And Jischke made a huge and long anticipated announcement. At an event in Loeb Playhouse, where he had been introduced as the incoming president just eighteen months earlier, he wrapped three big announcements into one. First, the University had launched a Campaign for Purdue with a goal of $1.3 billion;[7] second, nearly half that amount had already been raised, $615 million; and third, the University had received the largest private gift in its history, $52.5 million, from Indianapolis businessman and civic leader William Bindley. The announcement came a day after Purdue celebrated its largest corporate gift ever—$116 million in software from General Motors Corporation, Sun Microsystems, and EDS.

Jischke beamed as he made the announcements in Loeb Playhouse with "$1.3 billion" projected on a screen behind him. "Our alumni and the corporations that hire them are as excited as we are about our strategic plan's vision to move this University to the next level," Jischke said, and then he defined the next level—"preeminence." It was the largest public university campaign in the history of Indiana.[8]

As the year 2002 came to an end, no one doubted Jischke and Blackwelder would succeed. In fact, two years later Jischke announced that the campaign was going very well. But instead of being satisfied that it would reach its mark early, he raised the goal. He announced that the University planned to raise $1.5 billion. And when the campaign ended on June 30, 2007, it had gone far beyond that.

R. Graham Cooks
(Photo by Mark Simons/
Purdue University)

R. Graham Cooks, the Purdue University Henry B. Hass . Distinguished Professor of Chemistry, came to a realization early in his career. "The answer to impossible questions in chemistry is mass spectrometry," he said. "I have spent my career practicing this philosophy."[9]

Cooks is among the most cited chemists worldwide and considered possibly the leading mass spectrometrist in the world. His work in the Aston Laboratory of Mass Spectrometry at Purdue has resulted in more than fifty patents and led to the launch of four companies.

Mass spectrometry (MS), he said, "is the method, par excellence, for measuring chemicals. Both we, and the world, are comprised of chemicals—molecules, elements, isotopes, and mixtures of these, especially complex mixtures as in biofluids, our tissues, environmental pollutants, constituents of steel, trace elements in semi-conductors, vitamins and toxic industrial chemicals, medicines and waste effluents. Mass spectrometry is based on forming ions (electrically charged molecules, elements, etc.) and then measuring their masses."[10]

During his nearly five decades at Purdue, Cooks, with others, has added two important capabilities to mass spectrometry: first, a method of making ions from complex mixtures in air termed DESI, and second, a method of measuring masses of individual constituents of complex mixtures by performing MS twice (MS/MS). The combination of DESI and MS/MS allows intricate measurements on real world materials in real time.

For example, a mass spectrometer tool has been developed by a Cooks-led team to help brain surgeons test and more precisely remove cancerous tissue. The tool sprays a microscopic stream of charged solvent onto the tissue surface to gather information about its molecular makeup and produces a color-coded image that reveals the location, nature, and concentration of tumor cells.

"In a matter of seconds this technique offers molecular information that can detect residual tumor that otherwise may have been left behind in the patient," Cooks said.[11]

Cooks is working to take miniature mass spectrometers from the laboratory to people in the real environment, where there are countless potential uses such as rapid, precise detection and measurement of drugs through blood tests, checking for bacteria in food in minutes rather than days, "and almost anything else you can imagine that is useful out there," Cooks said.[12]

Born in South Africa, Cooks received his first PhD from the University of Natal, South Africa. His second was awarded at Cambridge University in the United Kingdom in 1967.

At Purdue Cooks has guided 136 people through their PhD.

39

Mission Accomplished

2003 to 2007

Keep your dreams for Purdue alive. The future
needs them. And Purdue needs you.

—Martin Jischke

Martin Jischke looked particularly pleased and satisfied on April 3, 2003, when
as part of the Campaign for Purdue he announced a $200 million student
scholarship fund drive including $5.5 million for a need-based Purdue Opportunity
Awards Program. It would provide a complete financial package for one freshman
from each of the state's ninety-two counties, beginning in the fall of 2004. At the
onset, the award, including tuition and room and board, was for one year. It was later
expanded to two with the promise that students would be helped to find additional
assistance for their final two years. The program was launched with a $2 million lead
gift from trustee chairman J. Timothy McGinley and his wife, Jane.

There was progress on Purdue's regional and West Lafayette campuses. In May
2003 ground was broken at Indiana University–Purdue University Fort Wayne for
its first student housing complex. The $25 million project included seven residen-
tial apartment-style buildings with space for 568 students. The fall of the year was
marked by completion of the three-year, $70 million renovation of Ross-Ade Stadium.
The single item that received the loudest applause was when the athletic director,

Morgan Burke, discussed the improvements in the addition and expansion of women's restrooms.

Weldon School of Biomedical Engineering

On October 24, 2004, during a ceremony in the Union Ballrooms that included golden indoor fireworks and the Glee Club singing "Hail Purdue," Jischke announced a $10 million gift from Norman and Carol Weldon and their son Thomas. The funds were used to expand the Department of Biomedical Engineering into a school. Additional faculty were hired and programs were expanded to the undergraduate level. It marked a turning point in biomedical engineering at Purdue that dated to 1974 when Leslie Geddes was recruited from Baylor University to start the Hillenbrand Biomedical Engineering Center. The center became the Department of Biomedical Engineering in 1998. Geddes had retired seven years earlier but continued coming to his lab at 4:30 in the morning. A U.S. president would soon recognize his accomplishments.

In 2006, Purdue dedicated a $25 million biomedical engineering building, partially funded by the Whitaker Foundation and the State of Indiana. Upon Jischke's retirement, the building was named the Martin C. Jischke Hall of Biomedical Engineering and the street that runs in front of it, Martin Jischke Drive.

President Emeritus Martin C. Jischke at the dedication of Jischke Hall of Biomedical Engineering.

Statewide Perception of Purdue

New buildings on campus were dedicated with celebrations, and as Jischke carried the Purdue story around the state, the perception of the University changed. At the beginning of his presidency an assessment of the perception of Purdue in Indiana was authorized. The company doing the assessment talked to citizens, community leaders, corporate leaders, politicians—people from all walks of life. The message that came back was not good. Purdue was not well positioned in the state compared to other universities. "In some areas we were quite behind other universities in the state," Jischke said. "It was pretty disappointing to me, and to the board. We had a lot of work to do. We weren't seen [as strong] in economic development. Quality was seen as slipping. We developed our strategic plan and we got to work on it. We hired the same firm three or four years later and had them redo the survey. It was a total turnaround. In fact, the fellow who led the effort on behalf of the consultants

The leaders of the band: Martin and Patty Jischke leading the "All-American" Marching Band at Homecoming.

said it was the most dramatic change he had ever seen in the public's perception of an institution."[1]

2005 Regional Campus Development and Faculty Chairs

By the end of 2005 a College Bound full-scholarship program had been launched by Purdue North Central for students in La Porte and Michigan City schools, and the regional campus had been approved to offer MBA degrees. Purdue Calumet opened residences for students. Construction began on a new music building for the Fort Wayne campus.

In West Lafayette, Purdue launched its first University-wide honors program, and a Lilly Endowment challenge grant fostered private gifts that enabled the West Lafayette campus to create twenty-two new faculty chairs.

Students were being well fed. Newly constructed and renovated residence hall dining facilities in West Lafayette won several national awards. On-campus housing had reached its own preeminence level with the completion of its master plan for dining services and residential housing space. Students now ate in attractive, contemporary dining courts, offering a wide selection of choices at five locations and winning national awards for building design and food quality. Rooms in Cary Quadrangle were reconfigured, providing walk-in closets and shared bathrooms. And 70 percent of the on-campus housing space was air-conditioned. On-campus living was as popular as ever.

Chief Financial Officer Frederick Ford played a significant role in accomplishing the Housing Master Plan. The first new dining court, an $18 million facility on Stadium Avenue, was named for Ford and his wife, Mary. Students called it "Fred and Mary's." "Fred Ford's financial leadership at Purdue—especially his twenty-four years as executive vice president and treasurer—enabled us [to] build one of the largest and best residence hall systems in the country," Jischke said.[2]

Progress in 2006

In May 2006 Purdue opened a technology park in New Albany, Indiana, and expanded its College of Technology programs there.

October 2006 was an especially exciting month. A $4 million gift from the J. Willard and Alice S. Marriott Foundation made possible construction of a $12 million building for Purdue's top-ranked School of Hospitality and Tourism Management. Purdue trustee Susan Bulkeley Butler gave a $1 million gift to build the University's first archives documenting the lives of women, and another $1 million from Virginia Kelly Karnes went toward an archives and special collections library. In December 2006 another gift came from Lilly Endowment, this time $25 million to promote pharmacy, education, and outreach.

Patty Jischke

The role of a university president in the twenty-first century is all-encompassing, and Jischke had a strong and active partner in his wife, Patty, who was an attorney. Priscilla Hovde, Nancy Hansen, Jane Beering, and Patty Jischke were all very active on the Purdue campus and in the community representing the University. But the "First Ladies of Purdue" were not compensated for their work.

The board of trustees recognized the work Patty Jischke performed by granting her the title "Ambassador for the President"—for life. "Tim McGinley was the individual who announced it at one of the board meetings and it was very kind of him to do that," she said. "I was thinking, wait a second, that can't be for life. But it was very kind and in step with the way spouses of university presidents are handled or treated."[3] The title signified her role in organizing and planning events, as well as her service in voluntary leadership roles for several organizations.[4]

Patty Jischke served as hostess for thousands of events at Westwood, on campus, and throughout the nation and world. At dinners the Jischkes always sat at separate tables to interact with more people. Along with her husband, she was very active with students. They both taught a student leadership course at Westwood. They attended men's and women's athletic events. Patty Jischke was an adviser for Iron Key, a service project honor society of select students. She was involved with the nonpartisan Purdue Rock the Vote to register students. She served as a Faculty Fellow, giving students in residence halls an opportunity to interact with her and others. She helped develop a mulch trail around the intramural playing fields.

Patty Jischke was also very involved with the community and was especially active in child abuse prevention. She was active with the Purdue Women's Club and supported reading programs for children. She served on many boards, including the Community Foundation of Greater Lafayette, the Lafayette Community Development Corporation, and the Museum at Prophetstown. She helped start the Dog Park Association in Greater Lafayette. In 2008 the University's Early Care and Education Center, for which Patty had advocated, was named for her.

Jischke's Final Year

Another groundbreaking during the Jischke era. (Photo by Michael Heinz, Lafayette *Journal and Courier*)

In the fall of 2006 Jischke announced his plan to retire. He turned sixty-five in August 2006, and University policy at that time required that he retire by June 30, 2007. The board of trustees offered him an extension, but Jischke didn't want to be a caretaker for a year or two. It was time for a new person to lead the University in bold new areas.

As retirement approached, he did not slow down. During the winter of 2007 the Purdue Research Foundation announced plans for a $14.5 million facility for its high-tech business incubator complex, and the Mann Foundation for Biomedical Engineering announced that Purdue would receive $100 million to endow an institute designed to speed commercialization of innovative biomedical technologies.

Martin and Patty Jischke at campaign celebration, June 2000.

By April 2007 the University announced that the strategic plan goals were being met. Purdue was on track to finish hiring the last of the three hundred additional faculty and achieving $300 million in sponsored program funding. Student SAT scores were at an all-time high, and six-year graduation rates improved 3.9 percent for 2006–2007. The Campaign for Purdue had met its goals and then some—$1.7 billion: System-wide, the Campaign for Purdue contributed to new facilities completed, under construction, or in the planning, totaling more than $1 billion; Purdue more than doubled its endowed professorships; the research program doubled. Student financial aid increased 66 percent; applications for admissions, graduation rates, student diversity, study abroad, and freshman academic preparedness were all at record levels.

Presidential Search

Throughout late 2006 and early 2007 the trustees were busy looking for a new president. On Monday, May 7, it was learned that a woman would lead Purdue for the first time in its history.

The new president would not only be the first woman to lead Purdue but also the first with Hispanic heritage, the first to have been born in Paris, the first to come from California—and, as she pointed out—the first to have been a soccer mom.

With the selection made, it was time to say thanks to Martin and Patty Jischke, who would remain in the community, with a dinner on June 30, 2007.

"We have dreamed together. We have built together," Jischke said in his closing remarks. "These are my final words to you as president of this University. Keep your dreams for Purdue alive.

"The future needs them. And Purdue needs you."[5]

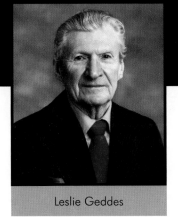

Leslie Geddes

Never one to withhold advice, Purdue University researcher Leslie Geddes offered President George W. Bush his three secrets to success as he received the nation's highest award for technological innovation in 2007.

Do something that other people consider important, he said. Don't make too many enemies. And, finally, live a long time. "You've got all three," Bush told Geddes, who was then eighty-six.[6]

During his career at Purdue, Geddes received more than thirty patents for biomedical devices that generated $15 million in royalties. He started the University's biomedical engineering program in 1974 with $250,000, and continued teaching and researching even after official retirement, arriving on campus at 4:30 a.m.

"He was incredibly curious and driven to test out ideas to solve special medical problems," said George Wodicka, head of Purdue's Weldon School of Biomedical Engineering. "He just loved getting into the medical laboratory with his students and figuring out what would work."[7]

Geddes's innovations included electrode design, tissue restoration, miniature blood pressure cuffs for infants in neonatal care, and pacemakers that can automatically increase a patient's heart rate during exercise.[8] He also did extensive work in cardiopulmonary resuscitation.

John DeFord, a student mentored by Geddes, said his research has impacted all people who have experienced cardiac arrest. "He was a pioneer in internal and external defibrillators," DeFord said.[9]

Geddes was born in Scotland in 1921 but his family moved to Canada. He studied electrical engineering at McGill University in Montreal and received his PhD in physiology from the Baylor University College of Medicine. As a graduate student at Baylor he helped develop monitoring systems for astronauts in the early days of the U.S. space program.[10]

He was recruited to Purdue in 1974 to start the University's biomedical engineering program and succeeded with limited funding by obtaining research grants and partnerships with industry. His initial biomedical research center grew into the Department of Biomedical Engineering and finally the Weldon School of Biomedical Engineering, with undergraduate and graduate programs.

The highlights of his career were in his laboratory, but the greatest honor Geddes received was at the White House, where Bush placed a red, white, and blue ribbon holding the National Medal of Technology around his neck.

Bush noted that Geddes said he would never retire. "I wouldn't know what else to do. I'm not done yet," Geddes said.

He was still arriving at his lab in the predawn hours until days before he died in 2009 at the age of eighty-eight.[11]

40

France Córdova:
A Résumé Out of This World

1947 to 2007

There are only a few campuses in the United States that have that kind
of a very special reputation. . . . Purdue is a magic name for me.

—France Córdova

Having been described as a "superstar," a "Renaissance person," and "out of this world," France A. Córdova stepped onto the stage to a standing ovation at Loeb Playhouse on May 7, 2007, and became the eleventh president of Purdue University.

"The opportunity to lead one of America's great universities is a wonderful privilege for me," she said. "I have tremendous respect for the achievements of Purdue under Martin Jischke's leadership, and I look forward to working with the faculty, staff and students here."[1]

As happy as Córdova looked wearing the gift of a brand-new Purdue letter jacket, the University trustees who elected her appeared even more pleased with the announcement and with her reception.

"Our trustees interviewed several very outstanding candidates who would have been excellent presidents, but France Córdova is the right person at the right time for Purdue," said board chair J. Timothy McGinley. "The breadth of her expertise, as we

Purdue president France Córdova.
(Photo by John Underwood/
Purdue University)

might expect from an astrophysicist, enables me to make a powerful statement. Dr. France Córdova's resume is truly out of this world. She has distinguished herself as a scientist, an administrator, and a creative writer. She is truly a Renaissance person."[2]

It was high praise. But the biggest applause, and laugh, of the morning came when a reporter asked Córdova if it would be challenging trying to fill the shoes of retiring president Martin Jischke. She didn't miss a beat. "If you saw my high heels," she said, "it would be a challenge for him to fill my shoes, too."[3]

It had all happened so fast that it was difficult for her to believe. An internationally recognized astrophysicist, Córdova served as chancellor at the University of California, Riverside, when a call came in 2007 from a company assisting Purdue in its presidential search. She loved California. She loved Riverside, its students, and its faculty. She had been chancellor and Distinguished Professor of Physics and Astronomy at Riverside since 2002. She was instrumental there in launching a medical school and a community–university art museum complex. Under her leadership the campus became a national model for the academic success of underrepresented students.

Córdova didn't want to leave. But the man calling her persisted, told her what a wonderful opportunity the Purdue presidency would be, and urged her to consider it.

In a very short time three trustees flew to California to meet with her, including John Hardin, co-chair of the search committee. They met at an airport restaurant for lunch. Trustee Susan Bulkeley Butler later made a separate trip to Riverside. In mid-April Córdova visited the Purdue campus. It was Spring Fest and there was snow and cricket spitting. The April snow surprised her.

Shortly after that visit she received a call from McGinley offering her the Purdue presidency. Students and faculty at Riverside urged her to stay, and there was talk of increasing her salary. Her fellow chancellors in the University of California system told her to stick with her West Coast roots. California governor Arnold Schwarzenegger asked her to stay. But she received another telephone call. Indiana governor Mitch Daniels reached out to Córdova and told her that Purdue was his "Pole Star," and he needed her to guide the University.

"Purdue has chosen the ideal person to continue its recent climb toward the front of the ranks of America's research universities," Daniels said after Córdova announced she would become a Boilermaker. "I was deeply impressed during my conversation with Dr. Córdova and found myself holding my breath in the hope that she would say 'yes' and come to Purdue."[4]

The arrival of Córdova at Purdue marked the beginning of an era at the University that she said would be "student-focused, innovative" and "reaching for the stars. . . . Purdue has always been a very special place to me because of its reputation in the sciences and engineering," she said. "There are only a few campuses in the United States that have that kind of a very special reputation in those areas—places like Caltech, MIT and Purdue. Purdue is a magic name for me."[5]

She was very familiar with Caltech and MIT. Now she would get to know Purdue.

Córdova was born in 1947 in Paris, France, where her father, Frederick, was chief of mission for CARE—a relief organization founded in 1945 at the end of World War II. He was born in Mexico and immigrated to the United States as a youth, and in 1946 he graduated from the U.S. Military Academy at West Point. Her mother, Joan, was born in New York and of Irish American heritage.

Her parents were expecting a boy in the days when no one knew the gender of an unborn child. The Córdovas had planned to name their son Frederick III, so Joan sewed an "F" into the unborn baby's clothing. When a girl was born, they christened her "Francoise." It fit the country where she was born and baptized, and also avoided having to redo all the letters sewn into the clothes. Her name would later be legally changed to "France."

Her family moved from Paris to Germany and finally to California, where her father started a company that imported and installed marble, granite, and onyx for large buildings.

"I was the oldest of twelve children, so I was the responsible child," Córdova said. "I babysat a lot. I scrubbed floors, folded diapers, and ironed uniforms. I was never off duty. We are still a close, tight-knit family. We have lots of family parties all year long celebrating all the things that families celebrate—births, first communions, weddings, graduations."[6]

Both her parents were very successful and served as role models, but her mother especially stood out for a young woman who grew up in the 1950s and 1960s and pursued a career path beginning in the 1970s. Opportunities for women were still a struggle heading into the 1970s. But, in addition to raising twelve children, her mother went back to school at age forty and graduated in the top ten of her class at California State University, Los Angeles. Then she started two companies.

Córdova's goal was to pursue a career in science, and her mother saved a diary in which the sixteen-year-old girl described that dream. Albert Einstein fascinated her. But in the early to mid-1960s her parents saw college as mainly an opportunity for a young woman to find a husband—common thinking for the time. So instead of being encouraged in science, she majored in English at Stanford University and graduated cum laude in three and a half years. She won a national competition to be a guest travel editor at *Mademoiselle* magazine in New York City. She also worked as a copy girl and copy editor at the *Los Angeles Times* and wrote a short book based on her college experience doing research in a Zapotec Indian pueblo in Oaxaca, Mexico.

The summer of 1969 was a turning point in her life. First, she was captivated by Neil Armstrong and his Apollo 11 mission to the moon in July of that year. Second, while living in Cambridge, Massachusetts, she watched a PBS television program about neutron stars. The two experiences coming so close together reignited her love for science, and she decided to take her life in that direction.

But her degree in English was not the background she needed. She would have to return to school. She got a job at the MIT Center for Space Research that paid almost nothing, but allowed her to do astrophysics research during the summer. Then, she was accepted into MIT graduate school. Family ties pulled her back to California, and she eventually transferred her graduate work to Caltech, immersing herself in X-ray astrophysics. Not surprisingly for the times, there was only one other woman in her graduate program in physics that year. One of her favorite photographs, which followed her from state to state and hung on her office walls, was a montage of the planets Jupiter and Saturn against the Pleiades, or "seven sisters" star cluster. She is one of seven sisters.

After she received her PhD from Caltech in 1979, Córdova took a staff position at the Los Alamos National Laboratory in New Mexico. In 1989, after nearly ten years there, she went to Penn State to head the department of astronomy and astrophysics. She was also the only woman faculty member in the department.

Opportunities continued to present themselves. From 1993 to 1996 she served as chief scientist for NASA—once again the first woman to hold that position. Her work allowed her close contact with NASA administrator Daniel Goldin, the longest tenured administrator of the agency. She worked on budgets and policies, testified before Congress, and gave speeches around the nation. She knew Neil Armstrong long before she arrived at Purdue. They met at a White House reception. She also worked with one of three Purdue female astronauts—Janice Voss.

When her term at NASA expired in 1996, Córdova became vice chancellor for research at the University of California, Santa Barbara, and in 2002 moved to Riverside as chancellor.

Among the many "firsts" Córdova brought to Purdue was a "First Gentleman," her husband, Chris Foster. They met at Los Alamos through their involvement in a rock-climbing club. He was a science teacher who worked summers at the Los Alamos National Laboratory when they met.

When they arrived at Purdue, their daughter, Anne-Catherine, was twenty-one and their son, Stephen, nineteen. Both children remained in college in California—Anne-Catherine at UC Riverside and Stephen at Stanford.

Foster selected the title "First Gentleman" for himself and played an active role on the campus. He became director of the Discovery Park K–12 Science, Technology, Math, and Engineering (STEM) programs in addition to the work he did assisting his wife.

When she arrived, Córdova spoke with enthusiasm about all aspects of the University—teaching, research, outreach, engineering, science, the liberal arts, and more. "Purdue is a university with a great reputation in research," she said. "This [the Purdue presidency] is an opportunity to speak on a national stage about issues that are really important to me; issues such as student success, innovative interdisciplinary research and the importance of students—especially women and minorities—going into science, technology, engineering and math careers. The coupling of science with engineering is essential for doing important research. The liberal arts are the foundation of a good education at our colleges and I intend to promote and enhance liberal arts and sciences at Purdue. I'd like to apply myself to more connections between the liberal arts and the sciences and engineering."[7]

But first she applied herself to serving everyone on campus some ice cream on a sunny summer day.

41

Progress Through Economic Recession

2007 to 2012

The recreation center is a shining component of the
Student Success Corridor taking shape along Third Street,
which will provide a multitude of ways for students
to find paths to success at Purdue and beyond.

—FRANCE CÓRDOVA

More than eighteen hundred students, faculty, and staff gathered in front of Hovde Hall on a beautiful eighty-five-degree summer day, July 16, 2007, to celebrate the official arrival of France Córdova as the eleventh president of Purdue University—and to enjoy some ice cream.

Wearing a sun hat with a wide brim to Purdue's first all-campus ice cream social, Córdova had to be urged away from a long line of people to deliver a brief address. She apologized to those she hadn't yet greeted and promised to talk with everyone. Her husband, Chris Foster, called the response from Purdue people "overwhelming."[1] Córdova had awakened at three o'clock that morning ready to start the new era at Purdue.

In five years at Purdue Córdova accomplished an impressive list of successes. By the end of her term, first-year students had the highest-ever SAT/ACT scores in University history; the first-year retention rate reached an all-time high of 90.2 percent; an Access and Success scholarship fund drive was on target to reach $304 million by 2014; more than $1.1 billion had been raised in private gifts; research funding had increased 40 percent; $564 million in facilities had been constructed; the University's annual impact on Indiana had reached $4.2 billion; the Purdue Research Park network's annual economic impact on Indiana had reached $1.3 billion; and annual research expenditures had doubled to $600 million.

In her last year as president, fund-raising hit $298.8 million, a 32 percent increase above the previous year, and at that time the second highest total in University history. She worked to develop a ten-year funding plan that sought to double Purdue's revenue and decrease its reliance on tuition and fees.

Victor Lechtenberg, Interim Provost

The first announcement of her presidency came at the ice cream social and it was a popular one. She named Victor Lechtenberg interim provost, replacing Sally Mason, who had left Purdue to become president of the University of Iowa. Lechtenberg was a popular person on the Purdue campus and across the state, and he was widely acknowledged as a strong administrator. Beginning in 2004 he had served as vice provost for engagement, leading the campus in its outreach efforts. Prior to that he had served ten years as the dean of agriculture. Córdova said she selected Lechtenberg because of the success he had in engagement and his support on campus. "Everybody thought he was the person to help me with this transition," she said.[2]

People recruited to Purdue from Iowa State University during the Jischke years continued in their positions with Córdova. Murray Blackwelder, vice president for advancement who had just finished a $1.7 billion campaign, was ready to start new fund-raising efforts. And Rabindra "Rab" Mukerjea remained as director of strategic planning and assessment.

Randy Woodson, Provost and Vice President

As he began his duties Lechtenberg announced he would not be a candidate for the position of provost, and Córdova planned a national search. But she didn't have to look very far. On May 1, 2008, Purdue dean of agriculture Randy Woodson became the new provost and vice president of academic affairs.

Randy Woodson, provost and vice president for academic affairs.

Jay Akridge, dean of agriculture and later provost and vice president for academic affairs.

Woodson had twenty-three years of experience on the campus. "Randy, being from inside of the University, can hit the ground running and that's very important with the new strategic plan coming," Córdova said when she announced the appointment. "He knows [the University's] strengths and he knows what needs fixing."[3] Lechtenberg returned to the position of vice provost for engagement. Córdova promoted Jay Akridge to replace Woodson as dean of agriculture.

A New Chairman of the Board of Trustees

A major change took place on the board of trustees during the Córdova years. In 2009 Tim McGinley retired from the board, ending twenty years of service, sixteen of them as chair. An alumnus, Keith Krach, of California, who had been appointed to the board three years earlier, was elected chair and remained until he retired in 2013. He took an active role on campus and for periods of time lived in the Union Club Hotel with his family.

New Synergies Strategic Plan and Recreational Sports Center

On June 20, 2008, eleven months after she started at Purdue, Córdova presented her strategic plan, New Synergies, to the board of trustees. The six-year plan focused on positioning the University to meet challenges facing humanity, creating and increasing opportunities for Indiana in the global economy, and enhancing student learning for success in a changing world. The three stated goals were launching tomorrow's leaders, discovery with delivery, and meeting global challenges.

At the same meeting the board approved planning for a $98 million expansion of the Recreational Sports Center—a long-delayed project. It was accomplished through a student fee that was endorsed by Purdue Student Government. The facility was named the France A. Córdova Recreational Sports Center in her honor. "She was vital in moving this project forward," said Howard Taylor, director of the facility. "In every step along the way, she was involved."[4]

The renovation was one of the highlights of her administration. "I'm elated to see this student dream fulfilled, and deeply honored by the naming," Córdova said when the building was dedicated. "This remarkable center is a tribute to the students' imagination, leadership and diligence. The recreation center is a shining component of the Student Success Corridor taking shape along Third Street, which will provide a multitude of ways for students to find paths to success at Purdue and beyond."[5] The facility was dedicated in October 2012.

Student Success Corridor

On April 8, 2011, the board of trustees approved construction of a $30 million student activities center and a $38 million apartment-style residence hall along Third Street near the Recreational Sports Center. They were part of what was called the Student Success Corridor. The student activities center opened in 2014 as the Krach Leadership Center, named for Keith Krach, who gave major funding and support for this project.

Global Economic Crises

Córdova's strategic plan and University finances hit a major challenge in the global economic crisis that began in 2008. Financial markets dropped rapidly, the country fell into a deep recession, and it would be years before the economy recovered. Among Córdova's accomplishments during her years at Purdue was that she successfully led the University through the financial crisis when state funds dried up.

"The recession caused a historic shakeup of all institutions, everywhere," Córdova said. "At Purdue that was really a tough time. All the states were trying to figure out what they were going to do; they were holding back money from public institutions. It was a real shock. Obviously we had a lot of really good people, and we just dug in and everyone rode it out together. We tried a lot of things and some didn't make people happy. That was really a rough time. Communication is the most important thing. It's something you have to work on every single minute. The biggest challenge was to get everyone to understand what was happening and that they had [a] role in trying to return to normalcy."[6]

She had help from many people, including Al Diaz, who was named Purdue's executive vice president for business and finance and treasurer July 1, 2009. Diaz came to Purdue from the University of California, Riverside, where he worked with Córdova as vice chancellor for administration. He had previously spent forty years with NASA, including service as director of the Goddard Space Flight Center.

College of Health and Human Sciences

In February 2010, the board of trustees approved a realignment creating the College of Health and Human Sciences and a separate, smaller College of Liberal Arts. (By this time, Purdue had chosen to call many of its umbrella academic institutions colleges instead of schools.) Córdova promoted the plan and said, "A college dedicated to health and human sciences would enhance student opportunities and promote faculty collaborations aimed at improving health and quality of life. The realignment could consolidate and elevate Purdue's reputation in the health and human sciences."[7]

The new college encompassed the preexisting departments of health and kinesiology, psychological sciences, speech, language and hearing sciences, child development and family studies, foods and nutrition, consumer sciences and retailing, and hospitality and tourism management. It also included the schools of nursing and health sciences. It became the second largest college at Purdue after engineering.

Woodson said the realignment was the work of a faculty-led task force. The nine academic units had been spread across three colleges, and that created challenges for student recruitment and success. "Combining them will provide an opportunity to develop new and strengthen existing interdisciplinary pre-medicine, pre-dental, pre-veterinary or other professional degree programs that are attractive to students," he said.[8]

The new college launched on July 1, 2010, with Chris Ladisch as dean. Ladisch had served as vice provost for academic affairs since July 2005 and as associate provost since 2001. She served as associate dean for academic affairs in the College of Consumer and Family Sciences from 1993 to 1999 and as department head of Consumer Sciences and Retailing from 1999 to 2001.

In 2010 Woodson left Purdue to become chancellor of North Carolina State University. Córdova named Tim Sands, director of the Birck Nanotechnology Center in Discovery Park, to replace him. Sands later went on to become, in 2014, president of Virginia Tech. Córdova also brought Julie Griffith to Purdue in 2011. Griffith had been an executive with Duke Energy in Indianapolis. At Purdue she was named vice president for public affairs as part of a reorganization that brought government and community relations and economic development efforts into one office.

Hanley Hall, Marriott Hall

The first new building in the College of Health and Human Sciences was the Bill and Sally Hanley Hall, dedicated in the fall of 2011. It was built with a $3 million gift from Bill and Sally Hanley and $1.5 million from Lilly Endowment. Hanley Hall was next to Fowler House, and together they became home to the Ben and Maxine Miller Child Development Laboratory School, the Department of Human Development and Family Studies, the Military Family Research Institute, and the Purdue Center on Aging and the Life Course.

Another Health and Human Sciences building was Marriott Hall for the School of Hospitality and Tourism Management. Dedicated in April 2012, the $13 million facility was made possible by a $5 million leadership gift from the J. Willard and Alice S. Marriott Foundation. Córdova called it "a cutting-edge environment in which to learn."[9] She was pleased to welcome Bill Marriott to the campus dedication of the new building. Marriott became the new home of the long-established John Purdue Room, the laboratory for students learning fine dining and service.

In addition to Hanley and Marriott Halls, the board of trustees credited Córdova with a number of other accomplishments during her tenure, including completion of the Hockmeyer Hall of Structural Biology, the Niswonger Aviation Technology Building, and the First Street Towers residence. She successfully acquired the Schowe House for the Global Policy Research Institute, and she was responsible for the Archer Daniels Midland Agricultural Innovation Center, the Hall for Discovery and Learning Research in Discovery Park, and the Roland G. Parrish Library of Management and Economics. Córdova said she relished the opportunity to thank the donors in person during the campus building dedications. "It was one of the highlights of my tenure," she said, "getting to experience the pride of Purdue's donors and their happiness at fulfilling their dreams."

The board also credited her with laying the foundation for the Center for Student Excellence and Leadership, Wang Hall for Electrical and Computer Engineering, Bailey Hall for Purdue Musical Organizations, the northwest site for baseball and improved soccer facilities, the Life and Health Sciences Park, Lyles-Porter Hall, the Drug Discovery Building and expansions to Herrick Labs, and the Bindley Bioscience Center.

Córdova said that among her greatest joys as president were her association with faculty and the opportunity to interact with students. She and First Gentleman Chris Foster held weekly freshmen leadership classes at Westwood. Chris was an adviser to Iron Key, a student honor society; he guided its members in building two iconic campus projects: the Purdue Unfinished "P" and a larger-than-life statue of a seated John Purdue. Córdova loved marching with the "All-American" band to Ross-Ade Stadium and sometimes sang and cheered with them during part of the game. She even learned how to twirl a baton and led the homecoming parade into the Ross-Ade Stadium with the Golden Girl and the Silver Twins.

The Unfinished "P" came to have two meanings. It was used to honor and remember students who died before finishing their degree. And it represented the fact that learning is a lifelong process—it is not finished upon graduation from Purdue.

It was an exciting time at Purdue. And October 6, 2010, was one of the most exciting days of all.

42

The 5 a.m. Telephone Call

2007 to 2012

Purdue has been around for most of our nation's history,
and it's wonderful to say you were part of it.

—France Córdova

E i-ichi Negishi received the call at five o'clock in the morning on October 6, 2010. His heart jumped as the telephone rang. He might have already been awake or at best his sleep was very light because he had been waiting years for this moment. The evening before, he told his wife, Sumire, that in the morning maybe it will happen, or maybe it will not. What is to be is to be.

When he hung up the telephone, he turned to his wife. She knew. "Now our fifty-year dream comes true," he said. "Finally it is a reality."[1]

Negishi, seventy-five, had just been told he was a 2010 recipient of the Nobel Prize in chemistry. It was the greatest honor of his career, the culmination of a life-time of work. But just a few hours after receiving the telephone call from Stockholm, Sweden, Negishi would excuse himself from an international news conference to walk across campus and teach a sophomore-level chemistry class.

Negishi, a Japanese citizen, came to the United States to study at the University of Pennsylvania after graduating from the University of Tokyo in 1958. In 1962 he attended a presentation by Herbert Brown, the Purdue University professor who would receive a Nobel Prize in 1979. He was fascinated.

Negishi received his PhD from Penn in 1963 and immediately moved to Purdue to work as a postdoctoral researcher under Brown. He stayed until 1972, when he went to Syracuse University as an assistant and then associate professor. He returned to Purdue in 1979.

Negishi said he predicted Brown would win a Nobel when the two men first met. "I always said how [Brown] will change the whole world of organic chemistry. That's why I came here [to Purdue]," he said.[2] In 1999 he was named the Herbert C. Brown Distinguished Professor of Chemistry.

He was co-recipient of the Nobel award with Richard Heck, of the University of Delaware, and Akira Suzuki, of Hokkaido University in Japan. Suzuki was also a protégé of Brown. Negishi, Suzuki, and Heck had discovered new ways to bond together carbon atoms, and their work had wide use in medicines, agriculture, and electronics.

As daylight dawned after the 5 a.m. Nobel telephone call, information about the award quickly spread through the Purdue campus and around the world. Within an hour Negishi's work and honor were making international news. Purdue provost Tim Sands, a neighbor, peeked out his window at a line of television trucks that filled their quiet West Lafayette Street.[3]

Purdue president France Córdova was among the first to arrive at Negishi's home to congratulate him. She called him "a new hero for science, for Purdue and for the world."[4]

Negishi knew people had been nominating him for the Nobel, and he admitted he was not surprised to receive it. "I'm extremely pleased," he said. "I'd be telling you a lie if I said this was totally unexpected." While he felt his work was worthy of the honor, he said he understood life does not always proceed as a person desires. "We had our own confidence in the value of what we had done," he said. "It became a realistic dream, but it was a dream. I was telling my wife we should never raise our hope high and these things should only be a consequence of your excellent work, which we should strive for."[5]

With media coming to campus to cover the story—including news outlets beaming reports by satellite to Japan—reporters and cameras quickly gathered in Dauch Alumni Center where a news conference was held. But it wasn't long before Negishi announced he had to leave to teach that class.

Reporters, photographers, and TV cameras hustled after him as he made his way to Wetherill Hall. Outside the lecture room, three hundred students had gathered and were preparing to enter. They had no idea what was going on. They asked reporters, TV camera people, and photographers what they were doing. "Your professor just won the Nobel Prize," they were told, and many stood in stunned silence for a moment. Inside the classroom the students took out their cell phones. Instead of taking notes on the lecture, they took photos of their professor. Many also called their parents and friends as the lecture continued. "Hey, Mom," one said. "My professor just won the Nobel Prize." Through the next eight years Negishi was very busy.

On March 13, 2018, he was driving his car with his wife, Sumire, in rural Ogle County, Illinois, south of Rockford near the Wisconsin border. His car became stuck in a ditch and Negishi, eighty-two, went for help, leaving Sumire, eighty, behind. Ogle County Police said they found Negishi at 5 a.m. wandering along a road suffering from "an acute state of confusion and shock." Upon discovering that the couple had been reported missing in Indiana at 8:13 p.m. the previous evening, police searched for the car and Sumire. When they found her she had died. According to a family statement, Sumire had been "near the end of her battle with Parkinson's."

Negishi told police he became lost while trying to find Chicago Rockford International Airport about eight miles from where the couple was found. The coroner's office ruled that Sumire died from exposure and hypothermia. Negishi was hospitalized in Rockford.

In a statement released by the University, Mitchell E. Daniels, Jr., who was president of Purdue at the time, praised Sumire for "a lifetime of love and loyalty."

"It appears that the Parkinson's disease from which she has been suffering and the mental confusion that age can bring to the most brilliant minds combined to produce the recent tragic events," Daniels said. "That these phenomena are so common does not make their consequences any less cruel. All Boilermakers everywhere join the Negishi family in sadness at the loss of Sumire, who made so many of her own contributions to her husband's life work and to the vitality of our community."

World Food Prizes

The years 2007 to 2010 during the presidency of Córdova marked a time of more huge honors. In 2007 and in 2009, two members of the Purdue faculty received the World Food Prize—considered the Nobel Prize in agriculture.

In May 2007 Purdue professor Philip Nelson, seventy-three, was trying to reach a class he was teaching when he received a telephone call. The call was from World Food Prize Foundation president Kenneth Quinn. "He asked me if I was familiar with the World Food Prize and explained what it was," Nelson said later. "I thought he was trying to sell me tickets. I wanted to get off the phone because I had a class to teach."[6]

When Quinn finally explained that he had won, Nelson said he wanted to stop everyone and tell them. But he couldn't until June 18, when the announcement would be

Purdue Distinguished Professor and World Food Prize Laureate Gebisa Ejeta stands with children in his native Ethiopia, across the street from the elementary school he attended. It is one of his favorite photos. "I was one of these kids," Ejeta said. "And if I can make it, they can, too." (Photo by Tom Campbell/Purdue University)

made at the U.S. State Department in Washington, D.C. He received the award October 17, 2007, in Des Moines, Iowa.

In the fall of 1960, Nelson had planned on entering Purdue's new School of Veterinary Medicine. But instead he accepted an assistantship position in the University's horticulture department. He went on to revolutionize large-scale storage and transportation of fresh fruits and vegetables. His discoveries have touched consumers everywhere in the form of fresh orange juice in cartons and much more. His work also benefited people in the developing world, preserving perishable food without refrigeration.

In 2009, the World Food Prize was presented to Purdue Distinguished Professor Gebisa Ejeta, who had designed hybrid sorghum seeds resistant to drought and weeds. It was estimated that his work had improved the food supply fivefold for hundreds of millions of people in sub-Saharan Africa. Ejeta came to the United States from Ethiopia.

"I grew up in the middle of poverty," he said when the award was announced. "So I was sensitive to the problems of poor farmers. That empathy has always been a part of me."[7] Córdova and her spouse later traveled to Ethiopia with Ejeta to increase collaborations between Purdue and universities there. "He was a rock star in that country," she said, "Everyone knew of Gebisa Ejeta and what he had accomplished."

In 2017 the Bill and Melinda Gates Foundation gave Ejeta a $5 million, four-year grant to develop hybrid grain seeds that could resist parasite weeds. It was the second grant he had received from the foundation.

Also in 2017 Purdue University alumnus Akinwumi Ayodeji Adesina, president of the African Development Bank Group, was named a World Food Prize laureate for his work as a reformer and leader of the agricultural sector in Africa.

National Medal of Technology and Innovation

During Córdova's tenure two members of the faculty received the National Medal of Technology and Innovation presented by the president of the United States at the White House. The recipients were Leslie Geddes, the Showalter Distinguished Professor Emeritus of Biomedical Engineering, and Rakesh Agrawal, the Winthrop E. Stone Distinguished Professor of Chemical Engineering. In later years the honor was also given to Nancy Ho, research professor emerita of chemical engineering. The medal recognizes people who have made lasting contributions to America's competitiveness, standard of living, and quality of life through technological innovation, and to recognize those who have made substantial contributions to strengthening the nation's technological workforce. Córdova was in attendance at all three White House ceremonies in which these distinguished faculty received their honors from the president.

Also during the Córdova years, a number of faculty were elected to the prestigious National Academies of Sciences, Engineering, and Medicine and the National Academy of Sciences.

On July 1, 2011, Córdova announced she would leave the University at the end of the fiscal year in mid-July 2012. "In thinking about the year ahead I wanted to remain true to my commitment to the trustees and stay for five years," she said.[8]

The previous year Purdue had climbed to eighteenth among *U.S. News and World Report* rankings. It had ranked twenty-sixth in 2008. "In science you do something right at the beginning. You set out your goals, and we did that in the strategic plan the first year I was here," Córdova said. "You make your biggest impact . . . and then you wrap it up."[9]

Five-year milestones included launching the Center for Faculty Success with a $4 million grant from the National Science Foundation; securing Purdue's largest research grant, $105 million from the National Science Foundation for a Network for Earthquake Engineering Simulation; launching an Honors College in 2013 and a Discovery College in 2014 for students who have yet to declare a major; and increasing diversity and launching the Native American Educational and Cultural Center.

Córdova went on to become chair of the Board of Regents of the Smithsonian Institution and then director of the National Science Foundation under presidents Barack Obama and Donald Trump. Her spouse, Chris, was honored by Purdue with the annual Christian J. Foster Award, which recognizes a faculty member on any Purdue campus "who has made demonstrable contributions to improving STEM teaching and learning in Indiana's K–12 schools."

Four Purdue presidents: (left to right) Martin Jischke,
France Córdova, Steven Beering, Arthur Hansen.

Before they left Purdue, they had one more event to celebrate. On June 30, 2012, the daughter of Córdova and Chris Foster married a Boilermaker. Anne-Catherine Foster and Derek Mauk were married beneath shade trees in the gardens at Westwood. It was the first marriage at the Westwood presidential home.

"I want to be remembered for my emphasis on students having a successful college experience and growing into mature leaders and thinkers," Córdova said. "I developed as a president. I had my fiscal challenges and I had many wonderful opportunities with this talented faculty. And we thrived together. We ended up better together. Purdue has been around for most of our nation's history, and it's an honor to say you were part of it. When I tell people now that I was president of Purdue, they pause. Then say, 'Oh, wow, Purdue!' It's like they're saying, 'That's something special.' Purdue is special. I feel privileged to have led a special institution."[10]

With Córdova's announcement that she was leaving, the board of trustees began searching for a new president. They thought they had a super candidate, until he told them he wasn't interested. But times change very quickly. People change.

And even a governor can change his mind.

GIANT LEAPS
IN RESEARCH

Rakesh Agrawal
*(Photo by Mark Simons/
Purdue University)*

Rakesh Agrawal, the Winthrop E. Stone Distinguished Professor in the Purdue University School of Chemical Engineering, holds 125 U.S. patents. Several years ago he stopped counting his international patents when the total passed 500.

For his "extraordinary record of innovations," in 2011 during a White House ceremony in the East Room, President Barak Obama awarded him the nation's highest honor for technological achievement, the National Medal of Technology and Innovation.

"It was totally surreal," his wife, Manju, said of the ceremony.[11]

Agrawal's extraordinary innovations include energy efficiency, improved electronic device manufacturing, liquefied gas production, and the supply of gases for diverse industries.

Born in northern India, Agrawal did his undergraduate work at the prestigious Indian Institute of Technology, Kanpur. He did his graduate studies in chemical engineering at the University of Delaware and the Massachusetts Institute of Technology. "I was fascinated by the world-class universities America has and wanted to learn how to do great research," he said.[12]

After receiving his doctorate he married Manju and joined Air Products and Chemicals in Allentown, Pennsylvania. His continued interest in research and different technologies led him to author articles published in professional journals that caught the attention of Purdue faculty, who invited him to join them.

"I was initially reluctant," Agrawal said. "But once I met the faculty and President Martin Jischke I became excited at the possibilities." "Purdue is among the most wonderful events that ever happened to me in my life," he said. "The faculty is very collaborative and supportive. The graduate students are bright, and a fair number of them are among the brightest the world has to offer. I immediately formed several collaborations to pursue the difficult and interdisciplinary subject of solar energy and its use to fulfill multiple needs in our daily life. Purdue has transcended my job into my hobby. It is here that I can dream big and begin to solve the food and solar energy challenge with a team of experts across the campus."[13]

His current passion is energy production, especially from renewable sources such as solar, fabricating low-cost solar cells based on nanotechnology and meeting essential human needs including transportation fuel, chemicals, and electricity, primarily from renewable energy sources.

"The question is as we move forward, how will the world look," he said. "The whole idea of my research is to paint that picture of the future and then find the transition pathways that will take us there in a harmonious path."[14]

Nancy W. Y. Ho
(Photo courtesy of Nancy Ho)

Nancy W. Y. Ho was born in China and grew up in Taiwan, where as a little girl her life was marked by illnesses that made her feel inferior to other children. She felt worthless and a burden to her family, and as a result she was determined to contribute to society.[15]

She succeeded beyond her dreams. None of the children Ho grew up with can say that based on their research accomplishments they were invited to sit with the First Lady while the president of the United States delivered his State of the Union message to a joint session of Congress. And none of those children were invited to the White House to receive the National Medal of Technology and Innovation from another president.

A professor emerita in the Purdue University School of Chemical Engineering and group leader emerita of the Molecular Genetics Group of Purdue's Laboratory of Renewable Resources Engineering (LORRE), Ho developed what is widely known as the Ho-Purdue yeast that can effectively produce ethanol from all types of cellulosic plant materials, such as corn stalks, wheat straw, wood, and grasses. In 2006 she founded Green Tech America in the Purdue Research Park to produce and market the yeast.

Ho received a bachelor's degree in chemical engineering from National Taiwan University. She continued her studies in the United States, earning a master's degree in organic chemistry from Temple University in 1960 and a PhD in molecular biology from Purdue in 1968. She began her research at Purdue in 1971.

In 2007 she was invited to Washington, D.C., to join First Lady Laura Bush in the House of Representatives gallery to hear President George W. Bush deliver his State of the Union address. Bush promoted cellulosic ethanol in his speech. Ho wondered how the president found out about her work. "We do research ourselves," a White House spokesperson said.[16]

President Barack Obama awarded her the nation's highest honor for scientific and technological achievement in 2016. At the age of eighty, Ho had recently retired. "There are few better examples for our young people to follow than the Americans we honor today," Obama said.[17]

Arvind Varma, the R. Games Slayter Distinguished Professor and Jay and Cynthia Ihlenfeld Head of the School of Chemical Engineering, called Ho, "a pioneering and visionary researcher who has worked tirelessly and creatively for more than 30 years to develop the most successful yeast currently available for the production of ethanol from cellulosic feedstock."[18]

43

A Governor for Purdue

2012 to 2018

When I got up this morning I was trying to identify the rather
singular feeling about the day ahead, and the only analogy
that came to mind was my wedding day. I knew that life was
about to change. I didn't know exactly what was on the other
side, but I was convinced that I picked the perfect mate.

—Purdue president Mitchell E. Daniels, Jr.

"My Man Mitch" was the campaign slogan for Mitchell E. Daniels, Jr., when
he ran for governor of Indiana in 2004 and again in 2008. It was painted
on his campaign RV and shouted by admiring crowds of people who greeted him. He
won with 53 percent of the vote in 2004 and more than 58 percent in 2008, receiving
more votes than any candidate for any office in state history. His approval rating in
2010 was 75 percent.

Daniels, a Republican, was the most popular political figure in Indiana. He
also had a large national following, and his second term as governor was filled with
speculation about what he would do next. Going back to 2010 and through the early
months of 2011 he had been widely encouraged to seek the Republican nomination
for president of the United States. Daniels did not discourage the speculation. He told
a *Washington Post* reporter in 2010 that he was open to the idea of running in 2012.
He polled well against the other potential Republican candidates.

But on May 21, 2011, he announced he was not a candidate. "The counsel and encouragement I received from important citizens like you caused me to think very deeply about becoming a national candidate," he said in a message to Indiana Republicans. "In the end, I was able to resolve every competing consideration but one, but that—the interests and wishes of my family—is the most important consideration of all. If I have disappointed you, I will always be sorry."[1] Some Republicans still urged Daniels to enter the race, but he continued to say he was not and would not be a candidate.

The decision left his future wide open upon leaving the office of governor in January 2013. Just two months after Daniels's 2011 announcement, Purdue president France Córdova made the announcement that she would leave the University the following year. So when the board of trustees formed a search committee for a new president, one name kept rising to the top: Daniels.

In addition to leading the state, he had experience in Washington, D.C., including as director of the Office of Management and Budget under President George W. Bush, who nicknamed him "my man Mitch." From 1997 to 2001 Daniels had worked in the private sector in a series of senior positions at Eli Lilly and Company.

Daniels seemed the perfect person to lead Purdue. An acting president could fill the gap until his term as governor ended and he was ready to move north to West Lafayette. But the search for a new Purdue leader continued.

Purdue's twelfth president, Mitch Daniels, talks with student.

Then on June 21, 2012, the Purdue trustees announced what during the previous twenty-four hours had become the worst-kept secret in the state: Mitch Daniels would become the twelfth president of Purdue. His appointment came amid great fanfare and loud applause in Loeb Playhouse on campus where Daniels was dressed in black and gold.

Purdue trustee Michael Berghoff, who led the presidential search and would become chair of the board in 2013, said the search committee narrowed a list of hundreds of candidates down to about ten. Daniels's name kept coming up again and again in committee discussions.

In reporting on Daniels's appointment, the *Indianapolis Star* said, "The search committee approached the governor several weeks ago about the job. Three interviews later Purdue decided to hire him."[2] There was speculation that the trustees had discussed the idea with Daniels much earlier but were told he was not interested.

"This is not a case where he sought the office. The office sought him," board of trustees chair Keith Krach said. Daniels said, "I gave it a lot of thought. I don't mind saying that for quite a long time I thought this was a wonderful opportunity for someone, but not the right thing for me. The more I learned, the more I listened, the more intrigued I got."[3]

Talking to the media and Purdue faithful at the announcement where the trustees voted unanimously to hire him, Daniels pushed aside comments he had heard that the state and the University should be run like a business by a businessman. "[The state] is not a business," he said. "There are checks and balances. That doesn't mean it can't be more businesslike. I have somewhat the same feeling about Purdue. It's completely different, too. Governance is very different. The faculty has got to be very involved. There are constituencies. I get all that. I have to figure out exactly how you do it, but I know it's very different. But underneath that, there's no reason we can't be more businesslike."[4]

In 1983 Steven Beering said he knew his life was changing forever when he gave up medicine to accept the Purdue presidency. Daniels had the same feeling. "When I got up this morning I was trying to identify the rather singular feeling about the day ahead, and the only analogy that came to mind was my wedding day," he said. "I knew that life was about to change. I didn't know exactly what was on the other side, but I was convinced that I picked the perfect mate. It's been a bit of a blur because there are so many things I want to learn and so many people I want to meet."[5]

Some faculty expressed concern that Daniels did not have an academic background. Others wondered about his age, which was sixty-three at the time of the announcement. Would he be required to retire at sixty-five? The trustees said no; because he was hired so close to his sixty-fifth birthday, the rule did not apply to him.

Daniels did not start at Purdue on July 15 when Córdova was set to leave. He served the rest of his term as governor while making thirteen visits to campus, where, often dressed in casual attire, he talked with administrators, staff, faculty, students, and community leaders. People who met with him said he did a lot of listening—more listening than talking.

Tim Sands, executive vice president for academic affairs and provost, was named acting president until Daniels arrived in January 2013.

The same January day that future U.S. vice president Mike Pence was sworn in as Indiana's fiftieth governor, the state's forty-ninth governor drove to West Lafayette and started work at Purdue. Daniels said that change was coming to higher education, "tsunamic change," and Purdue had to be prepared. "In ten years higher education won't look like [it] does today," he said. Purdue needed to "break out of the pack" and offer excellent education at a price students and families could afford.[6]

Randy Roberts, a Distinguished Professor of History at Purdue and a member of the search committee that recruited Daniels, said the president-elect's meetings with Purdue people went well. "Mitch Daniels is an absolute sponge," Roberts said. "He gets things very quickly. Education in today's world is not just for dilettantes who want to learn a lot about a lot of things. He believes education has to have value and I agree with him on that."[7]

Daniels was born April 7, 1949, in Monongahela, Pennsylvania, south of Pittsburgh and not too far from where John Purdue grew up. His father served in the U.S. Army and his mother was a registered nurse. He attended early elementary school in Tennessee and Georgia, and the family moved to Indianapolis in 1959. When Daniels graduated high school in 1967, he was named one of two Indiana Presidential Scholars, and he traveled to Washington, D.C., where he accepted the award from President Lyndon Johnson.

Daniels earned a bachelor's degree from the Woodrow Wilson School of Public and International Affairs at Princeton University and studied law first at the Indiana McKinney School of Law in Indianapolis and then at Georgetown in Washington, D.C. He received his law degree in 1979. He served as chief of staff for former Indiana senator Richard Lugar, was senior adviser to President Ronald Reagan, and was chief executive officer of the Hudson Institute before going to Lilly.

He wasted no time initiating measures to transform Purdue and set a new model for higher education nationally. The University had been searching for a president who could "think outside the box."

Daniels came from outside the box. And he brought new ideas with him.

GIANT LEAPS IN RESEARCH

Deborah Knapp

Researchers at the Purdue College of Veterinary Medicine Comparative Oncology Program made an interesting observation. The use of a nonsteroidal antiinflammatory drug had an impact on bladder cancer in dogs.

Across multiple studies, the treatment alone has caused dramatic cancer regression in 20 percent of dogs and the tumor to stop growing in 55 percent of dogs. The drug also improves the effect of chemotherapy.

These findings have led to the treatment annually of hundreds of thousands of dogs with bladder cancer, and the work has expanded to determine its use in humans.

The aim of the Purdue Comparative Oncology Program is to improve the outlook for pet animals and at the same time—because certain types of cancer are very similar in dogs and humans—research the implications of treatments on humans. The program centers on bladder cancer, brain cancer, and lymphoma.

Deborah Knapp, the Delores L. McCall Professor of Comparative Oncology, is director of the program and focuses on bladder cancer.

"The main focus is helping pet dogs that come to us with naturally occurring cancer," Knapp said.[8]

"We're studying how to prevent it, how to detect it early, and how to intervene to prevent it from becoming full-blown cancer," she said. "For those who have full-blown cancer, we're working to find out how we can treat it more effectively. We have done, in my opinion, the most rigorous work anywhere in characterizing dog models for human cancer. What we learn in dogs is helping us learn something about cancer in humans."[9]

The program works with the Purdue Center for Cancer Research, the Indiana University School of Medicine, and other universities such as Johns Hopkins and Duke.

Knapp did her undergraduate work at North Carolina State and studied veterinary medicine at Auburn. She had a private practice before coming to Purdue to combine her interests.

"Purdue has been very special to me," Knapp said. "It's been a place where I can marry my career goals. I love veterinary medicine, I enjoy teaching, and I have a passion for research. For comparative oncology, Purdue has consistently been one of the, if not the, strongest places for that kind of work in the world. I thought I'd be here for three years when I arrived in 1985. I'm still here. And the reason is because this is a very special community. The school is supportive, the University is supportive, and there are very few places where veterinary researchers can have the role in a human cancer research center like we have here. I feel very blessed."[10]

44

A New Model for Higher Education

2012 to 2018

Change is rarely risk-free, but who better to analyze
and calculate those risks than Boilermakers!

—Purdue president Mitchell E. Daniels, Jr.

Mitchell E. Daniels, Jr., promised change. And upon his arrival at Purdue in
January 2013, the first change was ending the traditional practice of annual
presidential "State of the University" speeches. Instead, he issued a widely distributed
and publicly available online "Open Letter to the People of Purdue." He continued
the tradition in the years that followed.

In his first letter he cautioned higher education was facing growing criticism,
change was coming, and Purdue needed to get out in front of it. "A growing literature
suggests that the operating model employed by Purdue and most American universities
is antiquated and soon to be displaced," he said. "My fundamental observation and
greatest source of excitement about the chance to enlist with you is that Purdue, already
a leader, has a chance to separate further from the pack. In a market now demanding
value for the education dollar, we plainly offer it. While others offer curriculum of weak
rigor and dubious relevance, we are a proud outlier. As I said to my fellow freshmen at
Boiler Gold Rush, 'You have chosen a tough school. Congratulations!'"[1]

Tuition Freeze

One of Daniels's first concerns was the rising cost of higher education. Since arriving on campus and extending at least through Purdue's 2019–2020 academic year, he froze tuition at the 2012 level for both in-state and out-of-state students. He announced the first freeze in March 2012, less than two months after arriving on campus. Initially it was for two years, and it was later extended several times. "Henceforth we will seek to adjust our spending to the budgets of students' families rather than require that they adjust their budgets to ours," Daniels said.

Annual jumps in tuition and fees at Purdue and other universities had become the norm, and Daniels said it was time to bring that to a halt. "I think we've reached the point where we ought to break that pattern," he said. "I am not saying this will be easy. It may make for a few differences. My prediction is it will be easier than most people think."[2]

At Purdue tuition had gone up every year since 1976 and had doubled in the prior twelve years. Daniels announced his freeze even before he knew how much money Purdue would receive from the state. The Indiana General Assembly had not yet finalized its next two-year budget. He believed it was time for action. "In this period of national economic stagnation, it's time for us to hit the pause button on tuition increases," Daniels said.[3]

The 2012 in-state tuition at Purdue was $9,900. It was $28,702 for out-of-state students and $30,702 for international students, with extra fees for some academic programs.

Along with the tuition freeze, the University cut room and board costs, adding to student savings, and in 2015 Amazon announced the grand opening of Amazon@ Purdue, its first-ever staffed customer order pickup and drop-off location. Daniels said the service resulted in an average 31 percent savings for students in the cost of their textbooks.

As a result of all of this, student borrowing at the University dropped 31 percent from the 2011–2012 academic year. Purdue consistently ranks among the best value degrees in the nation. At the University's sesquicentennial, the cost of attending Purdue was actually less than it was in 2012. The savings to Purdue families was about $465 million. In the fall of 2018 Purdue ranked fourth best in the nation for Best Value institutions in the United States, based on a survey by the *Wall Street Journal* and *Times Higher Education*.

In looking at holding the line on University expenses, Daniels started with his own pay. He asked for a cut in the president's base pay and negotiated a unique salary contract based on performance. His first-year base pay was set at $546,000, with the risk of losing a quarter of it if he did not meet performance incentives established by the board of trustees. The top amount he would be eligible to receive was less than Córdova made during her last year. In the first year, Daniels reached 88 percent of the set metrics, earning him a total pay of about $531,000. By the 2018–2019 academic year, one-third of his pay was at risk, tied to specific metrics in four broad categories: student affordability; student success, including graduation and retention rates; fundraising; and research productivity and reputation. Daniels also requested members of his senior leadership team adopt similar at-risk pay models, specifically the provost, the chief financial officer, and the executive vice president for research and partnerships.

Purdue Moves

In September 2013 another major announcement arrived when Daniels introduced a program called Purdue Moves, focused on what he called some of the greatest challenges facing higher education. The moves fell into four broad categories: science, technology, engineering, and math (STEM) leadership; world-changing research; transformative education; and affordability and accessibility.

The plans were transformative. Ten initiatives were identified: freezing tuition and cutting costs; expanding the College of Engineering; transforming the College of Technology that became the Purdue Polytechnic Institute; strengthening computer science; investing in drug discovery; advancing plant science research; changing how learning occurs; engaging students with international experiences; increasing success and value, living on campus; becoming a year-round university.

By 2018, 59 percent of Purdue students earned a STEM degree, a 15 percent increase from 2012. Retention increased to 91.9 percent in 2017. Four-year graduation rates increased to 60.3 percent for the 2014 incoming class.

In four years Purdue increased the number of students in engineering by about two thousand and added one hundred new faculty. The American Society for Engineering Education reported in the fall of 2017 that Purdue had the fifth most female graduates in engineering fields, and it ranked the University first for female graduates in engineering technology.

In transformative education Purdue began a process of "flipped classroom," where faculty lectures were available online for students and classroom time was spent on discussion and interactive projects. More than two hundred courses had been "flipped."

The plan focused on research commercialization. By 2015 the number of Purdue-affiliated start-up companies had increased by 400 percent, and the University had set a new record for patents.

Daniels said the University would progress on the major Purdue Moves plans as quickly as possible and that some of it could be accomplished in two or three years. "As we go ahead, I hope we'll find out we can move more quickly," he said. "[But] in some cases you may run into a barrier that says you can't go quite as far or you can't go quite as fast."[4] In 2018 five more initiatives were added: online learning; an Integrative Data Science Initiative to position Purdue at the forefront of advancing data science–enabled research; strengthening the interdisciplinary and comprehensive approach to addressing diseases and disorders that afflict families; increasing opportunities for degrees in three years; and ongoing development of Purdue Polytechnic High School in Indianapolis, aimed at bringing more minority and low-income students to Purdue.

There was more. State Street Redevelopment, a $120 million joint Purdue–West Lafayette project with private investment, transformed the main road through campus, and the Purdue Research Park Aerospace District was created on the west side. In March of 2017 Rolls-Royce officially moved into the Purdue Technology Center Aerospace, a 55,000-square-foot facility in the Purdue Research Park Aerospace District. In 2016 the Purdue Research Foundation and Browning Investment announced a partnership on a $1.2 billion Discovery Park District, including a hotel conference center, restaurants, retail, office, and business spaces, parks, research facilities, and industrial space. In January 2018 construction began on a private $86 million, four-story, 835-bed apartment complex in Discovery Park. The studio, two-bedroom, and four-bedroom apartments rent at prices comparable to or lower than others in the area. In September of 2018 Schweitzer Engineering Laboratories broke ground on SEL Purdue, a 100,000-square-foot facility for electric power research in the Discovery Park District. Research in the facility will support 300-plus high tech jobs. Company founder Edmund O. Schweitzer is a Purdue graduate.

Wilmeth Active Learning Center

A major feature of Purdue Moves arrived in August 2017 when the University opened the Wilmeth Active Learning Center. Located near Hovde Hall where the Engineering Administration Building and Power Plant were formerly located, it is a twenty-first-century library unlike anything in Purdue's history. The 164,000-square-foot, $79 million facility includes classroom and library space designed for active learning. Collaborative spaces are interspersed with the classrooms, and the center immediately became a destination for five thousand students daily.

"The Wilmeth Active Learning Center was designed to blend the learning experience of a classroom with those learning experiences that happen in the library,"

Mullins Reading Room in the Thomas S. and Harvey D. Wilmeth Active Learning Center. (Photo by Alex Kumar/Purdue University)

said James Mullins, Purdue dean of Libraries and the force behind creating the center. "Studies have shown that students retain information to a much greater extent when they learn by doing rather than by memorization, and the Wilmeth Active Learning Center is designed with that in mind."[5] Mullins retired from Purdue at the end of 2017.

Daniels called the center "a glimpse of the future. This innovative approach to classroom design and teaching is part of Purdue's commitment to transforming education and giving our students the best chance to truly learn and succeed," he said.[6]

He said he never wanted to look into the building and see a classroom with a professor lecturing students who were lined up at rows of desks taking notes.[7]

The Wilmeth Active Learning Center occupied a space where the Purdue smokestack once stood. After the dedication of the center, Purdue dean of students emerita Betty M. Nelson gave Mullins a commemorative brick from the smokestack wrapped with a gold ribbon. The reading room in the library was named for Mullins.

A large authorized copy of *Washington Crossing the Delaware* with his troops before a surprise attack during the Revolutionary War was placed in Wilmeth. German-born artist Emanuel Gottlieb Leutze painted the original *Washington Crossing the Delaware* in 1851. In 1969 Ann Hawkes Hutton, an author and civic activist who founded the Washington Crossing Foundation, commissioned the twelve-by-twenty-one-foot copy that hangs at Purdue. Hutton commissioned the painting in honor of her husband, Leon John Hutton, who was a 1929 graduate of Purdue's College of Science. The Washington Crossing Foundation owns the painting and has placed it on loan to the University.

Income Share Agreements

Looking for innovative, better ways to help students afford higher education, Daniels introduced income share agreements with the Back a Boiler Fund. The Back a Boiler Fund is a way for students to complete their education without having to worry about interest rates on loans. The program provides funding for students, who after graduation pay back a set percentage of their salary during an established number of years.

"The Back a Boiler program has begun to contribute meaningfully to the steady reduction in student debt replacing nearly $6 million that would otherwise have been borrowed by undergraduates," Daniels said.[8]

When taking out traditional loans, the students face risk if they can't afford to make their payments. In the Back a Boiler program student payments adjust according to income levels. The plan received national attention and more universities are offering it.

Purdue University Global

In April 2017 Daniels announced a new public university to further expand Purdue's land-grant mission of access to higher education. The initiative addressed changing dynamics in society. More and more working adults are in need of postsecondary education but cannot manage traditional classroom study time in their schedule that includes work and family. Online technologies provide opportunities to reach people in their own homes. Daniels said:

> Nearly 150 years ago, Purdue proudly accepted the land-grant mission to expand higher education beyond the wealthy and the elites of society. We cannot honor our land-grant mission in the twenty-first century without reaching out to the thirty-six million working adults, seven hundred fifty thousand of them in our state, who started but did not complete a college degree, and to the fifty-six million Americans with no college credit at all. None of us knows how fast or in what direction online higher education will evolve, but we know its role will grow, and we intend that Purdue be positioned to be a leader as that happens. A careful analysis made it clear that we are very ill equipped to build the necessary capabilities ourselves, and that the smart course would be to acquire them if we could. We were able to find exactly what we were looking for.[9]

To launch the new university, Purdue acquired Kaplan University and its institutional operations and assets, including its fifteen campuses and learning centers nationwide, thirty thousand students, fourteen learning centers, three thousand employees, and decades of experience in distance education. All existing Kaplan University students and faculty transitioned to the new university. Under the terms of the agreement, Purdue paid one dollar for Kaplan in exchange for a lengthy contract for services.

The new university, named Purdue University Global, is distinct from others in the Purdue system, relying only on tuition and fund-raising to cover operating expenses. No state appropriations are utilized. It operates primarily online. Indiana resident students receive about a 45 percent tuition discount and there is free tuition for Purdue employees.

Three-Year Degree

In the fall of 2017, ninety-three years after then president Edward C. Elliott first said it was necessary, Purdue launched its first three-year degree program. It was created in the College of Liberal Arts.

"The notion that it requires four years to complete an undergraduate degree is really little more than a matter of tradition, a uniquely American tradition," Daniels said. "That this college is moving ahead to offer students the opportunity to accelerate their education and do it in an economical way is a great testament to the efforts of the entire College of Liberal Arts faculty and its leaders. All of this, it's important to note, is without reducing the requirements or the quality of the degree. That its degrees will now be attainable in three years sets our College of Liberal Arts apart from its counterparts and marks it as a leader nationally."[10]

The College of Education also launched a three-year accelerated degree program for special education students beginning in the fall of 2018. It was the first accelerated special education degree in Indiana and it addressed the shortage of special education teachers in the state and nationwide. By the sesquicentennial, Health and Human Sciences majors had three-year track programs from retail management to kinesiology to public health to hospitality and tourism management, and all seven departments in the College of Science offered three-year degree degrees, including actuarial science and data science.

The three-year degree saves Indiana students more than $9,000 in tuition and fees. Out-of-state students save more than $18,000 and international students more than $19,000.

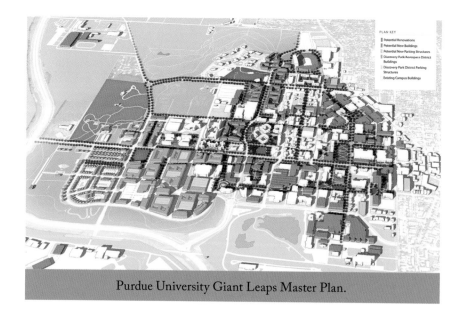

Purdue University Giant Leaps Master Plan.

Patents and Companies Started on Purdue Research

Daniels encouraged commercialization of research. By 2017 Purdue ranked twelfth in the world in U.S. utility patents. A utility patent covers creation of a new or improved product, process, or machine. Among universities without a medical school, Purdue follows only MIT and Caltech in utility patents One hundred sixty-five companies started on the campus from 2013 to 2017, "a record achieved by only a few universities anywhere in the world," Daniels said.[11] Sixteen new drugs from Purdue research were in human trials with forty more in the pipeline, and more than $28 million had been invested to help feed a rapidly growing world population.

New Master Plan

In the fall of 2018 Purdue Physical Facilities unveiled a new West Lafayette campus master plan. The Giant Leaps Master Plan builds on Purdue's 150-year history and looks to the future to imagine how the campus's physical space will set the stage for the next 50 years of giant leaps.

There were five key goals for the new master plan: invest in teaching, research and collaborative spaces; prioritize strategic renovations; focus housing and dining investments; enhance open space connectivity and campus circulation; and strengthen campus identity and gateways.

Intercollegiate Athletics

In August 2016, Mike Bobinski was named Purdue athletic director, replacing Morgan Burke, who retired. Bobinksi became the ninth full-time athletic director at the University. He had most recently been athletic director at Georgia Tech. In addition to his experience in athletics, Bobinski is a certified public accountant and has a bachelor's degree in business administration from the University of Notre Dame, where he graduated magna cum laude in 1979 and played four years on the baseball team.

"There is a lot to like about Mike," said Michael Berghoff, chairman of the Purdue University Board of Trustees. "He is competitive and expects to win. He was a student-athlete, a coach, and a CPA with career experience in both industry and universities with lofty academic and athletic expectations." Daniels said, "We went looking for the most competitive, determined, proven winner we could find, [and] someone who shares our values of academics and standards of conduct. Someone who wants to win, but in the right way."[12]

In December 2016, Bobinski hired Jeff Brohm as head football coach. In his first season Brohm finished the regular schedule 6–6 followed by a victory in the Foster Farms Bowl. It was the University's first bowl game since the 2012 season.

In Bobinski's first year at Purdue two teams won Big Ten Conference championships—men's basketball and women's outdoor track and field. Six teams finished in the top twenty-five at their respective NCAA championships.

Among student athletes, 100 percent participate in community service and a leadership activity every year. Purdue student athletes regularly perform equal to or better than the student body grade point average, maintaining a GPA of 3.0 or better.

Change Is Coming to Higher Education

In an open letter to Purdue people in January 2017, Daniels said higher education was changing as he had predicted and the University needed to respond. "Not so long ago, there was essentially zero risk in public higher education," he said. "Dollars flowed in steadily growing amounts from state treasuries, and were augmented by an apparently limitless ability to raise tuition and fees every year. No one questioned the value of a college diploma, nor whether too many young people were pursuing one. That's all over. During 2016, the skeptical trends of recent years intensified. Almost half of Americans now doubt that a higher education is a good investment, and nearly sixty percent say a college education is no longer necessary to a person's success. Escalating costs are the main driver of these concerns: seventy-five to ninety percent of freshmen nationally say that cost was a major factor in their college choice, and one-quarter to one-third said they could not afford their first choice. Forty-four percent of U.S. citizens label colleges 'wasteful and inefficient.'"[13]

He said alternatives to traditional higher education were continuing to expand and that some critics were saying the traditional residential college campus experience was heading for extinction. "At Purdue, we do not discount these alarming possibilities, but we do not accept their inevitability," he said. "We view the current challenges to higher education as real and the criticisms . . . legitimate. We believe that a posture of denial, obstinately insisting on the superiority of the model just as it is, would be both irresponsible and dangerous. But neither do we accept that the residential university experience is destined for the . . . bone yard. Modernized and enhanced, we believe strongly that Purdue and its sister schools can still offer a compelling case to ambitious and talented young people decades from now. That case must be built around provable, unquestionable value that cannot be replicated in even the most advanced alternative modes. Change is rarely risk-free," he said. "But who better to analyze and calculate those risks than Boilermakers!"[14]

Capital Campaign

When Daniels was named incoming president, a new capital campaign began. Amy Noah was named vice president for development to help Daniels lead the campaign that will include all donations from July 1, 2012, through June 30, 2019. Named Ever True: The Campaign for Purdue University, the goal is $2.019 billion, the largest capital campaign in Purdue history. The name of the campaign comes from a line in the Purdue fight song, "Hail Purdue" ("Ever grateful, ever true . . ."). It has three priorities: Place students first, build on the University's strengths, and champion research and innovation. The title of this book is tied to the name of the campaign.

As of June 30, 2018, the campaign had raised $1.964 billion. It was already the largest fund-raising effort in the University's history and well on its way to success.

It was set for completion June 30, 2019, just weeks after Purdue would mark its 150th anniversary in May 2019.

Purdue continues to celebrate its role in aviation and space. In April of 2014 it hosted another reunion of its astronauts. And with the fiftieth anniversary of Neil Armstrong's *Apollo 11* moon landing coinciding with the University's sesquicentennial year in 2019, more celebrations were planned.

There was much more to celebrate. At its sesquicentennial Purdue was at the center of unprecedented state and national attention, not only in rankings but in media reports recognizing its leadership in bucking national trends in higher education with its tuition freeze, enrollment growth, increasing student quality and success, online courses through the acquisition of Kaplan University, record research and commercialization, and focus on creating a new model for universities prepared for the challenges and opportunities of the twenty-first century and beyond.

GIANT LEAPS IN RESEARCH

Philip S. Low
(Photo courtesy of Philip Low)

Philip S. Low returned to his hometown of West Lafayette and Purdue in 1976 shortly after leaving the University of California, San Diego, with a PhD in biochemistry.

Having done his undergraduate work at Brigham Young University, he was happy to be back in the town where he grew up, played on the high school basketball team, and was the son of an internationally known professor of agriculture and a mother who was a widely respected musician.

But he had no idea where his work at Purdue would lead: cancer research, tools to improve surgeries, development of drugs to cure diseases and speed recovery from broken bones, and much more.

In 2018 Low, Purdue's Ralph C. Corley Distinguished Professor of Chemistry, had eight drugs in human clinical trials, which might be twice as many as any other entire university in the nation. He is the founder of four companies, one of which, Endocyte, a biopharmaceutical firm that develops targeted therapies for the treatment of cancer and inflammatory diseases, is the first Purdue spin-off company in the Purdue Research Park to have a market capitalization exceeding $1.5 billion. It was acquired by Novartis for $2.1 million in 2018.

It all started with research on plants and an interesting discovery. Low wondered if the findings had applications in humans. They did. He discovered he could target individual human cancer cells without damaging healthy surrounding tissue, leading to the formation of Endocyte in 1995.

"Now we've jumped into imaging," Low said. "We've moved into florescence-guided surgeries that should change the whole practice of surgery during the next five or six years. We anticipate that everyone will be using our cancer-targeted florescence dyes" that selectively light up malignant tumors.[15]

His drugs in clinical human trials include treatments for cancer, malaria, sickle cell disease, and rheumatoid arthritis. Teamed with his youngest son, who is doing postdoctoral research in his lab, he has formed a company, Novosteo, that received a $1.7 million National Science Foundation grant to fast-track an injectable drug that will accelerate and improve recovery from bone fractures, which currently impact 6.3 million Americans per year, some fatally.

"It would have been difficult to see as much success at almost any other university," Low said. "Purdue's willingness to encourage both my basic science and entrepreneurial interests was well ahead of its time when I founded Endocyte. Even today, most other universities are trying to figure out how Purdue manages these diverse activities so well.

"I think you have to step back every time you make a discovery and ask, 'How can I use this new information to help mankind?' If you stop and think about it, there are opportunities everywhere."[16]

45

Ever True

Purdue is not its buildings, or even its wonderful
past and traditions. Purdue at any point in time is
its faculty, its students, and the magic that happens
when they are brought together effectively.

—PURDUE PRESIDENT MITCHELL E. DANIELS, JR.

On December 1, 1870, Purdue University existed only in the thoughts and dreams
of a select few people when its board of trustees met and considered what the
new school would become. The trustees had spent the previous eighteen months
arguing among themselves. Their disagreements would continue as they struggled
with different philosophies and the endless details of creating a university that had
no past, no traditions, no students, no faculty—and no magic.

On that cold December day they gathered in downtown Lafayette at the Lahr
House, where John Purdue kept his residence in upstairs rooms. If they couldn't
agree on much that day, following a motion by Martin Peirce, a kindly, astute, and
successful Lafayette businessman, the trustees came to unanimous agreement on one
point: "Resolved, that in the judgment of this board the legitimate and proper work
of Purdue University is to furnish instruction of the highest order in those branches
of science which pertain to any profession or industrial pursuit in life, in preference
to the study of those branches pursued in the high school, academy or college."[1]

The University they created succeeded beyond their expectations and ultimately
beyond their dreams. With thirteen colleges and schools, the curriculum in the
twenty-first century has expanded far beyond the Purdue being planned in 1870.

But always the University has maintained its focus on education for careers, preparing young people not only for their future, but also preparing them to shape that future.

In 2016 as Purdue approached its sesquicentennial, the *Wall Street Journal* and *Times Higher Education* ranked it among the nation's top five public universities along with Michigan, UCLA, North Carolina, and the University of California, Berkeley.

To Purdue president Mitchell E. Daniels, Jr., that was a point of great pride. "As far as I'm concerned," he said in his 2017 letter to Purdue people, "those schools should be thrilled to be listed with us." Purdue ranked sixth among public universities in the 2018 report.

In 150 years Purdue and its people have impacted the world in countless ways. The profiles in this book of Purdue people who have accomplished giant leaps in life and research are just a short sampling of the University's many leaders and researchers. The list could go on and on:

- Adel Halasa, who received his doctorate in chemistry from Purdue in 1964, is credited with developing the tire tread polymers for the Goodyear Aquatred tire. He has more than 250 patents.
- Charles Pankow, who graduated with a civil engineering degree in 1947, is credited with helping develop a concrete building frame that is exceedingly resistant to earthquakes and launched a construction company responsible for more than two hundred buildings.
- R. Games Slayter, who graduated in chemical engineering in 1921, was a tuba player in Purdue's Marching Band and created the coarse fibers that led to the production of Fiberglass.
- Ward Cunningham, a 1978 Purdue graduate and computer programmer, wrote the first wiki application.
- Ralph S. Johnson, who graduated in 1930 with a degree in aeronautical and mechanical engineering, is credited with innovations that helped develop aviation.
- Bruce Rogers, Purdue 1890, took the publishing world by storm with a stylized typeface he dubbed Centaur.
- Riyi Shi, Purdue professor of neuroscience and biomedical engineering, is a medical scientist specializing in uncovering the mechanisms of central nervous system trauma and diseases, and instituting new treatments through innovative and pioneering research and strategies.
- Stephen Bechtel, who received a bachelor of science degree in civil engineering from Purdue in 1946, headed Bechtel Corporation, the nation's largest construction and civil engineering company.

- Suresh V. Garimella, who led four straight record years for Purdue University research funding, was appointed to the National Science Board in the fall of 2018. Garimella, Purdue's executive vice president for research and partnerships and the Goodson Distinguished Professor of Mechanical Engineering, was one of seven appointments to the board by President Donald Trump. Members are selected for their eminence in research, education, and records of distinguished service and are appointed by the U.S. president.

In 2018 ten Purdue faculty members were named 150th Anniversary Professors, recognizing their excellence in teaching and mentorship. "These outstanding faculty members are extraordinary educators, truly connecting with their students and inspiring them to achieve," Provost and executive vice president for academic affairs Jay Akridge said. "They exemplify everything a great teacher and mentor should be."

The faculty who received the distinction were:

- Erica Carlson, professor of physics and astronomy
- David Eichinger, associate professor of science education
- Christine Hrycyna, professor of biochemistry
- Charles Krousgrill, professor of mechanical engineering
- Suzanne Nielsen, professor of food science
- William Oakes, professor of engineering education
- David Rollock, professor of psychological sciences
- Christian Oseto, professor of entomology
- Randy Roberts, professor of history
- Kathleen Salisbury, professor of small animal surgery

Faculty who received the distinction obtained an annual discretionary allocation of $25,000 to support their teaching/scholarly program, funded by the Provost's Office until donors are found to endow the named professorships.

* * *

Anniversaries are as much about looking forward as recalling the past. Just as we celebrate all that Purdue and its people have accomplished in 150 years, we continue to prepare for a future that holds tremendous potential as well as challenges.

In recalling the University's history, the message that comes through is that Purdue people have always built "one brick higher." In triumph or in setback, they always worked to exceed everything that came before. And they always will.

In 1988 when the University published its last comprehensive history book, senior vice president emeritus John Hicks noted, "Perhaps the most fascinating and heartening aspect of Purdue's history is that it is a University still on the way up."[2] At its sesquicentennial, Purdue far surpasses all that it was in 1988. And in the years ahead it will exceed all that it is today. It will always be a University on the way up.

"Boilermakers believe in continuous improvement, always reaching one 'brick higher,'" Daniels said in his 2018 letter to the People of Purdue. "As our 150th birthday approaches what better way to celebrate than with another year of even greater achievements, and a growth of Purdue's service to state and nation by bringing quality education to additional thousands of our fellow citizens."[3]

Purdue president emeritus Steven Beering once noted that no history book on Purdue will ever conclude with the words "The End." The final words will always read "To be continued."[4]

The future lies in the lifelong bonds created when Purdue faculty and students come together effectively.

It is magic. It is to be continued.

And it is Ever True.

Notes

Chapter 1

1. John Purdue's size is estimated from a pair of his 1862 trousers at the Tippecanoe County Historical Association. The waist of the pants is 42 inches and the inseam, 30 inches.
2. Chase Osborn, "Purdue's Genesis—My Exodus," *Purdue Alumnus* 6, no. 9 (July 1919): 12, 13.
3. *Lafayette Daily Courier,* June 17, 1875.
4. Ibid.
5. Purdue University *Debris* yearbook, 1889. (Page numbers were not used. The description is near the back, under the headline "Purdue, Country, Chapter One.")
6. *Lafayette Daily Courier,* October 24, 1874, a four-page supplement.
7. Ibid.
8. 1875–1876 Lafayette Directory.
9. George Munro, *John Purdue and Purdue University* (gray book), 202–3, Purdue University Virginia Kelly Karnes Archives and Special Collections Research Center.
10. George Munro, *The New Purdue,* Purdue University Virginia Kelly Karnes Archives and Special Collections Research Center.
11. "Purdue University," *Lafayette Daily Journal,* November 11, 1874, 3.
12. Munro, *The New Purdue* (1946), Purdue University Virginia Kelly Karnes Archives and Special Collections Research Center.
13. *Lafayette Daily Courier,* June 17, 1875.
14. Ibid.
15. Ibid.
16. Ibid. The newspaper summarized Purdue's speech, but it's likely the reporter wrote it in his notes almost exactly as it had been spoken.
17. Ibid.

18. Munro, *John Purdue and Purdue University,* a study of relations between them from its organization to his death, 187, Purdue University Virginia Kelly Karnes Archives and Special Collections Research Center.

19. Ibid., 201.

20. "His Monument, the Zuni Dam," *Los Angeles Times*, March 29, 1908, Part 2, 2.

21. Osborn, "Purdue's Genesis—My Exodus."

22. David M. Hovde, "John Bradford Harper: The Graduating Class of 1875," manuscript unpublished as of writing of this book.

23. "His Monument, the Zuni Dam," 3.

24. Hovde, "John Bradford Harper: The Graduating Class of 1875."

Chapter 2

1. Irena McCammon Scott, *Uncle: My Journey with John Purdue* (West Lafayette, IN: Purdue University Press, 2008), 19.

2. Ibid., 31.

3. Robert C. Kriebel, *The Midas of the Wabash: A Biography of John Purdue* (West Lafayette, IN: Purdue University Press, 2002), 66, 67.

4. Ibid., 38.

5. Ibid., 19.

6. Robert Topping, *A Century and Beyond: The History of Purdue University* (West Lafayette, IN: Purdue University Press, 1988), 40.

7. James H. Madison, *Hoosiers: A New History of Indiana* (Bloomington and Indianapolis: Indiana University Press, Indiana Historical Society Press, 2014), 83, 84.

8. Ibid., 84.

9. Topping, *A Century and Beyond*, 43, 44.

10. Lafayette *Journal and Courier*, September 1, 1921.

11. Fowler House History, www.fowlerhouse.org.

12. William Murray Hepburn and Louis Martin Sears, *Purdue University: Fifty Years of Progress* (Indianapolis: Hollenbeck Press, 1925), 47.

13. *Purdue Exponent*, October 29, 1916.

14. Kriebel, *The Midas of the Wabash*, 35–36.

15. Harvey Wiley, *Harvey W. Wiley: An Autobiography* (Indianapolis: Bobbs-Merrill, 1930), 123, 124.

16. Madison, *Hoosiers: A New History of Indiana* 112–13.

17. Topping, *A Century and Beyond*, 46.

18. Kriebel, *Midas of the Wabash*, 56.

19. *Lafayette Daily Courier*, February 2, 1851.

Chapter 3

1. Journal of the House of Representatives of the State of Indiana, January 8, 1869, 69.
2. *Justin Morrill, Land for Learning,* Vermont PBS documentary, 1997.
3. Justin Morrill, "Bill Granting Lands for Agricultural Colleges," speech in the U.S. House of Representatives, April 20, 1858.
4. *Justin Morrill, Land for Learning.*
5. Journal of the House of Representatives of the State of Indiana, January 8, 1869, 68.
6. Ibid., 69 and 70.
7. Journal of the Indiana State Senate of 1869, March 1, 616.
8. *Lafayette Daily Journal,* March 25, 1869.
9. Ibid., April 3, 1869.
10. Ibid., April 7, 1869.
11. Ibid.
12. Journal of the Indiana Senate in Special Session, April 20, 1869, 90, 91.
13. Kriebel, *Midas of the Wabash,* 100.
14. Thomas D. Clark, *Indiana University, Midwestern Pioneer,* Vol. 1, 116.

Chapter 4

1. Chauncey Town Record, Purdue University Virginia Kelly Karnes Archives and Special Collections Research Center.
2. The University was not located in Lafayette, but it used the city as its address until June 10, 1972, when the Purdue University Board of Trustees unanimously changed its official location to West Lafayette. The change came at the request of West Lafayette mayor Joe Dienhart, who had worked for many years in Purdue intercollegiate athletics. "It's official; Purdue is in WL," Lafayette *Journal and Courier,* June 12, 1972, 6.
3. Munro, *John Purdue and Purdue University,* 82.
4. *New York Times,* May 5, 1872, 2.
5. Purdue University Board of Trustees minutes, August 13, 1872.
6. Richard Owen, Box 1, Diaries and Papers, photocopy of Report to Purdue Trustees, 1873, Purdue University Virginia Kelly Karnes Archives and Special Collections Research Center.
7. Ibid.
8. Ibid.
9. Ibid.
10. "Professor Owen and the Purdue University," *Lafayette Daily Courier,* January 8, 1874.

11. Richard Owen, Diary, Purdue University Virginia Kelly Karnes Archives and Special Collections Research Center.

Chapter 5

1. *Lafayette Daily Journal,* March 4, 1874.
2. Ibid., March 12, 1874.
3. Purdue University Board of Trustees minutes, April 16, 1875.
4. *Lafayette Daily Journal,* June 13, 1874.
5. Munro, *John Purdue and Purdue University,* 171.
6. Hepburn and Sears, *Purdue University: Fifty Years of Progress,* 52.
7. John Norberg, *Wings of Their Dreams: Purdue in Flight* (West Lafayette, IN: Purdue University Press, 2003), 13.
8. Purdue Annual Register, 1884–1885, 75, 76.
9. Munro, *The New Purdue.*
10. *Purdue Alumnus* 7, no. 1 (October 1919): 12.
11. *Debris,* 1900.
12. George Munro, *John Purdue and Purdue University,* unpublished manuscript, 1946.
13. Ibid.
14. Purdue University Board of Trustees minutes, December 8, 1875, 1.
15. H. B. Knoll, *The Story of Purdue Engineering* (West Lafayette, IN: Purdue University Studies, 1963), 21.
16. Topping, *A Century and Beyond,* 89.
17. Purdue Annual Register, 1874–1875, 27.
18. *Purdue Alumnus,* February 1920.
19. Hepburn and Sears, *Purdue University: Fifty Years of Progress,* 173–74.
20. "Purdue's First Co-ed Tells of Trials and Tribulations," Lafayette *Journal and Courier,* May 1, 1924, 12.
21. Ibid.
22. Ibid.
23. "City Mourns as M. W. Miller Is Taken by Death," Lafayette *Journal and Courier,* September 11, 1933, 1, 7.
24. Ibid.
25. *Daily Gazette,* Berkeley, July 13, 1936, 6.
26. Judy Porta of the Christian Science Society, Berkeley, interview with the author, August 2018.
27. *Daily Gazette,* Berkeley, November 20, 1939, 15.
28. Christian Science Society, Berkeley, California, home page, http://www.christian sciencesoci

29. Ernie Pyle, *Home Country* (New York: William Sloane Associates, 1947).
30. "Governor Chase Osborn," River of History Museum, Sault Ste. Marie, Michigan.
31. Carleton Angell, "The Laird of Duck Island," *Quarterly Review of University of Michigan Alumnus*, December 10, 1938.
32. "Governor Chase Osborn."
33. "Soo, Michigan First in Chase Osborn's Affections," *Evening News*, Sault Ste. Marie, Michigan, April 12, 1949.
34. Chase Osborn, "Purdue's Genesis—My Exodus," *Purdue Alumnus*, July 1919, 12–13.
35. Ibid.
36. "Soo, Michigan First in Chase Osborn's Affections."
37. George Pratt, interview with the author, July 15, 2018.
38. Osborn, "Purdue's Genesis—My Exodus."

Chapter 6

1. *Lafayette Daily Journal,* May 15, 1872.
2. Ibid., March 4, 1873.
3. *Lafayette Daily Courier,* March 6, 1876.
4. *Lafayette Daily Journal,* September 11, 1876.
5. *Lafayette Daily Courier,* September 12, 1876.
6. Ibid.
7. Ibid., September 13, 1876.
8. "Judge Purdue's Mental Condition," *Lafayette Daily Courier,* September 15, 1876.
9. Ibid.
10. Ibid.

Chapter 7

1. Wiley, *An Autobiography,* 156.
2. Knoll, *The Story of Purdue Engineering,* 14.
3. Wiley, *An Autobiography,* 157.
4. Ibid.
5. Ibid., 157, 158.
6. *Purdue Alumnus* 7, no. 5 (February 1920): 7.
7. Wiley, *An Autobiography,* 127.
8. Ibid., 126.
9. Ibid., 128.
10. Ibid., 130–31.
11. Topping, *A Century and Beyond,* 173.

12. Frederick Whitford and Andrew Martin, *The Grand Old Man of Purdue University and Indiana Agriculture: A Biography of William Carroll Latta* (West Lafayette, IN: Purdue University Press, 2005).

13. Fred C. Kelly, *George Ade: Warmhearted Satirist* (Indianapolis: Bobbs-Merrill, 1947), 52.

14. Topping, *A Century and Beyond*, 110.

15. "Purdue—Early Impressions by Harvey Wiley," *Purdue Alumnus* 7, no. 5 (February 1920): 8.

16. Wiley, *An Autobiography*, 160.

17. Oscar Anderson, *Health of a Nation: Harvey Wiley and the Fight for Pure Food* (Chicago: University of Chicago Press, 1958).

18. "Annie Peck Dies, 84; Mountain Climber," *New York Times*, July 19, 1935, 17.

19. Hannah Kimberley, *A Woman's Place Is at the Top: A Biography of Annie Smith Peck, Queen of the Climbers* (New York: St. Martin's Press, 2017), 251.

20. Ibid., 112.

21. Ibid., 46.

22. Ibid., 40

23. Ibid., 58–59

24. "Annie Peck Dies, 84; Mountain Climber."

Chapter 8

1. George Ade biography, Purdue University Virginia Kelly Karnes Archives and Special Collections Research Center.

2. Knoll, *The Story of Purdue Engineering*, 22.

3. Kelly, *George Ade*, 50, 54.

4. Ibid., 51, 52.

5. Frederick Whitford, Andrew Martin, and Phyllis Mattheis, *The Queen of American Agriculture, a Biography of Virginia Claypool Meredith* (West Lafayette, IN: Purdue University Press, 2008), 50.

6. Ibid., 53.

7. Emma Ewing biography, Purdue University Susan Bulkeley Butler Women's Archives.

8. Knoll, *The Story of Purdue Engineering*, 25.

9. John Coulter, *The Dean: An Account of His Career and of His Conviction* (John Coulter, 1940), 88, 89.

10. *Debris*, 1924, 133–136.

11. Bernice Nelson, "The Purdue of Yesterday," L. Murray Grant and Bernice Nelson

Grant papers, folder 4, 1903–1905, Purdue University Virginia Kelly Karnes Archives and Special Collections Research Center.

12. Emma McRae papers and biography, Purdue University Virginia Kelly Karnes Archives and Special Collections Research Center.

13. John McCutcheon, *Drawn from Memory: An Autobiography* (Indianapolis: Bobbs-Merrill, 1950), 42.

14. Ibid., 55.

15. Purdue Reamer Club, *A University of Tradition: The Spirit of Purdue,* 2nd ed. (West Lafayette, IN: Purdue University Press, 2013).

16. Knoll, *The Story of Purdue Engineering,* 26.

17. Hepburn and Sears, *Purdue University,* 85.

18. Ibid.

19. Ibid.

20. Knoll, *The Story of Purdue Engineering,* 33–34.

21. Goss, W. F. M., Purdue University Libraries and Special Collections biography.

22. Ibid.

23. Knoll, *The Story of Purdue Engineering,* 191.

24. Goss, Purdue University Libraries and Special Collections biography.

Chapter 9

1. Knoll, *The Story of Purdue Engineering,* 34.

2. Ibid.

3. Editorial, "Purdue's Benefactor," *Purdue Exponent,* November 1, 1892, 5.

4. Amos Heavilon Diary, October 31, 1892, Purdue University Virginia Kelly Karnes Archives and Special Collections Research Center.

5. *Purdue Exponent.*

6. Amos Heavilon Diary, September 22, Purdue University Virginia Kelly Karnes Archives and Special Collections Research Center.

7. Ibid., October 26, 1892.

8. Ibid., December 31, 1892.

9. Hepburn and Sears, *Purdue University,* 87.

10. *Morning Journal,* January 20, 1894, 1.

11. Ibid.

12. "Purdue in Mourning," *Lafayette Courier,* January 24, 1894, 1.

13. *Morning Journal,* January 24, 1894, 1.

14. Ibid.

15. "Purdue in Mourning," *Lafayette Courier,* January 24, 1894, 1.

16. John Norberg, ed., *Full Steam Ahead: Purdue Mechanical Engineering Yesterday, Today and Tomorrow* (West Lafayette, IN: Purdue University Press, 2013).

17. *Purdue Exponent* 11, no. 20 (March 1, 1900): 13.

18. *Debris*, 1924, 139.

19. *Purdue Engineer*, 26–27 (November 1930): 55.

20. Ibid.

21. "Tribute Paid Mead by the President," *New York Times*, January 28, 1936, 20.

Chapter 10

1. Whitford, Martin, and Mattheis, *The Queen of American Agriculture*, 134.

2. *Debris*, 1899, 98–99.

3. Topping, *A Century and Beyond*, 129, 130.

4. Purdue University Semi-Centennial Alumni Record, published by Purdue University, 1924, 311. George Washington Lacey's last known address was 3761 State Street, Chicago. He was listed as having a lost current address.

5. *Debris*, 1890.

6. Caitlyn Marie Stypa, "Purdue Girls: The Female Experience at a Land Grant University, 1887–1913" (graduate paper for a master of arts degree from Indiana University, July 2013), 5.

7. Ibid., 15.

8. Ibid., 30.

9. Ibid., 39.

10. *Lafayette Journal*, February 22, 1900, 1.

11. Winthrop Stone, speech, February 26, 1900, Stone papers, Box 1, folder 1, Purdue University Virginia Kelly Karnes Archives and Special Collections Research Center.

12. Ibid.

Chapter 11

1. Winthrop Stone, Stone papers, Box 1, Purdue University Virginia Kelly Karnes Archives and Special Collections Research Center.

2. "Dr. Stone Selected, Purdue's Popular Vice President to Succeed Dr. Smart," July 6, 1900, *Lafayette Courier*, 1.

3. *Purdue Alumnus* 7, no. 5 (February 1920).

4. Knoll, *The Story of Purdue Engineering*, 44.

5. Topping, *A Century and Beyond*, 151, 152.

6. David M. Hovde, "A Manly Spectacle, Purdue University's Tank Scrap," *Traces of Indiana and Midwestern History, Indiana Historical Society*, Winter 2014, 14.
7. Robert Kriebel, "Yearbook Notes Historic 'Water Tank Scraps,'" Old Lafayette column, Lafayette *Journal and Courier*, September 19, 1993, E6.
8. Winthrop Stone papers, speech, September 22, 1904, Box 1, Purdue University Virginia Kelly Karnes Archives and Special Collections Research Center.
9. Dave O. Thompson Sr. *Fifty Years of Cooperative Extension Service in Indiana* (1962), 22, 29.
10. H. S. Chamberlain, "How Indiana Has Forged to the Front as Producer of Best Corn in World," *Indianapolis Star*, Sunday, May 23, 1909, 52.
11. Ibid.
12. Ibid.
13. *Debris*, 1902, 264.
14. Ibid., 1904, 185.
15. Ibid., 110.
16. Purdue University Alumni Record and Campus Encyclopedia, published by Purdue University, 1929, 315.
17. *Debris*, 140.
18. Purdue University Alumni Record and Campus Encyclopedia, 1929, 350.
19. Ibid., 112.
20. *Debris*, 1913.
21. Purdue University Alumni Record and Campus Encyclopedia, 330.
22. Alexandra Cornelius, "Evolution of the Black Presence at Purdue University," Purdue African American Students, Alumni, and Faculty Collection, Box 1, Purdue University Virginia Kelly Karnes Archives and Special Collections Research Center.
23. *Debris*, 1905, 21.
24. Ibid., 1906, 107.
25. Knoll, *The Story of Purdue Engineering*, 168.
26. "Katherine Bitting Dies; Was Famous Purdue Graduate," Lafayette *Journal and Courier*, October 16, 1937, 11.
27. *Oxford Encyclopedia of Food and Drink in America* (2012), s.v. "Bitting, Katherine," p. 170.

Chapter 12

1. *Lafayette Morning Journal*, October 31, 1903.
2. Ibid.

3. Ibid., November 1, 1903.
4. *Lafayette Courier,* October 31, 1903.
5. "Speed to Scene of Wreck Like Madmen," *Indianapolis Star,* November 1, 1903, 7.
6. *Courier,* October 31, 1903.
7. *Indianapolis Journal,* October 31, 1903.
8. Ibid., November 1, 1903.
9. Ibid.
10. Winthrop Stone papers, Purdue University Virginia Kelly Karnes Archives and Special Collections Research Center.
11. Ibid.
12. Ibid.
13. *Purdue Exponent* 15, no. 3 (November 11, 1903): 9.
14. Winthrop Stone address at Memorial Service for those killed in October 31, 1903, Purdue train wreck, Winthrop Stone papers, Box 13, Purdue University Virginia Kelly Karnes Archives and Special Collections Research Center.
15. "Grand Jury's Report on the Purdue Wreck," *Indianapolis News,* December 24, 1903, 4.
16. Winthrop Stone papers, talk at dedication of the Memorial Gymnasium, May 30, 1908, Purdue University Virginia Kelly Karnes Archives and Special Collections Research Center.

Chapter 13

1. Winthrop Stone papers, Purdue University Virginia Kelly Karnes Archives and Special Collections Research Center.
2. Hepburn and Sears, *Purdue University,* 117.
3. Ibid., 117–18.
4. *Debris,* 1906, 48.
5. The separate paths but linked goals of Mary Matthews and Lella Gaddis are told in Angie Klink, *Divided Paths, Common Ground: The Story of Mary Matthews and Lella Gaddis, Pioneering Purdue Women Who Introduced Science into the Home* (West Lafayette, IN: Purdue University Press, 2011).
6. Angie Klink, *The Deans' Bible: Five Purdue Women and Their Quest for Equality* (West Lafayette, IN: Purdue University Press, 2014), 7.
7. Topping, *A Century and Beyond,* 175.
8. Ibid.
9. *Debris,* 1911, 138
10. Ibid., 139.
11. *Debris,* 1918, 150.

Chapter 14

1. Thomas D. Clark, *Indiana University, Midwestern Pioneer,* Vol. II, In Mid-Passage (Bloomington: Indiana University Press, 1973), 65.
2. Walter J. Daly, "The Origins of President Bryan's Medical School," *Indiana Magazine of History* 98, no. 4 (2002): 265–84.
3. William Lowe Bryan to Board of Trustees, April 3, 1905, Vol. III, President's Report.
4. "Offer to Give Away a College," *Indianapolis Star,* Sunday, June 21, 1903, 1.
5. "Waiting for Action by State University," *Indianapolis News,* July 21, 1903, 1.
6. "Colleges Do Not Agree on Merger," *Indianapolis Star,* July 28, 1903, 1.
7. "Medical College May Join the University," *Indianapolis News,* November 24, 1904.
8. Clark, *Indiana University, Midwestern Pioneer,* 70.
9. Ibid., 67.
10. Purdue University Board of Trustees minutes, September 1, 1905, 1–4.
11. "An Educational Merger," *Indianapolis Star,* September 2, 1905, 6.
12. *Indianapolis Star,* Saturday, September 2, 1905, 6.
13. *Indianapolis News,* Friday, September 1, 1905, 6
14. "Ft. Wayne School Here to Join Medical College," *Indianapolis Star,* October 10, 1905, 3.
15. Clark, *Indiana University, Midwestern Pioneer,* 70.
16. Ibid., 72.
17. Ibid., 73.
18. Robert Kriebel, Lafayette *Journal and Courier,* January 26, 1997, C-3.
19. *Indianapolis Star,* January 17, 1907.
20. "Senate Thronged for a Medical Fight," *Indianapolis Star,* January 23, 1907, 1.
21. Ibid.
22. Clark, *Indiana University, Midwestern Pioneer,* 87.
23. Ibid., 92.
24. Purdue University Virginia Kelly Karnes Archives and Special Collections Research Center, Purdue School of Medicine.
25. Clark, *Indiana University, Midwestern Pioneer,* 90.
26. Winthrop Stone papers, Box 1, Purdue University Virginia Kelly Karnes Archives and Special Collections Research Center.
27. "Basis of Trouble Between Purdue and Indiana Formally Set Forth," *Indianapolis Star,* December 30, 1906, 23.
28. "Renew Purdue and Indiana Athletics," *Indianapolis Star,* January 14, 1908, 10.
29. *Purdue Exponent,* "President Stone Speaks to a Large Number of Students and Faculty," January 10, 1907, 1.

Chapter 15

1. Winthrop Ellsworth Stone diary, June 15, 1911, Purdue University Virginia Kelly Karnes Archives and Special Collections Research Center.
2. "India Cult Causes Divorce, Dr. Stone of Purdue University," *New York Times*, June 20, 1911.
3. Ibid.
4. Winthrop Ellsworth Stone, interview with the author, October 2017.
5. *Lafayette Morning Journal*, July 11, 1912, 7; *Lafayette Courier*, June 10, 1912, 1.
6. Stone, interview with the author.
7. Ibid.
8. Ibid.
9. Ibid.
10. Winthrop Stone Papers, Purdue University Virginia Kelly Karnes Archives and Special Collections Research Center.
11. Christopher Klein, *How Teddy Roosevelt Saved Football*, The History Channel, September 6, 2012.
12. Terence Tobin, ed., *Letters of George Ade* (West Lafayette, IN: Purdue Research Foundation, 1973), 62.
13. Ibid.
14. President Stone, "A Real University Problem," *Purdue Alumnus* 7, no. 3 (December 1919): 12, 13.
15. George Ade, "Observations at Random," *Purdue Alumnus* 7, no. 4 (January 1920): 9.
16. Tobin, *Letters of George Ade*, 71–72.

Chapter 16

1. Winthrop Stone papers, 1917 diary, Purdue University Virginia Kelly Karnes Archives and Special Collections Research Center.
2. Ibid.
3. Ibid., speech May 27, 1917, "Patriotism Through Sacrifice."
4. Ibid., speech May 30, 1917, "Address Before the University."
5. Ibid.
6. Hepburn and Sears, *Purdue University*, 143.
7. Klink, *Divided Paths, Common Ground*, 95.
8. Ibid., 104.
9. *Debris*, 1918, 142.
10. Ibid., 1919, 55.

11. *Purdue Alumnus* 6, no. 2 (December 1918).
12. Lafayette *Journal and Courier,* November 12, 1918, 1.
13. Ibid., 4.

Chapter 17

1. Winthrop Stone papers, 1921 diary, Purdue University Virginia Kelly Karnes Archives and Special Collections Research Center.
2. Lafayette *Journal and Courier,* page 1 story by the Associated Press datelined Spray Falls, Alberta, July 30, 1921.
3. Lafayette *Journal and Courier,* Wednesday, July 27, 1921, Associated Press story with local input.
4. "Details of Tragedy," Lafayette *Journal and Courier,* July 31, 1921, 1.
5. "Eloquent Addresses Delivered at Stone Memorial," Lafayette *Journal and Courier,* October 13, 1921, 7.

Chapter 18

1. Frank Burrin, *Edward Charles Elliott, Educator* (West Lafayette, IN: Purdue Research Foundation, 1970), 16.
2. Joseph L. Bennett, *Boilermaker Music Makers* (West Lafayette, IN: Purdue Research Foundation, 1986), 50.
3. Burrin, *Edward Charles Elliott, Educator,* 4.
4. Knoll, *The Story of Purdue Engineering,* 69–70.
5. Ibid., 78.
6. Ibid.
7. Edward Elliott papers, Memorandum 22, February 6, 1924, Purdue University Virginia Kelly Karnes Archives and Special Collections Research Center.
8. Ibid., Memorandum 85.
9. Purdue Reamer Club, *A University of Tradition,* 48.
10. Topping, *A Century and Beyond,* 199.
11. Jeanne Norberg, "Voices of Purdue, Marking 90 Years of WBAA on Air," *Purdue Alumnus,* January/February 2012, 29.
12. Indiana Journalism Hall of Fame article on WBAA broadcaster John DeCamp.
13. Two of the seventeen new buildings were for instruction: Stanley Coulter Hall, completed in 1917, and the Armory, 1918. When Elliott arrived the Home Economics Building (Matthews Hall) and the Recitation Building were under construction. One hundred forty-two of the 302 faculty members had an advanced degree. During his

twenty-three years as president, Elliott would oversee construction of twenty-eight major buildings. Even during the depths of the Great Depression from 1929 to 1939, Purdue's assets tripled, according to the Purdue Reamer Club, *A University of Tradition*, 35.

14. Purdue Reamer Club, *A University of Tradition*, 74.
15. Knoll, *The Story of Purdue Engineering*, 91.
16. Purdue Reamer Club, *A University of Tradition*, 75.
17. The framed four-leaf clover is now preserved at the Purdue University Virginia Kelly Karnes Archives and Special Collections Research Center.
18. Whitford, Martin, and Mattheis, *The Queen of American Agriculture*, 275.
19. Fred C. Kelly and David Ross, *Modern Pioneer: A Biography* (New York: Alfred A. Knopf, 1946), 111.
20. Ibid., 177.
21. Ruth Freehafer, *R. B. Stewart and Purdue University* (West Lafayette, IN: Purdue Research Foundation, 1983), inside cover.
22. Edward C. Elliott papers, scrapbook, Purdue University Virginia Kelly Karnes Archives and Special Collections Research Center.
23. Klink, *Divided Paths, Common Ground*, 115.
24. Knoll, *The Story of Purdue Engineering*, 82–83.
25. Ibid.
26. Freehafer, *R. B. Stewart and Purdue University*, 30.
27. Ibid.
28. Ibid.
29. Klink, *Divided Paths, Common Ground*, 119.
30. "Mrs. Virginia Meredith," Lafayette *Journal and Courier*, December 12, 1936, 6.
31. Freehafer, *R. B. Stewart and Purdue University*, 28.
32. At the University's sesquicentennial there were five men's cooperative houses (Chauncey, Circle Pines, Fairway, Gemini, and Marwood) and seven for women (Ann Tweedale, Devonshire, Glenwood, Maclure, Shoemaker, Stewart, and Twin Pines).
33. Freehafer, *R. B. Stewart and Purdue University*, 187.
34. Ibid., 165.
35. Ibid., 166.
36. "Purdue Jet Center Director Retires," Lafayette *Journal and Courier*, December 9, 1965, 9.
37. Ellis Island passenger and crew list, *RMS Saxonia*.
38. "Purdue Jet Center Director Retires."
39. John Norberg, *Full Steam Ahead: Purdue Mechanical Engineering Yesterday, Today, and Tomorrow* (West Lafayette, IN: Purdue University Press, 2013), 49–50.

40. Maurice Zucrow papers, correspondences, Purdue University Virginia Kelly Karnes Archives and Special Collections Research Center

Chapter 19

1. Robert B. Eckles, *The Dean: A Biography of A. A. Potter* (West Lafayette, IN: Purdue University, 1974), 78.
2. Ibid., 8.
3. Harry H. Hirschl, "Andrey Abraham Potter: The Man for All Reasons," *Indiana Jewish History* 41, 2015, 31.
4. Ibid.; Jules Janick, "A Jewish History of Purdue 1920–1940," *Indiana Jewish History* 41, 2015, with quotes from Rabbi Gedalyah Engel, 19–21.
5. Eckles, *The Dean*, 7, 8.
6. Ibid., 6.
7. Ibid., 1.
8. Richard Grace, "Freshman Engineering at Purdue University: A Memoir," Purdue Engineering, December 1, 2010, 2.
9. Knoll, *The Story of Purdue Engineering*, 99.
10. Eckles, *The Dean*, 83.
11. Ibid.
12. Ibid., 84.
13. Knoll, *The Story of Purdue Engineering*, 399.
14. Ibid., 399–400.
15. Grace Hackel Lebow, interview with the author, April 2017.
16. Robert Topping, *Just Call Me Orville: The Story of Orville Redenbacher* (West Lafayette, IN: Purdue University Press, 2011), 6.
17. Ibid., 14.
18. "Popcorn King Dies," Lafayette *Journal and Courier*, September 20, 1995, 10.
19. Ibid., 10.

Chapter 20

1. Robert Kriebel, *Ross-Ade: Their Purdue Stories, Stadium, and Legacies* (West Lafayette, IN: Purdue University Press, 2009), 142–43.
2. Ibid., 159.
3. Ibid., 174.
4. Ibid., 174, 175.
5. Ibid., 176.

6. Purdue Reamer Club, *A University of Tradition,* 177–78.
7. Interview between Maurice G. Knoy and Robert B. Eckles, Purdue professor of history, July 11, 1972, 24–25, Purdue Office of Publications Oral History Program, Purdue University Virginia Kelly Karnes Archives and Special Collections Research Center.
8. Eckles, *The Dean,* 81.
9. Ibid.
10. Speeches of Edward C. Elliott, Purdue University Virginia Kelly Karnes Archives and Special Collections Research Center.
11. Klink, *The Deans' Bible,* 29.
12. Ibid., 32.
13. Norberg, *Wings of Their Dreams,* 174.
14. Topping, *A Century and Beyond,* 234.
15. Norberg, *Wings of Their Dreams,* 182.
16. Ibid., 180.
17. Story told to the author by Dorothy Stratton.
18. Klink, *The Deans' Bible,* 66.
19. Barron Hilton Flight Archives, Amelia Earhart Collection, "Lockheed Aircraft Corporation Order Blank," March 20, 1936, Purdue University Virginia Kelly Karnes Archives and Special Collections Research Center.
20. Amelia Earhart papers, Purdue University Virginia Kelly Karnes Archives and Special Collections Research Center.
21. George Palmer Putnam letter to Meikle, July 27, 1936.
22. Amelia Earhart papers, Purdue University Virginia Kelly Karnes Archives and Special Collections Research Center.
23. Bennett, *Boilermaker Music Makers,* 50.
24. Sonny Beck, interview with the author, July 2018.
25. Ibid.

Chapter 21

1. Edward C. Elliott papers, scrapbooks, remarks at the dedication of the Hall of Music, Purdue University Virginia Kelly Karnes Archives and Special Collections Research Center.
2. Burrin, *Edward Charles Elliott, Educator,* 111.
3. Ibid.
4. Ibid., 111–12.
5. Ibid., 112.

6. *Debris*, 1923, 457, 479.

7. *The Building of a Red Brick Campus: The Growth of Purdue as Recalled by Walter Scholer,* Tippecanoe County Historical Association, 1983, 24, 26.

8. Ibid.

9. "Administration Building to Be Built with P.W.A. Funds," Lafayette *Journal and Courier*, October 4, 1935, 7.

10. Ibid., "Rushing Work on Purdue Building," August 29, 1936, 4.

11. Physical Facilities Construction and Construction Documents 1930s–1940s, Book 3, letter from J. Andre Fouilhoux, consulting engineer with W. K. Harrison and J. A. Fouilhoux, Architects, 45 Rockefeller Plaza, New York, March 10, 1939, Purdue University Virginia Kelly Karnes Archives and Special Collections Research Center.

12. The consulting engineer for acoustic was F. R. Wason from the University of Illinois. J. Andre Fouilhoux, designer of New York's Radio City Music Hall, advised Scholer.

13. Edward C. Elliott papers, scrapbooks, Purdue University Virginia Kelly Karnes Archives and Special Collections Research Center.

14. William Meiners, cover story, *Purdue Engineering Extrapolations*, Summer 2001, 6–11.

15. Ibid., 9.

16. "Former Professor at Purdue Expires," Lafayette *Journal and Courier*, August 23, 1949, 1.

Chapter 22

1. Edward C. Elliott papers, scrapbooks, Purdue University Virginia Kelly Karnes Archives and Special Collections Research Center.

2. Ibid.

3. Ibid.

4. Burrin, *Edward Charles Elliott, Educator*, 128.

5. Robert Topping, *The Hovde Years: A Biography of Frederick L. Hovde* (West Lafayette, IN: Purdue Research Foundation, 1980), 171.

6. Klink, *The Deans' Bible*, 113.

7. Ibid., 112.

8. Ibid., 117–18.

9. Ibid., 119.

10. Knoll, *The Story of Purdue Engineering*, 107.

11. Ibid., 108.

12. Ibid., 109.

13. Topping, *A Century and Beyond*, 242.
14. Klink, *The Deans' Bible*, 136–37.
15. Klink, *The Deans' Bible*, 138.
16. Bennett, *Boilermaker Music Makers*, 1986.
17. Topping, *A Century and Beyond*, 242–43.
18. "Noisy Celebration Greet Peace News; 2-Day Holiday Here," Lafayette *Journal and Courier*, August 15, 1945, 1.
19. Scott R. Sanders, *Burger Chef*, Images of America (Mt. Pleasant, SC: Arcadia Publishing, 2009), 8.
20. Ibid.
21. Ibid., 10
22. "Agreement Made for Burger Chef," *Indianapolis Star*, October 17, 1967.
23. "Taos's Frank and Jill Thomas Celebrate 65 Years of Marriage," *Taos News*, Taos, New Mexico, November 21, 2007, C-4.
24. Connecticut Veterans Hall of Fame.
25. James S. Peters II, *Getting Over While Living Black: The Art of Making It* (Bloomington, IN: Xlibris, 2003), 52, 53.
26. Jeff Kelly Lowenstein, "Alumnus James Peters, a Charge for Integration," *ITT Magazine*, Illinois Institute of Technology, Winter 2008.
27. James S. Peters II, *The Saga of Black Navy Veterans of World War II: An American Triumph* (Bethesda, MD: International Scholars Publications, 1996), 96.

Chapter 23

1. Topping, *A Century and Beyond*, 246.
2. *Vision for Tomorrow: The Beering Years*, videotape, Purdue University, 2002.
3. Topping, *A Century and Beyond*, 259.
4. Ibid.
5. Richard Grace, interview with the author, 2017.
6. *Debris*, 1947, 57.
7. John Norberg, *A Force for Change: The Class of 1950* (West Lafayette, IN: Purdue Research Foundation, 1995), 181.
8. Ibid., 26.
9. Ibid., 63, 65.
10. Ibid., 60.
11. Ibid., 347.
12. Interview between Frederick L. Hovde and Purdue history professor Robert B. Eckles, recorded May 16–August 31, 1972, 155–57, Purdue Office of Publications

Oral History Program, Purdue University Virginia Kelly Karnes Archives and Special Collections Research Center.

13. *Purdue Exponent,* June 14, 1950.
14. Angie Klink, *Divided Paths, Common Ground: The Story of Mary Matthews and Lella Gaddis, Pioneering Purdue Women Who Introduced Science into the Home* (West Lafayette, IN: Purdue University Press, 2011), 202.
15. Jean Margaret Billings, "A Basic Mix for Household Use," January 29, 1947, Gertrude Sunderlin papers, Purdue University Virginia Kelly Ann Karnes Archives and Special Collections Research Center.
16. "Dr. Sunderlin Retiring," Lafayette *Journal and Courier,* June 18, 1954, 11.
17. Margalit Fox, "Ruth M. Siems, Inventor of Stuffing, Dies at 74," *New York Times,* November 23, 2005, https://www.nytimes.com/2005/11/23/us/ruth-m-siems-inventor-of-stuffing-dies-at-74.html.
18. Ibid.
19. Laura Shapiro, *All Things Considered,* National Public Radio, November 23, 2005.
20. Fox, "Ruth M. Siems, Inventor of Stuffing, Dies at 74."

Chapter 24

1. Norberg, *A Force for Change,* 280–81.
2. Jim Hitch, interview with the author, 2017.
3. Norberg, *A Force for Change,* 280.
4. Ibid.
5. Candace Galik, "They Call Him the Edison of the Poultry Industry," *Ithaca Journal,* Ithaca, New York, December 5, 1989, 23.
6. Mark Blitz, "The Cornell Professor Who Gave Us the Chicken Nugget," August 19, 2014, Today I Found Out, Feed Your Brain website, http://www.todayifoundout.com/index.php/2014/08/cornell-professor-gave-us-chicken-nugget/.
7. Associated Press, "Robert Baker, 84, Cornell Professor of Food Science," obituary, March 18, 2006.
8. Blitz, "The Cornell Professor Who Gave Us the Chicken Nugget."
9. Michael Baker, "How 'Barbecue Bob' Transformed Chicken," *Ezra,* Summer 2012, https://ezramagazine.cornell.edu/SUMMER12/People.html.
10. Blitz, "The Cornell Professor Who Gave Us the Chicken Nugget."
11. Barbara Mulligan, "Inventor Turned Up Energy Savings by Dimming the Lights," *National Geographic News,* June 6, 2010.
12. Spencer Soper, "Lutron Founder Joel Spira's Illuminating Career," *Morning Call,* Allentown, Penn., October 8, 2012.

13. Soper, "Lutron Founder Joel Spira's Illuminating Career."

Chapter 25

1. West Lafayette neighborhood covenant.
2. "A. Leon Higginbotham Jr., Federal Judge, Is Dead at 70," *New York Times*, December 15, 1998.
3. Ibid.
4. Alexandra Cornelius, "Evolution of the Black Presence at Purdue University," Purdue African American Students, Alumni, and Faculty Collection, Box 1, Purdue University Virginia Kelly Karnes Archives and Special Collections Research Center.
5. *The Negro Motorist Green Book,* 1949.
6. Social Action Committee, West Lafayette, Indiana, report, February 22, 1947, copy in possession of the author.
7. Frederick Parker, "An Account of Our Efforts to Secure Housing in the Women's Dormitory Halls at Purdue University," September 22, 1946, Amherst College Archives.
8. Ibid.
9. Purdue University Board of Trustees minutes, April 17, 18, 1946, Purdue University Virginia Kelly Karnes Archives and Special Collections Research Center.
10. Parker, "An Account of Our Efforts to Secure Housing."
11. Ibid.
12. Ibid.
13. Ibid.
14. Ibid.
15. Ibid.
16. Ralph Gates, letter to Frederick Parker, March 24, 1947.
17. Frederick Hovde, letter to Frederick Parker, December 16, 1946.
18. Ibid.
19. Ibid.
20. Frank Brown Jr., interview with the author, September 2018.
21. Ibid.
22. Ibid.
23. Ibid.
24. Ibid.
25. Ibid.
26. Forest Farmer, interview with the author, September 2018.
27. Ibid.

28. Ibid.
29. Ibid.

Chapter 26

1. Interview between Emanuel Weiler and Purdue history professor Robert B. Eckles, recorded 1970, Purdue Office of Publications Oral History Program, Purdue University Virginia Kelly Karnes Archives and Special Collections Research Center.
2. Topping, *A Century and Beyond*, 317.
3. John Norberg, *History of the Krannert School of Management, Purdue University*, unpublished manuscript, Purdue University Virginia Kelly Karnes Archives and Special Collections Research Center, 103–4.
4. Ibid., 106.
5. Ibid., 110.
6. Angela Roberts, *Celebrating a Continuum of Excellence, Purdue University School of Veterinary Medicine, 1959–2009* (West Lafayette, IN: Purdue University Press, 2009), 1.
7. Ibid., 6, 7.
8. Ibid., 7.
9. Interview between Frederick L. Hovde and Purdue history professor Robert B. Eckles, recorded May 16–August 31, 1972, 160, Purdue Office of Publications Oral History Program, Purdue University Virginia Kelly Karnes Archives and Special Collections Research Center.
10. Roberts, *Celebrating a Continuum of Excellence*, 8, 9.
11. Klink, *The Deans' Bible*, 202.
12. Klink, *The Deans' Bible*, 80.
13. Norberg, *Wings of Their Dreams*, 245.
14. "Earl Butz, Felled by Racial Remark, Is Dead at 98," *New York Times*, February 4, 2008.
15. Interview with Dr. Rita Colwell by Katherine Markee, January 12, 2009, 3, 4, Oral Histories, Purdue University Virginia Kelly Karnes Archives and Special Collections Research Center.
16. Purdue Reamer Club, *A University of Tradition*, 174.
17. Ibid., 168–73.
18. Ibid., 173, 174.
19. Tom Schott, *Purdue University Football Vault: The History of the Boilermakers* (Atlanta: Whitman Publishing, 2008), 66.
20. Brian Lamb, interview with the author, July 16, 2018.
21. "Family Values—Delon Hampton Honors Mother's Memory with Generous

Contribution to Civil Engineering," *Imprints*, Purdue College of Engineering, 2012, https://engineering.purdue.edu/Engr/Giving/Imprints/2012/family-values.

22. Ibid.
23. Delon Hampton with Bob Keefe, *A Life Constructed: Reflections on Breaking Barriers and Building Opportunities* (West Lafayette, IN: Purdue University Press, 2013), inside cover book jacket.
24. Jules Janick, interview with the author, August 2018.
25. "Outstanding Alumni Awards," *CALS News*, Cornell University College of Agriculture and Life Sciences, Fall 2010.
26. Jules Janick, interview with the author, August 2018.
27. Jules Janick and Arthur O. Tucker, *Unraveling the Voynich Codex* (Chad, Switzerland: Springer Nature, 2018).

Chapter 27

1. Klink, *The Deans' Bible*, 171–72.
2. Ibid., 174–75.
3. David Livingstone, "Christian Youth Hears Noted Integration Leader," Lafayette *Journal and Courier*, August 22, 1958, 10.
4. Eric McCaskill, interview with the author, 2018.
5. Eugenia Nixon, "Quiet Protest at Purdue," Lafayette *Journal and Courier*, May 16, 1968, 1.
6. Leroy Keyes, interview with the author, 2018.
7. Topping, *A Century and Beyond*, 330.
8. "Cheerleader Defends Black Man's Salute," Lafayette *Journal and Courier*, December 13, 1968, 14.
9. Betty M. Nelson, dean of students emerita, interview with the author, 2017.
10. Eric McCaskill, interview with the author, 2018.
11. Ibid.
12. Ibid.
13. Robert Kriebel, "No Charges Filed in Wake of Bomb Hoax at Purdue Airport," Lafayette *Journal and Courier*, April 14, 1969, 7.
14. Larry Schumpert, "Purdue," Lafayette *Journal and Courier*, April 21, 1969, 8, continued from 1.
15. Dean of students emerita Betty M. Nelson. Geoff Emerson, Evelyn Emerson's son, organized Purdue's first Earth Day program.
16. Klink, *The Deans' Bible*, 297.
17. Betty M. Nelson, interview with the author, 2017.

18. Lafayette *Journal and Courier*, "60 State Troopers Clear Purdue's Exec Building," May 7, 1969.

19. Betty M. Nelson, interview with the author, 2017.

20. Ibid.

21. Interview between John Hicks and Purdue history professor Robert B. Eckles, recorded July 5, 1972, 54, 61, 62, 63, Purdue Office of Publications Oral History Program, Purdue University Virginia Kelly Karnes Archives and Special Collections Research Center.

22. Interview between Maurice Knoy and Purdue history professor Robert B. Eckles, recorded July 11, 1972, 36–37, Purdue Office of Publications Oral History Program, Purdue University Virginia Kelly Karnes Archives and Special Collections Research Center.

23. Richard Nixon, remarks at a dinner in Indianapolis, June 24, 1971, honoring Dr. Hovde on his retirement as president of Purdue University, Weekly Compilation of Presidential Documents.

24. *Perspective*, Purdue University, 1983.

25. Roland Parrish, interview with the author, September 2018.

26. Ibid.

27. Ibid.

28. Diann Jordan, *Sisters in Science: Conversations with Black Women Scientists about Race* (West Lafayette, IN: Purdue University Press, 2006), 194.

29. Ibid.

30. "Dolores Shockley: In Her Own Words," Purdue University College of Pharmacy, interview conducted during the summer of 2009, https://www.pharmacy.purdue.edu /dolores-shockley-her-own-words.

31. Jordan, *Sisters in Science*, 196.

32. "Dolores Shockley: In Her Own Words."

33. Ibid.

Chapter 28

1. John Norberg, *Wings of Their Dreams*, 341.

2. Knoll, *The Story of Purdue Engineering*, 344–45.

3. Norberg, *Wings of Their Dreams*, 101–2.

4. Ibid., 334.

5. Ibid., 338.

6. Ibid., 392.

7. "Man Sets Foot on Moon," Lafayette *Journal and Courier*, July 21, 1969, 1.

8. Ibid., 4.

9. Phil H. Shook, "Frankly Speaking: A Profile of Dr. Frank Bass, Director of the PhD Programs," *The School of Management, The University of Texas at Dallas*, 1(2), 1998.

10. Ibid.

11. Ibid.

12. Ibid.

13. Ibid.

14. Ibid.

15. Scott Niswonger, interview with the author, September 11, 2018.

16. Ibid.

17. Ibid.

Chapter 29

1. Carolyn Gery, interview with the author, 2017.

2. Topping, *A Century and Beyond*, 338.

3. Ibid., 339.

4. Ibid., 340.

5. National Academy of Engineering, *Memorial Tributes: Volume 16* (Washington, DC, National Academies Press, 2012, 117), https://doi.org/10.17226/13338.

6. Jeanne Norberg, "Arthur G. Hansen, the Students' President," *Purdue Alumnus*, September/October 2010, 35.

7. Topping, *A Century and Beyond*, 337.

8. *Atlanta Constitution*, April 28, 1971, 4A.

9. Margaret Hansen, a professor of nursing at Georgia State University, died of cancer in July 1974.

10. Angelyn Rizzo, "Purdue's Hansen to Wed," Lafayette *Journal and Courier*, July 10, 1972, 1.

11. Topping, *A Century and Beyond*, 343.

12. Ibid., 346.

13. Ibid., 348.

14. Purdue University News Service, "Purdue Profiles: Hansen Legacy Built on Openness, Intellectual Drive," September 7, 2010.

15. D. William Moreau Jr., interview with the author, 2017.

16. Ibid. G. A. Ross was a 1916 graduate. The award is given to a senior man for academic achievement, outstanding leadership, strength of character, and contribution to Purdue.

17. Ibid.

18. Ibid.

19. Ibid.
20. Interview between John Hicks and Purdue history professor Robert B. Eckles, recorded July 5, 1972, 41, 42, Purdue Office of Publications Oral History Program, Purdue University Virginia Kelly Karnes Archives and Special Collections Research Center.
21. Klink, *The Deans' Bible*, 329.
22. Purdue University News Service, "Purdue Profiles: Hansen Legacy Built on Openness, Intellectual Drive," September 7, 2010.
23. Betty M. Nelson, interview with the author, 2017.
24. Klink, *The Deans' Bible*, 357.
25. Interview between John Hicks and Purdue history professor Robert B. Eckles, recorded July 5, 1972, 48, 49, Purdue Office of Publications Oral History Program, Purdue University Virginia Kelly Karnes Archives and Special Collections Research Center.
26. Carolyn Gery, interview with the author, June 24, 2017.
27. Ibid.
28. "Plan for the Eighties," speech at campaign kickoff at the Purdue Memorial Union, March 7, 1980, Arthur G. Hansen papers, Box 10, folder 20, Purdue University Virginia Kelly Karnes Archives and Special Collections Research Center.
29. H. Aspaturian, interview with Seymour Benzer, Oral History Project, California Institute of Technology Archives, 1991, http://oralhistories.library.caltech.edu/27.
30. HowStuffWorks, Seymour Benzer, https://science.howstuffworks.com/dictionary /famous-scientists/biologists/seymour-benzer-info.htm.
31. Louis Sherman, interview with the author, September 2018.
32. Gruber Foundation, "Seymour Benzer," https://gruber.yale.edu/neuroscience/sey mour-benzer.
33. Frederick C. Neidhardt, "H. Edwin Umbarger," Biographical Memoirs, National Academy of Sciences, 2014, http://www.nasonline.org/publications/biographical -memoirs/memoir-pdfs/umbarger-h-edwin.pdf.
34. Ibid.
35. Louis Sullivan, interview with the author, September 2018.
36. Arthur Aronson, interview with the author, September 2018.

Chapter 30

1. "Black Cultural Center Opens Today at Purdue," Lafayette *Journal and Courier*, December 4, 1970, 9.
2. Floyd Hayes III, "Zamora Leaves Rich Legacy to Community," Lafayette *Journal and Courier*, February 19, 1995.

3. Tony Zamora, interview with the author, 2017.

4. Ibid.

5. Senator Birch Bayh, interview with the author, 2018.

6. Kate Cruikshank, "The Art of Leadership: A Companion to an Exhibition from the Senatorial Papers of Birch Bayh, United States Senator from Indiana, 1963–1980," Trustees of Indiana University.

7. 118th Congress, Congressional Record, 5804 and 5808, 1972.

8. Cruikshank, "The Art of Leadership."

9. Betty M. Nelson, interview with the author, 2017.

10. Birch Bayh, interview with the author, 2018.

11. Kevin Cullen, "Purdue to Give Honor to Black It Wronged," Lafayette *Journal and Courier*, March 17, 1979, 1.

12. Beth Brooke-Marciniak, interview with the author, September 2018.

13. Ibid.

14. Ibid.

15. Carolyn Pottenger, interview with the author, September 2018.

16. Ibid.

17. Ibid.

18. Ibid.

19. Anthony Harris, interview with the author, August 28, 2018.

20. Ibid.

21. Ibid.

22. "NSBE History," National Society of Black Engineers website, http://www.nsbe.org /About-Us/NSBE-History.aspx#.XArAxmhKh9M.

23. Anthony Harris, interview with the author, August 28, 2018.

Chapter 31

1. Kevin Cullen, "Nobel Laureate Says Lab Thrill Tops," Lafayette *Journal and Courier*, October 17, 1979.

2. Ibid.

3. In addition to Brown, nine other people with Purdue connections have received the Nobel Prize. Ben Mottelson received a bachelor's degree from Purdue University in 1947. He won the 1975 Nobel Prize in physics for his work on the nonspherical geometry of atomic nuclei. Edward Purcell received his BSEE in electrical engineering from Purdue and shared the 1952 Nobel Prize in physics for his independent discovery of nuclear magnetic resonance in liquids and in solids. John B. Fenn was an American research professor of analytical chemistry who was awarded a share of

the Nobel Prize in chemistry in 2002 for work in mass spectrometry. Fenn earned his bachelor's degree from Berea College and took a summer classes in physical chemistry at Purdue. Julian Schwinger received a Nobel in 1965. He is best known for his work on the theory of quantum electrodynamics.

His first regular academic appointment was at Purdue in 1941. While on leave from Purdue during World War II, he worked at the Radiation Laboratory at MIT. He left Purdue after the war. Purdue Professor Ei-ichi Negishi received the Nobel in 2010. (See part 3, chapter 42.) Vernon Smith shared the 2002 Nobel Memorial Prize in economic sciences. He was a professor at Purdue from 1955 to 1967. Clinton Joseph Davisson was an American physicist who shared the 1937 Nobel Prize in physics for his discovery of electron diffraction. Wolfgang Ernst Pauli received the Nobel Prize in 1945 for his discovery of a new law of nature. Akira Suzuki shared the 2010 Nobel with Negishi. Suzuki was a postdoctoral researcher at Purdue with Brown.

4. Ibid.
5. "Purdue Chemist Wins Nobel," Lafayette *Journal and Courier,* October 15, 1979.
6. Interview with Herbert Brown, Distinguished Professor of Chemistry, by R. B. Eckles of the Department of History, September 22, 1970, 4, Oral Histories, Purdue University Virginia Kelly Karnes Archives and Special Collections Research Center.
7. Ibid., 5.
8. Ibid., 7, 9.
9. Jeanne Norberg, "Arthur G. Hansen, the Students' President," *Purdue Alumnus,* September/October 2010, 34.
10. Donald Powers, interview with the author, January 2008.
11. Laurie Jensen, "Hansen Announced Retirement," Lafayette *Journal and Courier,* November 13, 1981.
12. Joseph L. Bennett, interview with the author, January 6, 2009.
13. Ibid.
14. Michael Rossmann, interview with the author, September 2018.
15. Ibid.
16. Ibid.
17. Amy Raley and Susan Gaidos, "Anatomy of a Master," Purdue University News Service, 2002.
18. Michael Rossmann, interview with the author, September 2018.
19. Oral History Interviews, Michael Rossmann conducted by Katherine Markee, March 31, 2008, Purdue University Virginia Kelly Karnes Archives and Special Collections Research Center.
20. Carolyn Woo, interview with the author, September 2018.

21. "FP Power Map," *Foreign Policy*, no. 200 (May/June 2013): 76.
22. Carolyn Woo, interview with the author, September 2018.
23. Ibid.
24. Ibid.
25. Ibid.

Chapter 32

1. Steven C. Beering, interview with the author, January 2009.
2. Marvin Diskin, *A Vision for Tomorrow: The Beering Years,* Purdue University, 2002.
3. Steven C. Beering, interview with the author, January 2009.
4. The manifest for the *Britannic,* which arrived at Ellis Island July 24, 1948, lists a family of four: Alice and Stefan Bieringer with their sons Klaus and Hans Bieringer. Ellis Island immigration records list Klaus as "Klaus S. Bieringer."
5. Some sources say Stefan Bieringer was on the German Olympic team for the 1936 Berlin Games and competed in the breast stroke.
6. Michael Mauer, *Nineteen Stars of Indiana, Exceptional Hoosier Men,* IBJ Media in association with the Indiana Historical Society, Indianapolis, 2010, 88.
7. Steven C. Beering interview conducted by Katherine Markee, November 15, 2006, Purdue University Virginia Kelly Karnes Archives and Special Collections Research Center.
8. The *Pittsburgh Post-Gazette* obituary for Alice Beering states she had a "mass of Christian burial in St. Bede (Roman Catholic) Church, in Pittsburg," July 11, 1994, 14.
9. Mauer, *Nineteen Stars of Indiana,* 90.
10. Steven C. Beering, interview with the author, 2009.
11. Ibid.
12. Ibid.
13. Liz and Don Thompson interview with Gerry Robinos, Purdue University Development, September 2018, Footprints video, purdue.edu/footprints.
14. Ibid.

Chapter 33

1. Donald Powers, interview with the author, January 2008.
2. Jane Beering, interview with the author, 2009.
3. Katherine Markee interview with Bob Jesse, December 6, 2006, Purdue University Virginia Kelly Karnes Archives and Special Collections Research Center.

4. Karla Hudecek, "Purdue's New President to Take Office on July 1," Lafayette *Journal and Courier,* February 4, 1983.

5. Joseph L. Bennett, interview with the author.

6. Ibid.

7. Topping, *A Century and Beyond,* 378.

8. Steven C. Beering, interview with the author, February 2009.

9. Joseph L. Bennett, interview with the author.

10. Steven C. Beering, interview with the author.

11. The number of media requests Neil Armstrong received was overwhelming, so he generally did media events only on significant anniversaries of the moon landing. He did talk with space historians. In 1985 he signed autographs for Purdue people, especially students. Boilermakers cherished theirs. But he would later stop signing autographs altogether when he realized many of them from other events were being sold.

12. Steven C. Beering, interview with the author.

13. Lafayette *Journal and Courier,* October 1, 1987.

14. Margaret Rowe, interview with the author, September 2009.

15. Lafayette *Journal and Courier,* October 20, 1987.

16. Margaret Rowe, interview with the author.

17. Steven C. Beering, interview with the author.

18. Joseph L. Bennett, interview with the author.

19. Frederick Ford, interview with the author, December 2008.

20. Michael Perry, "Coach Couldn't Leave Purdue Tradition," Lafayette *Journal and Courier,* March 15, 1989.

21. Steven C. Beering, interview with the author.

22. Michael Perry, Lafayette *Journal and Courier,* March 15, 1989.

23. Gene Keady, interview with the author, 2017.

24. Lyrics to the hymn are: "Close by the Wabash, In famed Hoosier land, Stands old Purdue, Serene and Grand. Cherished in Memory, By all her sons and daughters true, Fair Alma Mater All Hail Purdue. Fairest in all the land Our own Purdue. Fairest in all the land Our own Purdue."

25. J. Timothy McGinley, interview with the author, 2017.

26. Morgan Burke, interview with the author.

27. Ibid.

28. Ibid.

29. Steven C. Beering, interview with the author.

30. Randy Brameir, "Annual Purdue Campus Run," Lafayette *Journal and Courier,* January 23, 1985, 3.

31. John Norberg, "Nude Olympian Shows Cold Endurance," Lafayette *Journal and Courier*, February 5, 1981, 3.

32. Tom Paczolt, interview with the author, 2017.

Chapter 34

1. Some of Peirce's original pine trees remain north of the Psychological Sciences building and on the south side of Stanley Coulter Hall. Chinquapin oak trees south of Windsor Hall predate the original construction of the campus. In 1984 astronaut and alumnus Charles Walker took two hundred sweet gum tree seeds into space, and they germinated before returning to earth. Five were planted on the campus beside the Forestry Products Building, Grissom Hall, the Electrical Engineering Building, Forney Hall of Chemical Engineering, and in Pickett Park. In 1990 astronaut Jerry Ross and his wife Karen, both Purdue graduates, planted the "space tree," a sycamore in front of Lilly Hall. Germinated in space during Ross's flight on the space shuttle *Atlantis* in 1988, the tree recognizes Purdue Extension, including 4-H.

2. John Collier, interview with the author, 2017.

3. Frederick Ford, interview with the author.

4. Steven C. Beering, interview with the author, February 2009.

5. The fountain from 1959 was a gift from Lafayette merchants Bert and June Loeb to honor Solomon Loeb, who founded their prominent department store in Lafayette.

6. Kathe Schuckel, "Crowd Gushes over New Fountain," Lafayette *Journal and Courier*, July 18, 1989.

7. The octagonal lobby with surrounding rooms was based on a traditional extended-family village. The "keyhole"-shaped walk-through outside the main entrance was similar to the entrance portal of an African village. The lattice design on the cast stone medallions embedded in the portal is a typical African ornamental motif that can be found in wood, basketry, textiles, wall mats, and even tattoos. The geometric brick pattern on the four elevations of the building was based on a design used in wall mats found in Zaire. Also in keeping with traditional African architecture, the facility is comprised of many circular spaces and domes, and its window openings are few, simple, and work within the overall pattern of the walls.

8. "Black Cultural Center Officially Opens," Purdue University News Release, October 1, 1999.

9. Frederick Ford, interview with the author.

10. Ibid.

11. Steven C. Beering, interview with the author.

12. Frederick Ford, interview with the author.

Chapter 35

1. *Purdue Exponent,* "Exposing the Beerings," March 31, 2000.
2. Frederick Ford, interview with the author.
3. Klink, *The Deans' Bible,* 383.
4. J. Timothy McGinley, interview with the author, 2008.
5. Ibid., 2017.
6. Lafayette *Journal and Courier,* May 8, 2000.

Chapter 36

1. J. Timothy McGinley, interview with the author, 2008.
2. Ibid.
3. Ibid.
4. Ibid.
5. Martin C. Jischke, interview with the author, 2009.
6. "Jischke Crowned as University's 10th President," Lafayette *Journal and Courier,* May 24, 2000.
7. Ibid.
8. Martin C. Jischke, interview with the author
9. Ibid.
10. Ibid.
11. Ibid.
12. Ibid.
13. Oral History Interviews, Martin Jischke conducted by Katherine Markee and Valerie Yazza, December 21, 2006, Purdue University Virginia Kelly Karnes Archives and Special Collections Research Center.
14. Ibid.
15. Martin C. Jischke, interview with the author, 2009.
16. Oral History Interviews, Martin Jischke conducted by Katherine Markee and Valerie Yazza, December 21, 2006, Purdue University Virginia Kelly Karnes Archives and Special Collections Research Center.
17. Ibid.
18. Ibid.
19. Ibid.
20. Ibid.

21. Ibid.

22. Ibid.

23. Ibid.

24. John Sautter, interview with the author, 2017.

25. "Vernon L. Smith—Facts," NobelPrize.org, Nobel Media AB 2018, https://www .nobelprize.org/prizes/economic-sciences/2002/smith/facts/.

26. Matt Holsapple, "Nobel Laureate's Research Has West Lafayette Roots," Lafayette *Journal and Courier*, October 10, 2002.

27. John Norberg, "History of Krannert School of Management, Purdue University," (unpublished manuscript, 1999, p. 15), Purdue University Virginia Kelly Karnes Archives and Special Collections Research Center.

28. Ibid., 18, 19.

29. Ibid.

Chapter 37

1. Martin C. Jischke, interview with the the author, 2009.

2. Ibid.

3. Ibid.

4. Drew Brees was the Super Bowl XLIV MVP on February 7, 2010, when he led the Saints to victory against the Indianapolis Colts. Brees and his Purdue college sweetheart and wife became generous philanthropists, including making gifts to Purdue and the New Orleans community.

5. Tom Campbell, "A Magic Moment," *Gold and Black Illustrated*, November/December 2017, 8.

6. Martin C. Jischke, interview with the author.

7. Adam Kovac and Kevin Cullen, "Post Game Damage Exceeds $50,000," Lafayette *Journal and Courier*, April 3, 2001.

8. Kevin Cullen, "Jischke: No Tolerance for Post-game Behavior," Lafayette *Journal and Courier*, April 5, 2001.

9. Kenneth Burns, "Fountain Fences: Dousing Danger or Is Purdue All Wet," Lafayette *Journal and Courier*, April 12, 2001.

10. Kevin Cullen, "Plan Would Enclose Lower Portion of Water Jet," Lafayette *Journal and Courier*, April 14, 2001.

11. Robert Ringel spent nearly forty years at Purdue as a professor, a dean, and finally executive vice president for academic affairs. He died in 2006.

12. Adam Kovac, "Purdue's Newest Grads Enjoy Ceremony, Family and Fountain," Lafayette *Journal and Courier*, May 14, 2001.

13. Blackwelder had been at Iowa State since 1991 and had served as vice president for external affairs since 1996.
14. Murray Blackwelder, interview with the author, 2001.
15. Elizabeth Garner, "Research Leads to First Treatment for Drug-Resistant HIV," Purdue University News Release, August 2, 2006.
16. Arun K. Ghosh, interview with the author, September 2018.
17. Ibid.
18. Ibid.
19. Ibid.

Chapter 38

1. Sally Mason, interview with the author, 2008.
2. Ibid.
3. Kevin Cullen, "100 Million Research Facility Is Unveiled," Lafayette *Journal and Courier*, September 8, 2001.
4. Martin C. Jischke, interview with the author, 2009.
5. The money was earmarked: $3 million for hiring new faculty to reduce dependence on teaching assistants, $2.6 million for new student aid, $1.2 million for faculty salaries to keep the best people at Purdue, $1 million for classroom technology, $200,000 for diversity programs, and $200,000 for study abroad and experiential learning.
6. Kevin Cullen, "Students Bracing for Tuition Hike," Lafayette *Journal and Courier*, July 7, 2002, 10.
7. Jischke believed in accountability, and from the beginning he accounted for how the money would be invested. The $1.3 billion goal by the end of June 2007 included: $200 million for student scholarships and fellowships, $200 million for faculty support, $600 million for new and improved facilities and equipment, $200 million for programs and centers, and $100 million in unrestricted donations.
8. Joe Thomas, "University Sets Sights High—$1.3 Billion," Lafayette *Journal and Courier*, September 28, 2002.
9. Tanya Brown, "Purdue Professor to be Honored for His Work," Lafayette *Journal and Courier*, February 1, 2005, 8.
10. R. Graham Cooks, interview with the author, September 2018.
11. Elizabeth K. Gardner, "Purdue-Designed Tool Helps Guide Brain Cancer Surgery," Purdue University News Release, July 2, 2014, https://www.purdue.edu/newsroom/releases/2014/Q3/purdue-designed-tool-helps-guide-brain-cancer-surgery.html.
12. R. Graham Cooks, interview with the author, September 2018.

Chapter 39

1. Martin C. Jischke, interview with the author, 2009.
2. Martin C. Jischke, dedication ceremony, Ford Dining Court, 2004.
3. Patty Jischke interview conducted by Katherine Markee on January 29, 2007, 6, Purdue University Virginia Kelly Karnes Archives and Special Collections Research Center.
4. Patty Jischke, interview with the author.
5. John Norberg, "A Night That Redefined Heroes," *Purdue Alumnus*, September/October 2007, 17–19.
6. Maureen Groppe, Gannett News Service, "Geddes Bends the President's Ear," Lafayette *Journal and Courier*, July 28, 2007, B-1.
7. Eric Weddle, "Purdue Biomedical Professor Created More Than 30 Patents" Lafayette *Journal and Courier*, October 26, 2009, 6.
8. Groppe, "Geddes Bends the President's Ear."
9. Weddle, "Purdue Biomedical Professor Created More Than 30 Patents."
10. Kinam Park, *The Geddes Way*, The National Academy of Sciences, Engineering, Medicine, Memorial Tributes, Vol. 15 (Washington, DC: National Academies Press, 2011), 116–21.
11. Weddle, "Purdue Biomedical Professor Created More Than 30 Patents."

Chapter 40

1. "Purdue's New President 'Out of This World,'" Purdue University News Release, May 7, 2007.
2. Ibid.
3. Andy Gammill and Tim Evans, "She's the Boss," *Indianapolis Star*, May 8, 2007, 1–2.
4. Dan Shaw, "Governor Cheers Purdue's Choice," Lafayette *Journal and Courier*, May 8, 2007, 4.
5. John Norberg, "New Girl in Town," *Purdue Alumnus*, July/August 2007, 16–21.
6. Ibid.
7. Ibid.

Chapter 41

1. Andrea Thomas, "She Looks You Right in the Eyes," Lafayette *Journal and Courier*, July 17, 2007, 1.
2. Ibid., 6.

3. Brian Wallheimer, "New Provost Accompanied by 23 Years of Experience at Purdue," Lafayette *Journal and Courier,* March 27, 2007, 1.
4. Eric Weddle and Megan Banta, "Renovated Purdue Co-Rec to Be Named after Ex-President," Lafayette *Journal and Courier,* July 17, 2012, 1.
5. "Purdue to Dedicate France A. Córdova Rec Sports Center," Purdue University News Release, October 10, 2012.
6. France A. Córdova, interview with the author, 2017.
7. "Purdue Strengthens Health, Human Sciences with College Recruitment," Purdue University News Release, February 11, 2010.
8. Ibid.
9. "Purdue Dedicates Hospital and Tourism Management's Marriott Hall," Purdue University News Release, April 17, 2012.

Chapter 42

1. Eric Weddle, "A Dream Comes True," Lafayette *Journal and Courier,* October 7, 2010, 1.
2. Ibid., 8.
3. Ibid.
4. Ibid.
5. Ibid.
6. "How He Found Out," Lafayette *Journal and Courier,* Thursday, October 18, 2007, 9.
7. Eric Weddle, "Seeds of Hope for Africa," Lafayette *Journal and Courier,* June 12, 2009, 1.
8. Eric Weddle, "Córdova to Step Down in 2012," Lafayette *Journal and Courier,* July 2, 2011, A6.
9. Ibid.
10. France A. Córdova, interview with the author, 2017.
11. Maureen Groppe, "Purdue Professor Honored," Lafayette *Journal and Courier,* October 22, 2011, 12.
12. Rakesh Agrawal, interview with the author, September 2018.
13. Ibid.
14. Maureen Groppe, "Purdue Engineer Gets Top Innovation Honor," *Indianapolis Star,* September 28, 2011, 6.
15. Green Tech America, "Nancy Ho, a Problem Solver and a Person with Vision," 2013, http://www.greentechamerica.com/ARCHIVE/AboutNancy.html.
16. Dan Shaw, "Ethanol Researcher a Guest for Address," Lafayette *Journal and Courier,* January 24, 2007, 8.

17. Maureen Groppe, "Obama Honors Purdue Professor with Technology Medal," *IndyStar*, May 19, 2016, https://www.indystar.com/story/news/politics/2016/05/19/obama-honors-purdue-professor/84616570/.

18. "Purdue Engineer Receives Top Award from President Obama," Purdue University News Release, December 22, 2015, https://www.purdue.edu/newsroom/releases/2015/Q4/purdue-engineer-receives-top-award-from-president-obama.html.

Chapter 43

1. Chris Cillizza, "Daniels Won't Run for President in 2012," *Washington Post*, May 22, 2011.

2. Chris Sikich and Mary Beth Schneider, "Daniels Quickly Rose to Top of List," *Indianapolis Star*, June 22, 2012, B-3.

3. Ibid.

4. Mary Beth Schneider, "Daniels Rallies a New Crowd," *Indianapolis Star*, June 22, 2012, B-3.

5. Justin Mack, "The Only Analogy . . . My Wedding Day," Lafayette *Journal and Courier*, June 22, 2012, A-5.

6. Kathy Mayer, "Daniels Brings New Verse to 'Hail Purdue,'" *Purdue Alumnus*, January/February 2013, 18.

7. Ibid., 20.

8. Deborah Knapp, interview with the author, September 2018.

9. Ibid.

10. Ibid.

Chapter 44

1. Mitch Daniels, "Open Letter to the People of Purdue," January 2013.

2. Eric Weddle, "Purdue Tuition Freeze Breaks a Pattern," Lafayette *Journal and Courier*, March 2, 2013, 1.

3. Ibid., "Daniels, the Pause Button on Tuition," 5.

4. "Daniels," Lafayette *Journal and Courier*, September 10, 2013, A-6.

5. "Wilmeth Active Learning Center to Offer Advanced Learning Environment," Purdue University News Release, August 7, 2017, https://www.purdue.edu/newsroom/releases/2017/Q3/wilmeth-active-learning-center-to-offer-advanced-learning-environment.html.

6. Ibid.

7. "Wilmeth," Lafayette *Journal and Courier*, August 4, 2017, A-2.

8. Mitch Daniels, "Open Letter to the People of Purdue," January 2018.
9. "Purdue to Acquire Kaplan University, Increase Access for Millions," Purdue University News Release, April 27, 2017.
10. "Purdue Liberal Arts Launches Three-Year Degree Tracks," Purdue University News Release, September 5, 2017.
11. Mitch Daniels, "Open Letter to the People of Purdue," January 2018.
12. "Mike Bobinski," Purdue Sports, https://purduesports.com/staff.aspx?staff=690.
13. Mitch Daniels, "Open Letter to the People of Purdue," January 2017.
14. Ibid.
15. Philip Low, interview with the author, September 2018.
16. Ibid.

Chapter 45

1. Purdue University Board of Trustees minutes, December 1, 1870.
2. Topping, *A Century and Beyond,* ix.
3. Mitch Daniels, "Open Letter to the People of Purdue," January 2018.
4. Topping, *A Century and Beyond,* 382.

Bibliography

Anderson, Oscar. *Health of a Nation: Harvey Wiley and the Fight for Pure Food.* Chicago: University of Chicago Press, 1958.

Bennett, Joseph L. *Boilermaker Music Makers: Al Stewart and the Purdue Musical Organizations.* West Lafayette, IN: Purdue Research Foundation, 1986.

Burrin, Frank. *Edward Charles Elliott, Educator.* West Lafayette, IN: Purdue Research Foundation, 1970.

Butler, Susan. *East of the Dawn: The Life of Amelia Earhart.* Philadelphia: Da Capo Press, 1997.

Coulter, John. *The Dean: An Account of His Career and of His Conviction.* John Coulter, 1940.

Cross, Coy F., II. *Justin Smith Morrill, Father of the Land-Grant Colleges.* East Lansing: Michigan State University Press, 1999.

Eckles, Robert B. *The Dean: A Biography of A. A. Potter.* West Lafayette, IN: Purdue University, 1974.

———. *Purdue Pharmacy: The First Century.* West Lafayette, IN: Purdue University Press, 1979.

Fitzgerald, Francis J. *Greatest Moments in Purdue Football History.* West Lafayette, IN: Gold and Black Illustrated, 1999.

Freehafer, Ruth. *R. B. Stewart and Purdue University.* West Lafayette, IN: Purdue Research Foundation, 1983.

Geddes, L. A. *A Century of Progress: The History of Electrical Engineering at Purdue (1888–1988).* West Lafayette, IN: Purdue Research Foundation, 1988.

Grandt, A. F., Jr., W. A. Gustafson, and L. T. Cargnino. *One Small Step: The History of Aerospace Engineering at Purdue University,* 2nd ed. West Lafayette, IN: Purdue School of Aeronautics and Astronautics, 2010.

Gray, Ralph. *IUPUI: The Making of an Urban University.* Bloomington: Indiana University Press, 2003.

Hepburn, William Murray, and Louis Martin Sears. *Purdue University: Fifty Years of Progress*. Indianapolis: Hollenbeck Press, 1925.

Indiana Jewish History. Indiana Jewish Historical Society, 2015.

Kelly, Fred C. *George Ade: Warmhearted Satirist*. Indianapolis: Bobbs-Merrill, 1947.

———, and David Ross. *Modern Pioneer: A Biography*. New York: Alfred A. Knopf, 1946.

Kimberley, Hannah. *A Woman's Place Is at the Top: A Biography of Annie Smith Peck, Queen of the Climbers*. New York: St. Martin's Press, 2017.

Klink, Angie. *The Deans' Bible: Five Purdue Women and Their Quest for Equality*. West Lafayette, IN: Purdue University Press, 2014.

———. *Divided Paths, Common Ground: The Story of Mary Matthews and Lella Gaddis, Pioneering Purdue Women Who Introduced Science into the Home*. West Lafayette, IN: Purdue University Press, 2011.

Knoll, H. B. *The Story of Purdue Engineering*. West Lafayette, IN: Purdue University Studies, 1963.

Kriebel, Robert. *The Midas of the Wabash: A Biography of John Purdue*. West Lafayette, IN: Purdue University Press, 2002.

———. *One Hundred Fifty Years of Lafayette Newspapers*. Lafayette, IN: Tippecanoe County Historical Association, 1981.

———. *Ross-Ade: Their Purdue Stories, Stadium, and Legacies,* West Lafayette, IN: Purdue University Press, 2009.

Madison, James H. *Hoosiers: A New History of Indiana*. Bloomington: Indiana University Press, 2014.

McCutcheon, John. *Drawn from Memory: An Autobiography*. Indianapolis: Bobbs-Merrill, 1950.

Mullins, James L., ed. *A Purdue Icon: Creation, Life, and Legacy*. West Lafayette, IN: Purdue University Press, 2017.

Norberg, John. *A Force for Change: The Class of 1950*. West Lafayette, IN: Purdue Research Foundation, 1995.

———, ed. *Full Steam Ahead: Purdue Mechanical Engineering Yesterday, Today and Tomorrow*. West Lafayette, IN: Purdue University Press, 2013.

———. *Heartbeat of the University: 125 Years of Purdue Bands*. West Lafayette, IN: Purdue University Press, 2011.

———. *Three Tigers and Purdue*. West Lafayette, IN: Purdue Research Foundation, 1999.

———. *Wings of Their Dreams: Purdue in Flight*. West Lafayette, IN: Purdue University Press, 2003.

Purdue Reamer Club. *A University of Tradition: The Spirit of Purdue*. 2nd ed. West Lafayette, IN: Purdue University Press, 2013.

Roberts, Angela. *Celebrating a Continuum of Excellence: Purdue University School of Veterinary Medicine, 1959–2009*. West Lafayette, IN: Purdue University Press, 2009.

Scholer, Walter. *The Building of a Red Brick Campus*. Lafayette, IN: Tippecanoe County Historical Association, 1983.

Schott, Tom. *Purdue University Football Vault: The History of the Boilermakers*. Atlanta, GA: Whitman Publishing and Purdue University Press, 2008.

Scott, Irena McCammon. *Uncle: My Journey with John Purdue*. West Lafayette, IN: Purdue University Press, 2008.

Smith, Andrew F., ed. *The Oxford Encyclopedia of Food and Drink in America*, 2nd ed. New York: Oxford University Press, 2012.

Thompson, Dave O., Sr. *Fifty Years of Cooperative Extension Service in Indiana*. Purdue Agricultural Extension, 1962.

Tobin, Terence, ed. *Letters of George Ade*. West Lafayette, IN: Purdue Research Foundation, 1973.

Topping, Robert. *The Book of Trustees*. West Lafayette, IN: Purdue University Press, 2004.

———. *A Century and Beyond: The History of Purdue University*. West Lafayette, IN: Purdue University Press, 1988.

———. *The Hovde Years: A Biography of Frederick L. Hovde*. West Lafayette, IN: Purdue Research Foundation, 1980.

Twenty-Five Greatest Sports Stories in the History of Indiana. Evansville, IN: M. T. Publishing, 2017.

Whitford, Frederick. *For the Good of the Farmer, A Biography of John Harrison Skinner, Dean of Purdue Agriculture*. West Lafayette, IN: Purdue University Press, 2013.

———, and Andrew Martin. *The Grand Old Man of Purdue University and Indiana Agriculture: A Biography of William Carroll Latta*. West Lafayette, IN: Purdue University Press, 2005.

———, Andrew Martin, and Phyllis Mattheis. *The Queen of American Agriculture, a Biography of Virginia Claypool Meredith*. West Lafayette, IN: Purdue University Press, 2008.

Wiley, Harvey. *Harvey W. Wiley: An Autobiography*. Indianapolis: Bobbs-Merrill, 1930.

Woods, Paula, and Fern Martin. *The Best of Lafayette*. G. Bradley Publishing, 2000.

Index

Page numbers in italics refer to images.

A

B